Lecture Notes in Computer Science 11713

Andreas Holzinger · Peter Kieseberg ·
A Min Tjoa · Edgar Weippl (Eds.)

Machine Learning and Knowledge Extraction

Third IFIP TC 5, TC 12, WG 8.4, WG 8.9, WG 12.9
International Cross-Domain Conference, CD-MAKE 2019
Canterbury, UK, August 26–29, 2019
Proceedings

Springer

Editors
Andreas Holzinger 🆔
Medical University of Graz
Graz, Austria

A Min Tjoa 🆔
Vienna University of Technology
Vienna, Austria

Peter Kieseberg
St. Pölten University of Applied Sciences
St. Pölten, Austria

Edgar Weippl
SBA Research
Vienna, Austria

ISSN 0302-9743   ISSN 1611-3349   (electronic)
Lecture Notes in Computer Science
ISBN 978-3-030-29725-1   ISBN 978-3-030-29726-8   (eBook)
https://doi.org/10.1007/978-3-030-29726-8

LNCS Sublibrary: SL3 – Information Systems and Applications, incl. Internet/Web, and HCI

This Springer imprint is published by the registered company Springer Nature Switzerland AG
The registered company address is: Gewerbestrasse 11, 6330 Cham, Switzerland

Preface

The International Cross-Domain Conference for Machine Learning and Knowledge Extraction CD-MAKE, is a joint effort of IFIP TC 5, TC 12, IFIP WG 8.4, IFIP WG 8.9, and IFIP WG 12.9 and is held in conjunction with the International Conference on Availability, Reliability and Security (ARES). The third conference was organized at the University of Kent at Canterbury, UK. A few words about IFIP:

IFIP – the International Federation for Information Processing – is the leading multinational, non-governmental, apolitical organization in information and communications technologies and computer sciences, is recognized by the United Nations (UN), and was established in the year 1960 under the auspices of the UNESCO as an outcome of the first World Computer Congress held in Paris in 1959.

IFIP is incorporated in Austria by decree of the Austrian Foreign Ministry (20th September 1996, GZ 1055.170/120-I.2/96) granting IFIP the legal status of a non-governmental international organization under the Austrian Law on the Granting of Privileges to Non-Governmental International Organizations (Federal Law Gazette 1992/174).

IFIP brings together more than 3,500 scientists without boundaries from both academia and industry, organized in more than 100 Working Groups (WGs) and 13 Technical Committees (TCs).

CD stands for "cross-domain," and means the integration and appraisal of different fields and application domains to provide an atmosphere to foster different perspectives and opinions. The conference fosters an integrative machine learning approach, taking into account the importance of data science and visualization for the algorithmic pipeline with a strong emphasis on privacy, data protection, safety, and security. It is dedicated to offering an international platform for novel ideas and a fresh look on methodologies to put crazy ideas into business for the benefit of humans. Serendipity is a desired effect and should lead to the cross-fertilize of methodologies and the transfer of algorithmic developments.

The acronym MAKE stands for "MAchine Learning and Knowledge Extraction," a field of Artificial Intelligence (AI) that, while quite old in its fundamentals, has just recently begun to thrive based on both novel developments in the algorithmic area and the availability of vast computing resources at a comparatively low cost.

Machine learning studies algorithms which can learn from data to gain knowledge from experience and to generate decisions and predictions. A grand goal is in understanding intelligence for the design and development of algorithms that work autonomously (ideally without a human-in-the-loop) and can improve their learning behavior over time. The challenge is to discover relevant structural and/or temporal patterns ("knowledge") in data, which is often hidden in arbitrarily high dimensional spaces, and thus simply not accessible to humans. Knowledge extraction is one of the oldest fields in AI and sees a renaissance, particularly in the combination of statistical methods with classical ontological approaches. AI currently undergoes a kind of

Cambrian explosion and is the fastest growing field in computer science today thanks to the usable successes in machine learning. There are many application domains, e.g., in medicine, etc. with many use cases from our daily lives, e.g., recommender systems, speech recognition, autonomous driving, etc. The grand challenges lie in sense-making, in context understanding, and in decision-making under uncertainty. Our real world is full of uncertainties and probabilistic inference had an enormous influence on AI generally and machine learning specifically. Inverse probability allows to infer unknowns, to learn from data, and to make predictions to support decision-making. Whether in social networks, recommender systems, health applications or industrial applications, the increasingly complex data sets require a joint interdisciplinary effort, bringing the human-in-control to foster ethical and social issues, accountability, retractability, explainability, causability and privacy, as well as safety and security.

To acknowledge all those who contributed to the efforts and stimulating discussions is not possible in a preface with limited space like this one. Many people contributed to the development of this volume, either directly or indirectly, and it is impossible to list all of them here. We herewith thank all local, national and international colleagues, and friends for their positive and supportive encouragement. Finally, yet importantly, we thank the Springer management team and the Springer production team for their professional support.

Thank you to all! Let's MAKE it!

August 2019

Andreas Holzinger
Peter Kieseberg
A Min Tjoa
Edgar Weippl

Organization

International Cross-Domain Conference for Machine Learning and Knowledge Extraction (CD-MAKE 2019)

CD-MAKE Conference Organizers

Andreas Holzinger	Medical University and Graz University of Technology, Austria
Peter Kieseberg	SBA Research, Austria
Edgar Weippl (IFIP WG 8.4 Chair)	SBA Research, Austria
A Min Tjoa (IFIP WG 8.9. Chair, Honorary Secretary IFIP)	TU Vienna, Austria

Program Committee

Jose Maria Alonso	CiTiUS – University of Santiago de Compostela, Spain
Joel P. Arrais	Centre for Informatics and Systems, University of Coimbra, Portugal
Smaranda Belciug	University of Craiova, Romania
Elisa Bertino	Purdue University, USA
Tarek R. Besold	City University of London, UK
Chris Biemann	Language Technology Group, FB Informatik, Technische Universität Darmstadt, Germany
Malin Bradley	Health Information Privacy Lab, Health Data Science Center, Vanderbilt University, USA
Francesco Buccafurri	Security and Social Networks Group, Universita Mediterranea die Reggio Calabria, Italy
Mirko Cesarini	Università di Milano Bicocca, Italy
Ajay Chander	Stanford University, Fujitsu Labs of America, USA
Krzysztof J. Cios	Data Mining and Biomedical Informatics Lab, VCU, USA
Gloria Cerasela Crisan	Vasile Alecsandri University of Bacau, Romania
Alfredo Cuzzocrea	University of Trieste, Italy
Isao Echizen	Digital Content and Media Sciences Research Division, National Institute of Informatics, Japan
Kapetanios Epaminondas	University of Westminster, UK
Barbara Di Fabio	Universita di Bologna, Italy
Aldo Faisal	Brain & Behaviour Lab, Machine Learning Group, Imperial College London, UK
Massimo Ferri	University of Bologna, Italy

Hugo Gamboa	PLUX Wireless Biosensor, Universidade Nova de Lisboa, Portugal
Aryya Gangopadhyay	UMBC Center of Cybersecurity, University of Maryland, USA
Panagiotis Germanakos	University of Cyprus, Cyprus
Randy Goebel	Centre for Machine Learning, University of Alberta, Canada
Barbara Hammer	CITEC – Bielefeld University, Germany
Pim Haselager	Donders Institute for Brain, Cognition and Behaviour, Radbound University, The Netherlands
Barna Laszlo Iantovics	Petru Maior University, Romania
Beatriz De La Iglesia	Knowledge Discovery & Data Mining Group, University of East Anglia, UK
Xiaoqian Jiang	University of California San Diego, USA
Andreas Kerren	ISOVIS Group, Linnaeus University, Sweden
Peter Kieseberg	SBA Research, Austria
Robert S. Laramee	Data Visualization Group, Swansea University, UK
Freddy Lecue	Accenture Technology Labs, Ireland, and Inria Sophia Antipolis, France
Brian Y. Lim	National University of Singapore, Singapore
Luca Longo	Knowledge and Data Engineering Group, Trinity College Dublin, Ireland
Oswaldo Ludwig	Nuance Communications, Germany
Ljiljana Majnaric-Trtica	University of Osijek, Hungry
Vincenzo Manca	University of Verona, Italy
Sjouke Mauw	Security and Trust of Software Systems Group, University of Luxembourg, Luxembourg
Yoan Miche	Nokia Bell Labs, Finland
Marian Mrozek	Jagiellonian University, Poland
Daniel E. O'Leary	School of Business, University of Southern California, USA
Vasile Palade	School of Computing, Electronics and Mathematics, Coventry University, UK
Camelia–M. Pintea	Technical University of Cluj-Napoca, Romania
Irena Spasic	School of Computer Science & Informatics, Cardiff University, UK
Catagaj Turkay	City University London, UK
Jean Vanderdonckt	LSM, Université catholique de Louvain, Belgium
Pinar Yildirim	Okan University, Turkey
Jianlong Zhou	CSIRO, Australia
Christian Bauckhage	Fraunhofer Institute Intelligent Analysis and University of Bonn, Germany
Vaishak Belle	Belle Lab, Centre for Intelligent Systems and their Applications, School of Informatics, University of Edinburgh, UK
Ivan Bratko	University of Ljubljana, Slovenia

Frederico Cabitza	Università degli Studi di Milano-Bicocca, DISCO, Italy
David Evans	University of Virginia, USA
Benoit Frenay	Universite de Namur, Belgium
Ulrich Furbach	Universität Koblenz-Landau, Germany
Bryce Goodman	Oxford Internet Institute, UK, and San Francisco Bay Area, USA
Hani Hagras	Computational Intelligence Centre, School of Computer Science & Electronic Engineering, University of Essex, UK
Pitoyo Hartono	Chukyo University, Japan
Shujun Li	KirCCS, University of Kent, UK
Daniele Magazzeni	Trusted Autonomous Systems Hub, King's College London, UK
Timothy Miller	School of Computing and Information Systems, The University of Melbourne, Australia
Huamin Qu	HKUST HCI Initiative, Hong Kong University of Science and Technology, SAR China
Stephen Reed	Center for Research in Mathematics and Science Education, San Diego State University, USA
Gerhard Schurz	Düsseldorf Center for Logic and Philosophy of Science, University Düsseldorf, Germany
Marco Scutari	Instituto Dalle Molle di Studi sull'Intelligenza Artificiale, Switzerland
Sameer Singh	University of California Irvine, USA
Alison Smith	University of Maryland, USA
Ivan Stajduhar	University of Rijeka, Hungry
Isaac Triguero Velázquez	University of Nottingham, UK
Marco Tulio Ribeiro	Microsoft Research, Redmond, USA
Andrea Vedaldi	Visual Geometry Group, University of Oxford, UK

Contents

KANDINSKY Patterns as IQ-Test
for Machine Learning

Andreas Holzinger[1]([✉])(iD), Michael Kickmeier-Rust[2], and Heimo Müller[1]

[1] Medical University Graz, Auenbruggerplatz 2, 8036 Graz, Austria
{andreas.holzinger,heimo.mueller}@medunigraz.at
[2] University of Teacher Education, Notkerstrasse 27, 9000 St. Gallen, Switzerland
michael.kickmeier@phsg.ch

Abstract. AI follows the notion of human intelligence which is unfortunately not a clearly defined term. The most common definition given by cognitive science as mental capability, includes, among others, the ability to think abstract, to reason, and to solve problems from the real world. A hot topic in current AI/machine learning research is to find out whether and to what extent algorithms are able to learn abstract thinking and reasoning similarly as humans can do – or whether the learning outcome remains on purely statistical correlation. In this paper we provide some background on testing intelligence, report some preliminary results from 271 participants of our online study on explainability, and propose to use our Kandinsky Patterns as an IQ-Test for machines. Kandinsky Patterns are mathematically describable, simple, self-contained hence controllable test data sets for the development, validation and training of explainability in AI. Kandinsky Patterns are at the same time easily distinguishable from human observers. Consequently, controlled patterns can be described by *both* humans and computers. The results of our study show that the majority of human explanations was made based on the properties of individual elements in an image (i.e., shape, color, size) and the appearance of individual objects (number). Comparisons of elements (e.g., more, less, bigger, smaller, etc.) were significantly less likely and the location of objects, interestingly, played almost no role in the explanation of the images. The next step is to compare these explanations with machine explanations.

Keywords: Artificial intelligence · Human intelligence ·
Intelligence testing · IQ-Test · Explainable-AI ·
Interpretable machine learning

1 Introduction and Motivation

*"If you can't measure it, nor assign it an exact numerical value, nor express it
in numbers, then your knowledge is of a meager and unsatisfactory kind"*
(attributed to William Thomson (1824–1907), aka Lord Kelvin)

© IFIP International Federation for Information Processing 2019
Published by Springer Nature Switzerland AG 2019
A. Holzinger et al. (Eds.): CD-MAKE 2019, LNCS 11713, pp. 1–14, 2019.
https://doi.org/10.1007/978-3-030-29726-8_1

Impressive successes in artificial intelligence (AI) and machine learning (ML) have been achieved in the last two decades, including: (1) IBM Deep Blue [6] defeating the World Chess Champion Garry Kasparov in 1997, (2) the success of IBM Watson [10] in 2011 in defeating the Jeopardy players Brad Rutter and Ken Jennings, or (3) the sensation of DeepMind's Alpha Go [42] in defeating Go masters Fan Hui in 2015 and Lee Sedol in 2016.

Such successes are often seen as milestones for and "measurements" of AI. We argue that such successes are reached in very specific tasks and not appropriate for evaluating the "intelligence" of machines.

The development of intelligence, therefore, is the result of the incremental interplay between challenge/task, a conceptual change (physiological as well as mentally) of the system, and the assessment of the effects of the conceptual change. To advance AI, specifically in the direction of explainable AI, we suggest bridging the human strength and the human assessment methods with those of AI. In other words, we suggest introducing principles of human intelligence testing as an innovative benchmark for artificial systems.

The ML community is becoming now aware that human IQ-tests are a more robust approach to machine intelligence evaluation than such very specific tasks [9]. In this paper we provide (1) some background on testing intelligence, (2) report on some preliminary results from 271 participants of our online study on explainability[1], and (3) propose to use our Kandinsky Patterns [32][2] as an IQ-Test for machines.

2   Background

A fundamental problem for AI are often the vague and widely different definitions of the notion of intelligence and this is particularly acute when considering artificial systems which are significantly different to humans [28]. Consequently, intelligence testing for AI in general and ML in particular has generally not been in the focus of extensive research in the AI community. The evaluation of approaches and algorithms primarily occurred along certain benchmarks (cf. [33,34]).

The most popular approach is the one proposed by Alan Turing in 1950 [45], claiming that an algorithm can be considered intelligent (enough) for a certain kind of tasks if and only if it could finish all the possible tasks of its kind. The shortcoming of this approach, however, is that it is heavily task-centric and that it requires an a-priori knowledge of all possible tasks and the possibility to define these tasks. The latter, in turn, bears the problem of the granularity and precision of definitions. An indicative example is the evaluation, or in other terms, the "intelligence testing" for autonomously driving cars [29], or another example is CAPTCHA (completely automated public Turing test to tell computers and humans apart), which are simple for humans but hard for machines and therefore

---

[1] https://human-centered.ai/experiment-exai-patterns.
[2] https://human-centered.ai/project/kandinsky-patterns.

used for security applications [1]. Such CAPTCHAs use either text or images of different complexity and pose individual differences in cognitive processing [3].

In cognitive science, the testing of human aptitude – intelligence being a form of cognitive aptitude – has a very long tradition. Basically, the idea of psychological measurement stems from the general developments in 19th century science and particularly physics, which put substantial focus on the accurate measurement of variables.

This view was the beginning of so-called *anthropometry* [36] and subsequently the psychological measurement. The beginning of intelligence testing occurred around 1900 when the French government had passed a law requiring all French children to go to school. Consequently, the government regarded it as important to find a way to identify children who would not be capable to follow school education. Alfred Binet (1857–1911) [11] started the development of assessment questions to identify such children. Remarkably, Binet not only focused on aspects which were explicitly taught in schools but also on more general and perhaps more abstract capabilities, including attention span, memory, and problem solving skills. Binet and his colleagues found out that the children's capacity to answer the questions and solve the tasks was not necessarily a matter of physical age. Based on this observation, Binet proposed a "mental age" – which actually was the first intelligence measure [4]. The level of aptitude was seen relative to the average aptitude of the entire population. Charles Spearman (1863–1945) coined in 1904 [43], in this context, the term *g-factor*, a general, higher level of intelligence.

This very early example for an intelligence test already makes the fundamental difference to the task-centric evaluation of later AI very clear. Human intelligence was *not* seen as the capability to solve one particular task, such as a pure classification task, it was considered being a much wider construct. Moreover, human intelligence generally was not measured in an isolated way but always in relation to an underlying population. By the example of the self-driving cars, the question would be whether one car can drive "better" against all the other cars, or even whether and to what extent the car does better than human drivers. In the 1950s, the American psychologist David Wechsler (1896–1981) extended the ideas of Binet and colleagues and published the *Wechsler Adult Intelligence Scale* (WAIS), which, in its fourth revision, is a quasi standard test battery today [48]. The WAIS-IV contains essentially ten subtests and provides scores in four major areas of intelligence, that is, verbal comprehension, perceptual reasoning, working memory, and processing speed. Moreover, the test provides two broad scores that can be used as a summary of overall intelligence. The overall full-scale intelligence value (IQ was already coined by William Stern in 1912 for the German term Intelligenzquotient) uses the popular mean 100, standard deviation 15 metric.

In advancing Spearman's g-factor idea, Horn and Cattell [17] argued that intelligence is determined by about 100 interplaying factors and proposed two different levels of human intelligence, fluid and crystallized intelligence. The former includes general cognitive abilities such as pattern recognition, abstract reasoning, and problem solving. The latter is based on experience, learning, and acculturation; it includes general knowledge or the use of language. In addition to Wechsler's WAIS-IV, among the most commonly used tests, for example, is *Raven's Progressive Matrices* [37], which is a non-verbal multiple choice measures of the reasoning component of Spearman's g, more exactly, the two components (i) "thinking clearly" and "making sense of complexity", and (ii) the "ability to store and reproduce information". The test was originally developed by John Raven in 1936 [37]. The task is to continue a visual pattern (cf. Fig. 1). Other tests are the *Reynolds Intellectual Assessment Scales, the Multidimensional Aptitude Battery II, the Naglieri Nonverbal Ability Test* (cf. Urbina [46]), and in German speaking countries the *IST-2000R* [2] or the *Berlin Intelligence Structure Test* (BIS; [20]).

There exists a large amount of classifications and sub-classifications of sub-factors of intelligence, The Cattell-Horn [17] classification includes, for example:

- Quantitative knowledge (the ability to understand and work with mathematical concepts)
- Reading and writing
- Comprehension-Knowledge (the ability to understand and produce language)
- Fluid reasoning (incl. inductive and deductive reasoning and reasoning speed)
- Short term memory
- Long term storage and retrieval
- Visual processing (including closure of patterns and rotation of elements)
- Auditory processing (including musical capabilities)
- General processing speed.

An - at the first sight similar - classification was introduced by Gardner [12] based on his *theory of multiple intelligences*. As opposed to prior classification, his theory includes a much broader understanding of intelligence as human aptitude. Gardner's theory, therefore, was a starting point for an (often discussed as inflationary) increase of types of intelligence, for example in direction of emotional, social, and artistic "intelligence" [30]. Over the past 120 years, the 20th century ideas of human intelligence have been further developed and new models have been proposed. These new models tend to interpret general intelligence as an emergent construct reflecting the patterns of correlations between different test scores and not as a causal latent variable. The models aim to bridge correlational and experimental psychology and account for inter-individual differences in terms of intra-individual psychological processes and, therefore, the approaches look into neuronal correlates of performance [7]. One of these new approaches is, for example, *process overlap theory*, a novel sampling account, based upon cognitive process models, specifically models of working memory [22].

When explaining predictions of deep learning models we apply an explanation method, e.g. simple sensitivity analysis, to understand the prediction in terms of the input variables. The result of such an explainability method can be a heatmap. This visualization indicates which pixels need to be changed to make the image look (from the AI-systems perspective!) more or less like the predicted class [40]. On the other hand there are the corresponding human concepts and "contextual understanding" needs effective mapping of them both [24], and is among the future grand goal of human-centered AI [13].

For a detailed description of the KANDINSKY Patterns please refer to [32].

When talking about explainable AI it is important from the very beginning to differentiate between Explainability and Causability: under explainability we understand the property of the AI-system to generate machine explanations, whilst causability is the property of the human to understand the machine explanations [15]. Consequently, the key to effective human-AI interaction is an efficient mapping of explainability with causability. Compared to the map metaphor, this is about establishing connections and relations - not drawing a new map. It is about identifying the *same areas in two completely different maps*.

3 Related Work

Within the machine learning community there is an intensive debate if e.g. neural networks can learn abstract reasoning or whether they merely rely on pure correlation. In a recent paper the authors [41] propose a data set and a challenge to investigate abstract thinking inspired by a well-known human IQ test: the Raven test, or more specifically the Raven's Progressive Matrices (RPM) and Mill Hill Vocabulary Scales, which were developed 1936 for use in fundamental research into both the genetic and the environmental determinants of "intelligence" [37]. The premise behind RPMs is simple: one must reason about the relationships between perceptually obvious visual features – such as shape positions or line colors – to choose an image that completes the matrix. For example, perhaps the size of squares increases along the rows, and the correct image is that which adheres to this size relation (see Fig. 1). RPMs are strongly diagnostic of abstract verbal, spatial and mathematical reasoning ability. To succeed at the challenge, models must cope with various generalisation 'regimes' in which the training and test data differ in clearly-defined ways.

The amazingly advancing field of AI and ML technologies adds another dimension to the discourse of intelligence testing, that is, the evaluation of artificial intelligence as opposed to human intelligence. Human intelligence tends to focus on adapting to the environment based on various cognitive, neuronal processes. The field of AI, in turn, very much focuses on designing algorithms that can mimic human behavior (weak or narrow AI). This is specifically true in applied genres such as autonomously driving cars, robotics, or games. This also leads to distinct differences in what we consider intelligent. Humans have a consciousness, they can improvise, and the human physiology exhibits plasticity that leads to "real" learning by altering the brain itself. Although humans tend to make more errors, human intelligence as such is usually more reliable and robust against catastrophic errors, whereas AI is vulnerable against software, hardware and energy failures. Human intelligence develops based on infinite interactions with an infinite environment, while AI is limited to the small world of a particular task.

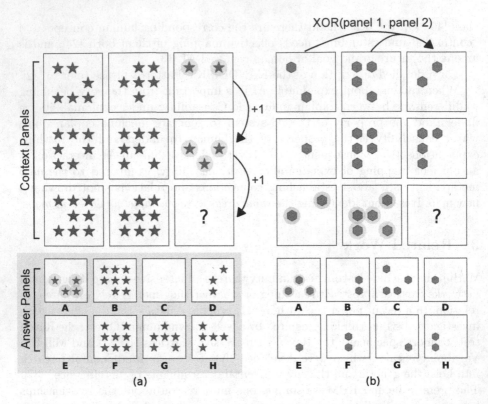

Fig. 1. Raven-style Progressive Matrices. In (a) the underlying abstract rule is an arithmetic progression on the number of shapes along the columns. In (b) there is an XOR relation on the shape positions along the rows (panel 3 = XOR(panel 1, panel 2)). Other features such as shape type do not factor in. **A** is the correct choice for both. Figure taken from [41].

The development of intelligence, therefore, is the result of the incremental interplay between challenge/task, a conceptual change (physiological as well as mentally) of the system, and the assessment of the effects of the conceptual change. To advance AI, specifically in the direction of explainable AI, we suggest bridging the human strength and the human assessment methods with those of AI. In other words, we suggest introducing principles of human intelligence testing as an innovative benchmark for artificial systems.

We want to exemplify this idea by the challenge of the identification and interpretation/explanation of visual patterns. In essence, this refers to the human ability to make sense of the world (e.g., by identifying the nature of a series of visual patterns that need to be continued). Sensemaking is an active processing of sensations to achieve an understanding of the outside world and involves the acquisition of information, learning about new domains, solving problems, acquiring situation awareness, and participating in social exchanges of knowledge

[35]. The ability can be applied to concrete domains such as various HCI acts [35] but also to abstract domains such as pattern recognition.

This topic was specifically in the focus of medical research. Kundel and Nodine [23], for example, investigated gaze paths in medical images (a sonogram, a tomogram, and two standard radiographic images). They were asked to summarize each of the images in one sentence. The results of this study revealed that correct interpretations of the images were related to attending the relevant areas of the images as opposed to attending visually dominant areas of the images. The authors also found a strong relation of explanations to experiences with images.

A fundamental principle in the perception and interpretation of visual patterns is the likelihood principle, originally formulated by Helmholtz, which states that the preferred perceptual organization of an abstract visual pattern is based on the likelihood of specific objects [27]. A, to a certain degree competing, explanation is the minimum principle, proposed by Gestalt psychology, which claims that humans perceive a visual pattern according the simplest possible interpretation. The role of experience is also reflected in studies in the context of the perception of abstract versus representative visual art; [47] demonstrated distinct differences in art experts and laymen in the perception and their preferences of visual art. Psychological research could demonstrate that the nature of perceiving and interpreting visual patterns, therefore, is a function of expectations [50]. On the one hand, this often leads to misinterpretations or premature interpretations, on the other hand, it increases the "explainability" of interpretations since the visual perception is determined by existing conceptualizations.

4 How Do Humans Explain? How Do Machines Explain?

In a recent online study [14], we asked (human) participants to explain random visual patterns (Fig. 2). We recorded and classified the free verbal explanations of in total 271 participants. Figure 3 summarizes the results. The results show that the majority of explanations was made based on the properties of individual elements in an image (i.e., shape, color, size) and the appearance of individual objects (number). Comparisons of elements (e.g., more, less, bigger, smaller, etc.) were significantly less likely and the location of objects, interestingly, played almost no role in the explanation of the images.

In a natural language statement about a Kandinsky Figure humans use a series of basic concepts which are combined through logical operators. The following (incomplete) examples illustrate some concepts of increasing complexity.

- Basic concepts given by the definition of a Kandinsky Figure: a set of *objects*, described by *shape, color, size* and *position*, see Fig. 4(A) for color and (B) for shapes.
- Existence, numbers, set-relations (*number, quantity* or *quantity ratios* of objects), e.g. *"a Kandinsky Figure contains 4 red triangles and more yellow objects than circles"* , see Fig. 4(C).

Fig. 2. Visual patterns to be explained by humans.

| | Explanatory Element | | | |
	1st	2nd	3rd	Total
Location	0	0	3	3
Number comparisons	1	2	0	3
Color comparisons	9	19	6	34
Size comparisons	16	14	18	48
Shape comparisons	17	13	18	48
Size	32	42	18	92
Color	46	41	45	132
Number	127	124	116	367
Shape	133	129	91	353

Fig. 3. Visual patterns to be explained by humans.

- Spatial concepts describing the arrangement of objects, either absolute (*upper, lower, left, right, . . .*) or relative (*below, above, on top, touching, . . .*), e.g. *"in a Kandinsky Figure red objects are on the left side, blue objects on the right side, and yellow objects are below blue squares"*, see Fig. 4(D).
- Gestalt concepts (see below) e.g. *closure, symmetry, continuity, proximity, similarity*, e.g. *"in a Kandinsky Figure objects are grouped in a circular manner"*, see Fig. 4(E).
- Domain concepts, e.g. *"a group of objects is perceived as a "flower""*, see Fig. 4(F).

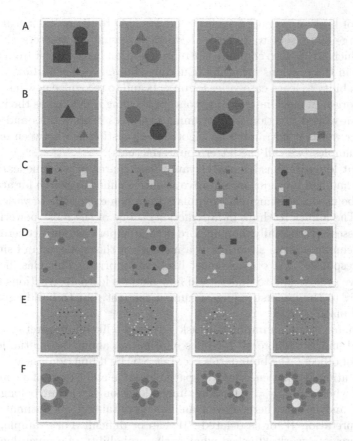

Fig. 4. Kandinsky Pattern showing concepts as color (A), shape (B), numeric relations (C), spatial relations (D), Gestalt concepts (E) and domain concepts (F) (Color figure online)

These basic concepts can be used to select groups of objects, e.g. 'all red circles in the upper left corner', and to further combine single objects and groups in a statement with logic operator, e.g. 'if there is a red circle in the upper left corner, there exists no blue object', or with complex domain specific rules, e.g. 'if the size of a red circle is smaller then the size of a yellow circle, red circles are arranged circular around yellow circles'.

In their experiments [18] discovered, among others, that the visual system builds an image from very simple stimuli into more complex representations. This inspired the neural network community to see their so-called "deep learning" models as a cascading model of cell types, which follows always similar simple rules: at first lines are learned, then shapes, then objects are formed, eventually leading to **concept representations**.

By use of back-propagation such a model is able to discover intricate structures in large data sets to indicate how the internal parameters should be adapted, which are used to compute the representation in each layer from the representation in the previous layer [26]. Building *concept representations* refers to the human ability to learn categories for objects and to recognize new instances of those categories. In machine learning, concept learning is defined as the inference of a Boolean-valued function from training examples of its inputs and outputs [31] in other words it is training an algorithm to distinguish between examples and non-examples (we call the latter counterfactuals).

Concept learning has been a relevant research area in machine learning for a long time and had it origins in cognitive science, defined as search for attributes which can be used to distinguish exemplars from non exemplars of various categories [5]. The ability to think in abstractions is one of the most powerful tools humans possess. Technically, humans order their experience into coherent categories by defining a given situation as a member of that collection of situations for which responses x, y, etc. are most likely appropriate. This classification is not a passive process and to understand how humans learn abstractions is essential not only to the understanding of human thought, but to building artificial intelligence machines [19].

In computer vision an important task is to find a likely interpretation W for an observed image I, where W includes information about the spatial location, the extent of objects, the boundaries etc. Let SW be a function associated with an interpretation W that encodes the spatial location and extent of a component of interest, where $SW_{(i,j)} = 1$ for each image location (i, j) that belongs to the component and 0 else-where. Given an image, obtaining an optimal or even likely interpretation W, or associated SW, can be difficult. For example, in edge detection previous work [8] asked what is the probability of a given location in a given image belonging to the component of interest.

[44] presented a model of concept learning that is both computationally grounded and able to fit to human behaviour. He argued that two apparently distinct modes of generalizing concepts – abstracting rules and computing similarity to exemplars – should both be seen as special cases of a more general *Bayesian learning framework*. Originally, Bayes (and more specific [25]) explained the specific workings of these two modes, i.e. which rules are abstracted, how similarity is measured, why generalization should appear in different situations. This analysis also suggests why the rules/similarity distinction, even if not computationally fundamental, may still be useful at the algorithmic level as part of a principled approximation to fully Bayesian learning.

Gestalt-Principles ("Gestalt" = German for shape) are a set of empirical laws describing how humans gain meaningful perceptions and make sense of chaotic stimuli of the real-world. As so-called Gestalt-cues they have been used in machine learning for a long time. Particularly, in learning classification models for segmentation, the task is to classify between "good" segmentations and "bad" segmentations and to use the Gestalt-cues as features (the priors) to train the learning model. Images segmented manually by humans are used as examples of "good" segmentations (ground truth), and "bad" segmentations are constructed by randomly matching a human segmentation to a different image [39]. Gestalt-principles [21] can be seen as rules, i.e. they discriminate competing segmentations only when everything else is equal, therefore we speak more generally as Gestalt-laws and one particular group of Gestalt-laws are the Gestalt-laws of grouping, called Prägnanz [49], which include the law of Proximity: objects that are close to one another appear to form groups, even if they are completely different, the Law of Similarity: similar objects are grouped together; or the law of Closure: objects can be perceived as such, even if they are incomplete or hidden by other objects.

Unfortunately, the currently best performing machine learning methods have a number of disadvantages, and one is of particular relevance: Neural networks ("deep learning") are difficult to interpret due to their complexity and are therefore considered as "black-box" models [16]. Image Classifiers operate on low-level features (e.g. lines, circles, etc.) rather than high-level concepts, and with domain concepts (e.g images with a storefront). This makes their inner workings difficult to interpret and understand. However, the "why" would often be much more useful than the simple classification result.

5   Conclusion

By comparing both the strengths of machine intelligence and human intelligence it is possible to solve problems where we are currently lacking appropriate methods. One grand general question is "How can we perform a task by exploiting knowledge extracted during solving previous tasks?" To answer such questions it is necessary to get insight into human behavior, but not with the goal of mimicking human behavior, rather to contrast human learning methods to machine learning methods. We hope that our Kandinsky Patterns challenge the international machine learning community and we are looking forward to receiving many comments and results. Updated information can be found at the accompanying Web page[3]. A single Kandinsky pattern may serve as an "intelligence (IQ) test" for an AI system. To make the step towards a more human-like and probably in-depth assessment of an AI system, we propose to apply the principles of human intelligence tests, as outlined in this paper. In relation to the Kandinsky patterns we suggest applying the principle of Raven's progressive matrices. This test is strongly related to the identification of a "meaning" in the complex visual patterns [38]. The underlying complex pattern, however, is not based on a single image, the meaning only arises from the sequential combination of multiple images. To assess AI, a set of Kandinsky patterns, each of which complex in itself, can be used. A "real" intelligent achievement would be identifying the concepts - and therefore the meaning ! - of sequences of multiple Kandinsky patterns. At the same time, the approach solves one key problem of testing "strong AI", the language component. With this approach it is not necessary to verbalize the insights of the AI system. Per definition, the identification of the right

---

[3] https://human-centered.ai/kandinksy-challenge.

visual pattern that "traverses" the Kandinsky patterns (analogous to Raven's matrices) indicates the identification of an underlying meaning. Much further experimental and theoretical work is needed here.

Acknowledgements. We are grateful for interesting discussions with our local and international colleagues and their encouragement. Parts of this project have been funded by the EU projects FeatureCloud, EOSC-Life, EJP-RD and the Austrian FWF Project "explainable AI", Grant No. P-32554.

References

1. von Ahn, L., Blum, M., Hopper, N.J., Langford, J.: CAPTCHA: using hard AI problems for security. In: Biham, E. (ed.) EUROCRYPT 2003. LNCS, vol. 2656, pp. 294–311. Springer, Heidelberg (2003). https://doi.org/10.1007/3-540-39200-9_18
2. Amthauer, R.: Intelligenz-Struktur-Test 2000 R: I-S-T 2000 R Manual, 2nd edn. Hogrefe, Göttingen (2001)
3. Belk, M., Germanakos, P., Fidas, C., Holzinger, A., Samaras, G.: Towards the personalization of CAPTCHA mechanisms based on individual differences in cognitive processing. In: Holzinger, A., Ziefle, M., Hitz, M., Debevc, M. (eds.) SouthCHI 2013. LNCS, vol. 7946, pp. 409–426. Springer, Heidelberg (2013). https://doi.org/10.1007/978-3-642-39062-3_26
4. Binet, A.: L'étude expérimentale de l'intelligence. Schleicher fréres & cie, Paris (1903)
5. Bruner, J.S.: Chapter 2: on attributes and concepts. In: Bruner, J.S., Goodnow, J.J., Austin, G.A. (eds.) A Study of Thinking, pp. 25–49. Wiley, Hoboken (1956)
6. Campbell, M., Hoane Jr., A.J., Hsu, F.: Deep blue. Artif. Intell. **134**(1–2), 57–83 (2002). https://doi.org/10.1016/S0004-3702(01)00129-1
7. Conway, A., Kovacs, K.: New and emerging models of human intelligence. Cogn. Sci. (2015). https://doi.org/10.1002/wcs.1356
8. Dollar, P., Tu, Z., Belongie, S.: Supervised learning of edges and object boundaries. In: IEEE Computer Society Conference on Computer Vision and Pattern Recognition (CVPR 2006), pp. 1964–1971. IEEE (2006). https://doi.org/10.1109/CVPR.2006.298
9. Dowe, D.L., Hernández-Orallo, J.: IQ tests are not for machines, yet. Intelligence **40**(2), 77–81 (2012). https://doi.org/10.1016/j.intell.2011.12.001
10. Ferrucci, D., Levas, A., Bagchi, S., Gondek, D., Mueller, E.T.: Watson: beyond jeopardy!. Artif. Intell. **199**, 93–105 (2013). https://doi.org/10.1016/j.artint.2012.06.009
11. Funke, J.: Handbook of Anthropometry. Springer, Heidelberg (2006)
12. Gardner, H.: Changing Minds: The Art and Science of Changing Our Own and Other People's Minds. Harvard Business School Press, Boston (2004)
13. Holzinger, A., Biemann, C., Pattichis, C.S., Kell, D.B.: What do we need to build explainable AI systems for the medical domain? arXiv:1712.09923 (2017)
14. Holzinger, A., Kickmeier-Rust, M.D., Müller, H.: Human explanation profiles for random visual patterns as a benchmark for explainable AI. In: Preparation (2019)
15. Holzinger, A., Langs, G., Denk, H., Zatloukal, K., Müller, H.: Causability and explainability of AI in medicine. Wiley Interdisc. Rev.: Data Min. Knowl. Discov. (2019). https://doi.org/10.1002/widm.1312

16. Holzinger, A., Plass, M., Holzinger, K., Crisan, G.C., Pintea, C.M., Palade, V.: A glass-box interactive machine learning approach for solving NP-hard problems with the human-in-the-loop. arXiv:1708.01104 (2017)
17. Horn, J.L., Cattell, R.B.: Refinement and test of the theory of fluid and crystallized general intelligences. J. Educ. Psychol. 57, 253–270 (1966)
18. Hubel, D.H., Wiesel, T.N.: Receptive fields, binocular interaction and functional architecture in the cat's visual cortex. J. Physiol. 160(1), 106–154 (1962). https://doi.org/10.1113/jphysiol.1962.sp006837
19. Hunt, E.B.: Concept Learning: An Information Processing Problem. Wiley, Hoboken (1962). https://doi.org/10.1037/13135-001
20. Jäger, A.O.: Validität von intelligenztests. Diagnostica 32, 272–289 (1986)
21. Koffka, K.: Principles of Gestalt Psychology. Harcourt, New York (1935)
22. Kovacs, K., Conway, A.: Process overlap theory: a unified account of human intelligence. Psychol. Inquiry 27, 151–177 (2016)
23. Kundel, H.I., Nodine, C.F.: A visual concept shapes image perception. Radiology 146(2), 363–368 (1983)
24. Lake, B.M., Salakhutdinov, R., Tenenbaum, J.B.: Human-level concept learning through probabilistic program induction. Science 350(6266), 1332–1338 (2015). https://doi.org/10.1126/science.aab3050
25. Laplace, P.S.: Mémoire sur les probabilités. Mémoires de l'Académie Royale des sciences de Paris 1778, 227–332 (1781)
26. LeCun, Y., Bengio, Y., Hinton, G.: Deep learning. Nature 521(7553), 436–444 (2015)
27. Leeuwenberg, E.L., Boselie, F.: Against the likelihood principle in visual form perception. Psychol. Rev. 95(4), 485–491 (1988)
28. Legg, S., Hutter, M.: Universal intelligence: a definition of machine intelligence. Minds Mach. 17(4), 391–444 (2007). https://doi.org/10.1007/s11023-007-9079-x
29. Li, L., et al.: Parallel testing of vehicle intelligence via virtual-real interaction. Sci. Robot. 4(eaaw4106), 1–3 (2019)
30. Locke, E.A.: Why emotional intelligence is an invalid concept. J. Organ. Behav. 26, 425–431 (2005)
31. Mitchell, T.M.: Machine Learning. McGraw Hill, New York (1997)
32. Müller, H., Holzinger, A.: Kandinsky patterns. arXiv:1906.00657 (2019). https://arxiv.org/abs/1906.00657
33. Nambiar, R.: Towards an industry standard for benchmarking artificial intelligence systems. In: 34th International Conference on Data Engineering (ICDE 2018). IEEE (2018). https://doi.org/10.1109/ICDE.2018.00212
34. Nambiar, R., Ghandeharizadeh, S., Little, G., Boden, C., Dholakia, A.: Industry panel on defining industry standards for benchmarking artificial intelligence. In: Nambiar, R., Poess, M. (eds.) TPCTC 2018. LNCS, vol. 11135, pp. 1–6. Springer, Cham (2019). https://doi.org/10.1007/978-3-030-11404-6_1
35. Pirolli, P., Russell, D.M.: Introduction to this special issue on sensemaking. Hum.-Comput. Interact. 26, 1–8 (2011)
36. Preed, V.R.: Handbook of Anthropometry. Springer, Heidelberg (2012). https://doi.org/10.1007/978-1-4419-1788-1
37. Raven, J.: The raven's progressive matrices: change and stability over culture and time. Cogn. Psychol. 41(1), 1–48 (2000). https://doi.org/10.1006/cogp.1999.0735
38. Raven, J.C., Court, J.H.: Raven's Progressive Matrices and Vocabulary Scales. Oxford Psychologists Press, Oxford (1998)

39. Ren, X., Malik, J.: Learning a classification model for segmentation. In: Ninth IEEE International Conference on Computer Vision (ICCV), pp. 10–17. IEEE (2003). https://doi.org/10.1109/ICCV.2003.1238308
40. Samek, W., Wiegand, T., Müller, K.R.: Explainable artificial intelligence: understanding, visualizing and interpreting deep learning models. arXiv:1708.08296 (2017)
41. Santoro, A., Hill, F., Barrett, D., Morcos, A., Lillicrap, T.: Measuring abstract reasoning in neural networks. In: 35th International Conference on Machine Learning, pp. 4477–4486. PMLR (2018)
42. Silver, D., et al.: Mastering the game of go with deep neural networks and tree search. Nature **529**(7587), 484–489 (2016). https://doi.org/10.1038/nature16961
43. Spearman, C.: "General intelligence", objectively determined and measured. Am. J. Psychol. **15**(2), 201–292 (1904)
44. Tenenbaum, J.B.: Bayesian modeling of human concept learning. In: Solla, S.A., Leen, T.K., Müller, K.R. (eds.) Advances in Neural Information Processing Systems (NIPS 1999), pp. 59–68. NIPS Foundation (1999)
45. Turing, A.M.: Computing machinery and intelligence. Mind **59**(236), 433–460 (1950). https://doi.org/10.1093/mind/LIX.236.433
46. Urbina, S.: Chapter 2: tests of intelligence. In: Kaufman, S.B. (ed.) The Cambridge Handbook of Intelligence. Cambridge University Press, Cambridge (2011)
47. Uusitalo, L., Simola, J., Kuisma, J.: Perception of abstract and representative visual art. In: TBA, January 2009
48. Wechsler, D.: The Measurement and Appraisal of Adult Intelligence, 4th edn. Williams and Witkins, Baltimore (1958)
49. Wertheimer, M.: Laws of organization in perceptual forms. In: Ellis, W.D. (ed.) A Source Book of Gestalt Psychology, pp. 71–88. Paul Kegan, London (1938). https://doi.org/10.1037/11496-005
50. Yanagisawa, H.: How does expectation affect sensory experience? A theory of relativity in perception. In: 5th International Symposium on Affective Science and Engineering, ISASE 2019, pp. 1–4. Japan Society of Kansei Engineering (2019). https://doi.org/10.5057/isase.2019-C000014

Machine Learning Explainability Through Comprehensible Decision Trees

Alberto Blanco-Justicia and Josep Domingo-Ferrer(✉)

Department of Computer Science and Mathematics,
CYBERCAT-Center for Cybersecurity Research of Catalonia,
UNESCO Chair in Data Privacy, Universitat Rovira i Virgili, Av. Països Catalans 26,
43007 Tarragona, Catalonia, Spain
{alberto.blanco,josep.domingo}@urv.cat

Abstract. The role of decisions made by machine learning algorithms in our lives is ever increasing. In reaction to this phenomenon, the European General Data Protection Regulation establishes that citizens have the right to receive an explanation on automated decisions affecting them. For explainability to be scalable, it should be possible to derive explanations in an automated way. A common approach is to use simpler, more intuitive decision algorithms to build a surrogate model of the black-box model (for example a deep learning algorithm) used to make a decision. Yet, there is a risk that the surrogate model is too large for it to be really comprehensible to humans. We focus on explaining black-box models by using decision trees of *limited size* as a surrogate model. Specifically, we propose an approach based on microaggregation to achieve a trade-off between *comprehensibility* and *representativeness* of the surrogate model on the one side and *privacy* of the subjects used for training the black-box model on the other side.

Keywords: Explainability · Machine learning · Data protection · Microaggregation · Privacy

1 Introduction

Since the turn of the century, big data are a reality. One of the main uses of this wealth of data is to train machine learning algorithms. Once trained, these algorithms make decisions, and a good number of decisions affect people: credit granting, insurance premiums, diagnosis, etc. While transparency measures are being implemented by public administrations worldwide, there is a risk of automated decisions becoming an omnipresent black box. This could result in formally transparent democracies operating in practice as computerized totalitarian societies.

To protect citizens, explainability requirements are starting to appear in legal regulations and ethics guidelines. For example, article 22 of the EU General Data Protection Regulation (GDPR, [6]) states the right of citizens to an explanation

© IFIP International Federation for Information Processing 2019
Published by Springer Nature Switzerland AG 2019
A. Holzinger et al. (Eds.): CD-MAKE 2019, LNCS 11713, pp. 15–26, 2019.
https://doi.org/10.1007/978-3-030-29726-8_2

on automated decisions. Also, the European Commission's Ethics Guidelines for Trustworthy AI [5] urge organizations making automated decisions to be ready to explain them on request of the affected citizens, whom we will call also subjects in what follows.

To be scalable, explanations must be automatically generated: even if a human auditor was able to produce a compelling explanation, one cannot assume that such an auditor will be available to explain every automated decision to the affected subject. Older machine learning models, based on rules, decision trees or linear models, are understandable by humans and are thus self-explanatory, as long as they are not very large (*i.e.* as long as the number of rules, the size of the decision trees or the number of explanatory attributes stay small). However, the appearance of deep learning has worsened matters: it is much easier to program an artificial neural network and train it than to understand why it yields a certain output for a certain input.

Contribution and Plan of this Paper

A usual strategy to generate explanations for decisions made by a black-box machine learning model, such as a deep learning model, is to build a surrogate model based on more expressive machine learning algorithms, such as the aforementioned decision rules [10,14], decision trees [1,12], or linear models [13]. The surrogate model is trained on the same data set as the black-box model to be explained or on new data points classified by that same model. Global surrogate models explain decisions on points in the whole domain, while local surrogate models build explanations that are relevant for a single point or a small region of the domain.

We present an approach that assumes that the party generating the explanations has unrestricted access to the black-box model and to the training data set. We will take as surrogate models decision trees trained on disjoint subsets of the training data set. We focus on the comprehensibility of the models which we measure as the inverse of the number of nodes of the trained decision trees. In general, the fewer the nodes of a decision tree, the easier it is to comprehend it.

Section 2 characterizes the type of explanations we seek to generate and the risks of generating them through straighforward release of surrogate models. Section 3 describes our microaggregation-based approach to generate explanations of limited size. Experimental results are provided in Sect. 4. Finally, Sect. 5 gathers conclusions and future research directions.

2 Explanations via Surrogate Models

2.1 Machine Learning Explanations

According to [9], an explanation for a black-box machine learning model should take into account the following properties:

Accuracy. This property refers to how well an explanation predicts unseen data. Low explanation accuracy can be fine only if the black-box model to be explained is also inaccurate.

Fidelity. The explanations ought to be close to the predictions of the explained model. Accuracy and fidelity are very related: if the black-box model is very accurate and the explanation has high fidelity, then the explanation has also high accuracy.

Consistency. Explanations should apply equally well to any model trained on the same data set.

Stability. When providing explanations to particular instances, similar instances should produce similar explanations.

Representativeness. A highly representative explanation is one that can be applied to several decisions on several instances.

Certainty. If the model at study provides a measure of confidence on its decisions, an explanation of this decision should reflect this.

Novelty. This property refers to the capability of the explanation mechanism to cover instances far from the training domain.

Degree of Importance. The explanation should pinpoint the important features.

Comprehensibility. Explanations should be understandable to humans. This depends on the target audience and has psychological and social implications, although short explanations generally go a long way towards comprehensibility.

Miller analyzes explainability from the social sciences perspective [8] and makes four important observations: (i) people prefer *contrastive* explanations, *i.e.* why the algorithm took a certain decision does not matter as much to us as why did it not take a different decision instead; (ii) people *select* only a few causes from the many causes that make up an explanation, and personal biases guide this selection; (iii) referring to probabilities or statistical connections is not as effective as referring to causes; and (iv) explanations are *social*, and thus should be part of a wider conversation, or an interaction between the explainer and the explainee.

In [7], the authors emphasize the importance of human field experts guiding the development of explanation mechanisms, given that current machine learning systems work on a statistical and/or model-free mode, and require context from human/scientific models to convey convincing explanations (especially for other field experts).

No single explanation model in the current literature is able to satisfy all the above properties (refer to [2,9] for extensive surveys on explainable artificial intelligence techniques). In what follows we will focus on accuracy, fidelity, stability, representativeness and comprehensibility, to which we will add privacy. See Sect. 2.2 about the privacy risks of explanations.

2.2   Risks of Surrogate Model Release

A common strategy to provide explanations satisfying the above properties is via a surrogate model based on intrinsically interpretable algorithms. However, care must be exercised to ensure that the surrogate model does not violate trade secret, privacy and explainability.

Trade Secret Risks. A very detailed surrogate model may reveal properties of the data set that was used to train the black-box model. This may be in conflict with *trade secret.* Indeed, training data are often the result of long-term corporate experience and reflect successes and failures. It takes time to accumulate good training data. Thus, organizations owning such data regard them as a valuable asset they do not want disclosed to competitors.

At the same time, too much detail in the released surrogate model may reveal more about the black-box model to be explained than its owner is willing to disclose. Traning complex models, like for example deep models, requires a costly process involving time and computing power. Hence, a well-trained black-box model is also a highly valued asset that organizations view as a trade secret.

Privacy Risks. If the released surrogate model leaks information on the training data and these contain personally-identifiable information, then we have a conflict with privacy legislation [6].

Comprehensibility Risks. A complex surrogate model, even if based on intrinsically interpretable algorithms, may fail to be comprehensible to humans. We illustrate this risk in Figs. 1 and 2.

Figure 1 shows a simple data set with two continuous attributes, represented by the two dimensions of the graph, and a binary class attribute, represented by the color of points in the graph. Thus, points represent the records in the data set. Figure 2 shows a surrogate model consisting of a decision tree trained on the example data set. With 303 nodes, this model is not very useful as an explanation to humans: it is very hard to comprehend it.

3   Microaggregation-Based Surrogate Models

To avert the risks identified in Sect. 2.2 while achieving as many of the properties listed in Sect. 2.1 as possible, we need a method to construct surrogate models that keep at bay leakage and complexity. To that end, we propose to provide subjects with partial or local explanations, that cover an area of the original training data set close to the subject (that is, attribute values similar to the subject's). Algorithm 1 describes a procedure for the owner of the training data and the black-box model to generate cluster-based explanations. Then, Protocol 1 shows how a subject obtains an explanation close to her. The fact that explanations are cluster-based favors *stability*: all instances in the cluster are similar and they are explained by the same interpretable model, so explanations can be expected to be similar.

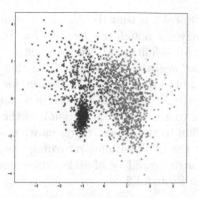

Fig. 1. Example data set

Fig. 2. Decision tree trained on the example data set

Algorithm 1: Generation of cluster-based explanations

Input: Training data set $\mathbf{X}$

1 Compute a clustering $C(\mathbf{X})$ for $\mathbf{X}$ based on all attributes except the class attribute

2 **for** *each cluster $C_i \in C(\mathbf{X})$* **do**

3 | Compute a representative, *e.g.* the centroid or average record $\tilde{c}_i$

4 **end**

5 **for** *each cluster $C_i \in C(\mathbf{X})$* **do**

6 | Train an interpretable model, such as a decision tree DT_i

7 **end**

Protocol 1 (Provision of explanations)

1. *A subject submits a query $\hat{x}$ to the black-box model.*
2. *The black-box model returns to the subject:*
   (a) *A decision $d = f(\hat{x})$;*
   (b) *The closest representative $\tilde{c}_x = \arg\min_{\tilde{c}_i} \operatorname{dist}(\tilde{c}_i, \hat{x})$ for some distance* dist*;*
   (c) *The interpretable model DT_x corresponding to the cluster represented by $\tilde{c}_x$.*

A shortcoming of Protocol 1 is that the decision output by the interpretable model DT_x on input $\hat{x}$ may not match the decision $d = f(\hat{x})$ made by the black-box model. This is bad for fidelity and can be fixed by returning the closest representative to $\hat{x}$ *whose decision tree yields* d. In this way, the explanation provision is *guided* by the black-box model. The search for a valid representative is restricted by a parameter N: if none of the decision trees associated with the N closest representatives to $\hat{x}$ matches the decision of the black-box model, the decision tree corresponding to the closest representative is returned. While this may hurt the fidelity of the explanations, returning the tree of an arbitrarily distant cluster representative would be of little explanatory power. The guided provision is formalized in Protocol 2.

Protocol 2 (Guided provision of explanations)

1. *A subject submits a query $\hat{x}$ to the black-box model.*
2. *The black-box model owner does:*
   (a) *Compute the decision $d = f(\hat{x})$ using the black-box model;*
   (b) *let U be the list of cluster representatives $\tilde{c}_i$ ordered by their distance to $\hat{x}$, being $\tilde{c}_1$ the closest representative;*
   (c) **let** $i = 1$;
   (d) **let** *found* $= 0$;
   (e) **repeat**
       i. *let DT_i be the interpretable model corresponding to the cluster represented by $\tilde{c}_i$;*
       ii. **if** $DT_i(\hat{x}) = d$ **then** *found=1* **else** $i = i + 1$;
       until *found* $= 1$ **or** $i > N$;
   (f) **if** *found* $= 1$ **then return** d, $\tilde{c}_i$ and DT_i
       else return d, $\tilde{c}_1$ and DT_1

We choose microaggregation [3,4] as the type of clustering in Algorithm 1, because it allows enforcing that clusters consist of at least a minimum number k of records. This minimum cardinality allows trading off *privacy* and *representativeness* for *comprehensibility* of explanations:

- Parameter k ensures that returning the representative $\tilde{c}_x$ in Protocol 1 is compatible with k-anonymity [4,11] for the subjects in the training data set. Indeed, the representative equally represents k subjects in the training data set. In this respect, the larger k, the more privacy.
- Additionally, large values of k result in clusters that contain larger parts of the domain, thus yielding explanations with higher representativeness.
- While choosing large values for k has a positive effect on privacy and representativeness, it does so at the expense of comprehensibility. A small k results in very local explanations, that have the advantage of consisting of simpler and thus more comprehensible surrogate models.

Specifically, we compute microaggregation clusters using MDAV (Mean Distance to Average Vector), a well-known microaggregation heuristic [4]. We recall it in Algorithm 2.

Algorithm 2: MDAV

Input: $\mathbf{X}$, k
Output: $\mathbf{C}$: set of clusters

1 $\mathbf{C} \leftarrow \emptyset$
2 **while** $|\mathbf{X}| \geq 3k$ **do**
3 $\quad$ $x_c \leftarrow$ **mean_record**($\mathbf{X}$)
4 $\quad$ $x_r \leftarrow \text{argmax}_{x_i}$ **distance**(x_i, x_c)
5 $\quad$ $x_s \leftarrow \text{argmax}_{x_i}$ **distance**(x_i, x_r)
6 $\quad$ $C_r \leftarrow$ **cluster**($x_r, k, \mathbf{X}$) // Algorithm 3
7 $\quad$ $C_s \leftarrow$ **cluster**($x_s, k, \mathbf{X}$)
8 $\quad$ $\mathbf{C} \leftarrow \mathbf{C} \cup \{C_r, C_s\}$
9 $\quad$ $\mathbf{X} \leftarrow \mathbf{X} \setminus C_r \setminus C_s$
10 **end**
11 **if** $2k \leq |\mathbf{X}| < 3k$ **then**
12 $\quad$ $x_c \leftarrow$ **mean_record**($\mathbf{X}$)
13 $\quad$ $x_r \leftarrow \text{argmax}_{x_i}$ **distance**(x_i, x_c)
14 $\quad$ $C_r \leftarrow$ **cluster**($x_r, k, \mathbf{X}$)
15 $\quad$ $\mathbf{C} \leftarrow \mathbf{C} \cup \{C_r\}$
16 $\quad$ $\mathbf{X} \leftarrow \mathbf{X} \setminus C_r$
17 **else**
18 $\quad$ $\mathbf{C} \leftarrow \mathbf{C} \cup \{\mathbf{X}\}$
19 **end**
20 **return** $\mathbf{C}$

Figure 3 depicts the representatives (centroids) of clusters computed by MDAV with $k = 200$ on the example data set of Fig. 1. The figure also shows the decision trees that are obtained as explanations for three of the clusters.

Algorithm 3: cluster

Input: x, k, $\mathbf{X}$
Output: C: cluster

1 $C \leftarrow \{x\}$
2 **while** $|C| < k$ **do**
3 $\quad$ $x_i \leftarrow \text{argmin}_{x_i}$ **distance**(x_i, x)
4 $\quad$ $C \leftarrow C \cup \{x_i\}$
5 $\quad$ $\mathbf{X} \leftarrow \mathbf{X} \setminus \{x_i\}$
6 **end**
7 **return** C

---

4   Empirical Work

We generated a data set consisting of 30,000 records, each with 10 numeric continuous attributes and a single binary class label using the `make_classification` method from Scikit-learn[1]. Out of the 30,000 records, we reserved 2/3 to train the models, and the remaining 1/3 to validate them. The code to generate the data set and conduct all the experiments reported in this section is available as a Jupyter notebook[2].

We took as a black-box model a neural network denoted by ANN with three hidden layers of 100 neurons each, which achieves 94.22% classification accuracy. We also trained a decision tree on the whole data set, to check its accuracy and its number of nodes. The classification accuracy of this global decision tree is 88.37%, and it has 2,935 nodes. We expected our local decision trees (trained on a single cluster) to achieve a similar accuracy on average, although it could happen that clusters containing points from a single class would produce more accurate classifiers.

Then, we tested our cluster-based mechanism for different values of k. As stated in Sect. 3, smaller values of k could be expected to produce *simpler* classifiers. Instead of directly choosing arbitrary values for k, we chose several percentages of the 20,000 records of the training data set that we wanted the clusters to contain, ranging from 0.1% to 50%; this translated to k values ranging from 20 to 10,000.

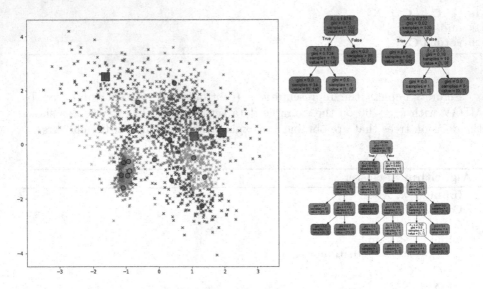

Fig. 3. Left, clusters produced by MDAV with $k = 200$ for the data set of Fig. 1; for each cluster, points are in a different color and the centroid is depicted. Right, decision tree-based explanations generated for the three clusters whose centroids have been marked with □ symbols on the left figure. (Color figure online)

---

[1] https://scikit-learn.org/stable/index.html.
[2] Download   address: https://www.dropbox.com/s/ex46twifl780fj4/MDAV-DT-Explainability.ipynb?dl=0.

The experiment was as follows. For each value of k, we used MDAV to obtain a clustering of the training data set. Then we computed the centroid representatives of clusters, and we trained a decision tree for each cluster. After that, we measured the classification accuracy and the fidelity of the explanations. Classification accuracy was computed in the usual manner, with the ground truth being the labels in the evaluation data set (1/3 of the original data set, that is, the 10,000 records not used for training). Fidelity was computed as the classification accuracy with respect to the decisions made by the black-box model.

Figure 4a shows the **accuracy** of our local explanations, which for all values of k is lower than the accuracy of the black-box model ANN by around 5% in the unguided approach (Protocol 1) and by only around 2% in the guided approach (Protocol 2, with $N = 3$). On the other hand, the accuracy of the unguided approach of Protocol 1 was basically the same as that of the global decision tree mentioned above, with the guided approach of Protocol 2 clearly outperforming both.

Moreover, it is important to note that accuracy is not very affected by the value of k, although very small values of k seem to produce slightly better results. This same behavior has been observed for several different generated data sets, so it cannot be attributed to randomness. In fact, even our unguided approach outperforms the global decision tree by around 5% in classification accuracy when trained on very small clusters (0.1% to 1%, or $k = 20$ to $k = 200$ for our data set size). The most plausible reason for this phenomenon is that for these small values of k, a substantial number of clusters are such that all records in the cluster have the same class attribute value. For these clusters, the decision tree is trivial. This hypothesis is further supported by Fig. 5, discussed in more detail below, where for small values of k we find decision trees containing 0 nodes: these must correspond to clusters whose records all belong to the same class. Whether this is beneficial from the point of view of explainability is to be further explored.

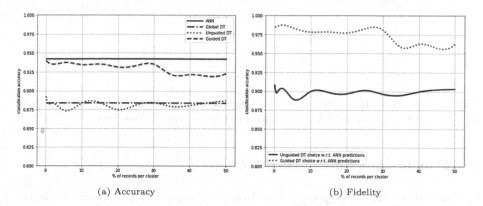

(a) Accuracy   (b) Fidelity

Fig. 4. Accuracy and fidelity of the decision trees for each value of k. For the guided approach $N = 3$ was used.

Figure 4b, on the other hand, depicts the **fidelity** of our explanations with respect to the black-box model. For Protocol 1 (unguided approach) the explanations coincide with the black-box model for 90% of the decisions. When using Protocol 2 (guided by the ANN with $N = 3$), these results improve to up to 97% coincidence, which demonstrates that our method achieves a high accuracy and fidelity with respect to the black-box model.

Figure 5 deals with the **comprehensibility** of the explanations, by depicting the number of nodes of the decision trees trained for each choice of k: since there is one decision tree per cluster, the box plot represents for each k the median and the upper and lower quartiles of the number of nodes per decision tree. We can see that for small k (up to 1% of the training set, in our case $k = 200$), the number of nodes per decision tree is well below 100. We argue that decision trees with 100 or more nodes are not very useful as explanations of a decision. Since according to Fig. 4a k does not significantly affect accuracy, *one should take the smallest k that is deemed sufficient for* **privacy** (explanations are best understood if trees have no more than 10 or 20 nodes).

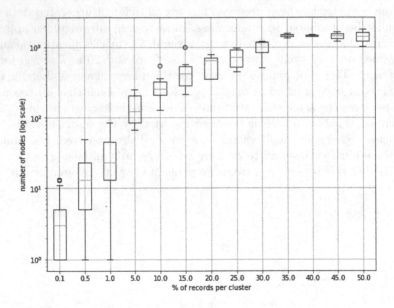

Fig. 5. Comprehensibility of explanations: the box plot represents for each k the median and lower and upper quartiles of the number of nodes per decision tree.

5   Conclusions and Future Research

We have presented an approach based on microaggregation that allows deriving explanations of machine learning decisions while controlling their accuracy, fidelity, representativeness, comprehensibility and privacy preservation. In addition, being based on clusters our explanations offer stability by design.

Future research will involve trying different distances in Protocols 1 and 2 and also in the microaggregation algorithm, in order to improve the trade-off between the above properties. Options to be explored include various semantic distances.

On the other hand, in this paper we have assumed that explanations are generated by the owner of the black-box model and the training data set. It is worth investigating the case in which a third party or even the subjects themselves generate the explanations. In this situation the black-box model owner may limit access to his model to protect his trade secrets. We will explore ways to generate microaggregation-based explanations that are compatible with such access restrictions. Possible strategies include cooperation between subjects and/or smart contracts between the generator of explanations and the owner of the black-box model.

Acknowledgments and Disclaimer. The following funding sources are gratefully acknowledged: European Commission (project H2020-700540 "CANVAS"), Government of Catalonia (ICREA Acadèmia Prize to J. Domingo-Ferrer and grant 2017 SGR 705) and Spanish Government (project RTI2018-095094-B-C21 "CONSENT"). The views in this paper are the authors' own and do not necessarily reflect the views of UNESCO or any of the funders.

References

1. Alonso, J.M., Ramos-Soto, A., Castiello, C., Mencar, C.: Hybrid data-expert explainable AI beer style classifier. In: IJCAI-18 Workshop on Explainable Artificial Intelligence (XAI 2018) (2018)
2. Biran, O., Cotton, C.: Explanation and justification in machine learning: a survey. In: IJCAI-17 Workshop on Explainable Artificial Intelligence (XAI 2017) (2017)
3. Domingo-Ferrer, J., Mateo-Sanz, J.M.: Practical data-oriented microaggregation for statistical disclosure control. IEEE Trans. Knowl. Data Eng. **14**(1), 189–201 (2002)
4. Domingo-Ferrer, J., Torra, V.: Ordinal, continuous and heterogeneous k-anonymity through microaggregation. Data Min. Knowl. Discov. **11**(2), 195–212 (2005)
5. European Comission's High-Level Expert Group on Artificial Intelligence: Draft Ethics Guidelines for Trustworthy AI (2018)
6. European Union: General Data Protection Regulation. Regulation (EU) 2016/679 (2016). https://gdpr-info.eu
7. Holzinger, A., Langs, G., Denk, H., Zatloukal, K., Müller, H.: Causability and explainabilty of artificial intelligence in medicine. In: Wiley Interdisciplinary Reviews: Data Mining and Knowledge Discovery, p. e1312 (2019)
8. Miller, T.: Explanation in artificial intelligence: insights from the social sciences. Artif. Intell. **267**, 1–38 (2019)
9. Molnar, C.: Interpretable machine learning: a guide for making black box models explainable. Leanpub (2018). https://christophm.github.io/interpretable-ml-book/
10. Ribeiro, M. T., Singh, S., Guestrin, C.: Anchors: high-precision model-agnostic explanations. In: 32nd AAAI Conference on Artificial Intelligence-AAAI 2018, pp. 1527–1535. AAAI (2018)

11. Samarati, P., Sweeney, L.: Protecting privacy when disclosing information: k-anonymity and its enforcement through generalization and suppression. SRI International Report (1998)
12. Singh, S., Ribeiro, M.T., Guestrin, C.: Programs as black-box explanations. arXiv preprint arXiv:1611.07579 (2016)
13. Strumbelj, E., Kononenko, I.: An efficient explanation of individual classifications using game theory. J. Mach. Learn. Res. **11**, 1–18 (2010)
14. Turner, R.: A model explanation system. In: IEEE International Workshop on Machine Learning for Signal Processing, MLSP 2016. IEEE (2016)

New Frontiers in Explainable AI: Understanding the GI to Interpret the GO

Federico Cabitza[1]($\boxtimes$), Andrea Campagner[1,2], and Davide Ciucci[1]

[1] Università degli Studi di Milano-Bicocca, Milan, Italy
federico.cabitza@unimib.it
[2] IRCCS Istituto Ortopedico Galeazzi, Milan, Italy

Abstract. In this paper we focus on the importance of interpreting the quality of the input of predictive models (potentially a GI, i.e., a Garbage In) to make sense of the reliability of their output (potentially a GO, a Garbage Out) in support of human decision making, especially in critical domains, like medicine. To this aim, we propose a framework where we distinguish between the Gold Standard (or Ground Truth) and the set of annotations from which this is derived, and a set of quality dimensions that help to assess and interpret the AI advice: fineness, trueness, representativeness, conformity, dryness. We then discuss implications for obtaining more informative training sets and for the design of more usable Decision Support Systems.

Keywords: Ground truth · Explainable AI · Reliability · Usable AI

1 Introduction

In the specialist literature around the topics of Fairness, Accountability, and Transparency in Machine Learning (FAT-ML), many approaches to make *AI explainable* (XAI) are proposed and discussed. A XAI system can be *intrinsically interpretable*, when it adopts a model whose internal functioning is immediately accessible to the decision maker, like in the case of linear or rule-based models (e.g., decision trees); or it can be made interpretable by focusing on two aspects: the model itself; or its output on one or more given cases. The former case of interpretability (also called *understandability* or *intelligibility*) regards "how the model works": this kind of model interpretability is pursued by providing the decision makers, i.e., the users of XAI systems, with indications about how the model produced a certain prediction, e.g., by plotting the loss function, or by visualizing the boundary region on a PCA-reduced space, or by telling what feature the model based more on to produce its prediction, as represented by feature relevance scores or saliency maps. In the latter case, when authors speak also of *post-hoc interpretations*, the focus is on output data, and the aim is "to explain the predictions without elucidating the mechanisms by which models

© IFIP International Federation for Information Processing 2019
Published by Springer Nature Switzerland AG 2019
A. Holzinger et al. (Eds.): CD-MAKE 2019, LNCS 11713, pp. 27–47, 2019.
https://doi.org/10.1007/978-3-030-29726-8_3

work" [31]. In this case, decision makers can be given counterfactual outputs (that is alternative outputs if the input case were different) or the rules or functional relationships that locally apply for the output of surrogate (and more interpretable, in the sense mentioned above) models. These models are intended to locally "simulate" the black-box model "at the terminals", and explain the original relationship between the prediction and the input instance more intuitively. This approach is also the basis for the only proposal, to our knowledge, to make the concept of interpretability fully formalized [29].

In this paper we want discuss a third, and still neglected, general approach: instead of focusing on either the model or its outputs, we aim to discuss *input explainability*, that is on ways to have the decision makers to get an idea of how much they should trust the single output prediction on the basis of the "quality" of the *ground truth* on which the model has been trained, that is on the basis of the input of the learning process that yielded the model.

2   First Things First: The Importance of Input

Ground Truth, or Gold Standard (as the reference data are commonly called in medicine, our reference domain), is assumed to be *true*, by definition: the ML model is then supposed to "learn" from it the hidden patterns actually lying in the complex and manifold relationships between the phenomenon's predictors (variables that express the phenomenon symbolically) and the target variable (seen as a sort of interpretation or further measure of the phenomenon). However, any data is but an approximation of reality, a mere representation of it: as obvious as it sounds, maps are not the territory, likewise, also our "truths" are more "map truth" rather than ground truths. However, scholars in the Machine Learning and AI communities seldom address the question of *how good their ground truth actually is*, that is how much "golden" their Gold Standard is (or, to adopt the jewellery jargon, what its *fineness* is).

Most works that compare machine and human performance in delicate tasks, like diagnostic ones in medical practice, assume ground truth good enough to yield reliable results but, at the same time, understand that relying on the interpretation of a single source or interpreter would be over-optimistic, hence lead to too inaccurate performance. For this reason, Gold Standard sets are usually built by gathering a number of observers (or raters, annotators) and asking them to observe a phenomenon of interest (i.e., a unit of observation, or case), judge it, rate it and annotate the sets of data that describe it with a value from a scale of measurement, which can be either scalar, ordinal or nominal in nature. In this later case, the raters annotate the case with a code, class or category, which best describes the case. The ML model is then aimed at associating the one best class with any new case extracted from the same reference population.

The multiplicity of ratings at the origin on the Gold Standard does not result only from multi-rater settings, but also when there is the necessity to "sample" a complex phenomenon with multiple measurements. For instance, a Gold Standard could regard the outcome of a medical intervention as it is perceived 3 months after the intervention; this outcome could be represented in terms of PROs (i.e., Patient Reported Outcome Measures), by asking the patient to report how they feel on an ordinal scale a number of times in the week occurring approximately a dozen of weeks after the intervention, and then averaging these measures [6].

Both in multi-rater and in single-rater settings, it is seldom considered whether the Gold Standard built from a set of annotations is reliable or not, i.e., whether each case were described by a sufficient number of rating, or whether the raters involved were expert or adequately committed to the task. For instance (to limit ourselves to some of the most relevant works in medical AI), the authors of [14] report to have used Gold Standard diagnoses "based on expert opinion (including dermatologists and dermatopathologists)" from open-source repositories, where yet the details on the number and expertise of the raters involved are not available. Also the supplementary materials related to the work by Haenssle and colleagues [21] do not provide any detail on the number of dermatologists involved. The dataset used in [22] was annotated by just three dermatologists. The data set used in [37] for the task to detect tumor cells was annotated by non-specialists. One of the studies that has involved more raters to date, i.e., the study mentioned in [19], involved 54 raters, and these were all either US-licensed ophthalmologists or ophthalmology trainees; however, we do not know the proportions of trainees, and inter-rater reliability was assessed for less than a third of the sample, as only 16 raters had graded a sufficient volume of repeat images; furthermore, agreement proportions were not adjusted for chance effects. Although these are only anecdotal mentions, we argue that current debate on accuracy (and explainability) of AI focuses primarily on the technology (i.e., the model), and not on the underlying data, whose production and validation still lies in the background.

Notwithstanding this relative lack of transparency on the number and skills of the original annotators involved in ground truthing, in various ambits – and especially in medicine – the phenomenon of observer variability has been known (and studied) since decades [3]: this phenomenon regards how different observers, who are called to annotate data can simply differ and disagree with each others. This observer variability affects the reliability of the resulting data set, what we call Diamond Standard (as it represents a multi-perspective view on, and a multi-facet record of, reality). In the context of observer variability assessment reliability is defined as the concordance of repeated measurements (the annotations of the multiple annotators) and is usually calculated by the intraclass correlation coefficient (ICC) [33], which estimates the average correlation among all possible orderings of data pairs. As ICC is sensitive to data range also standard error of measurement SEM is proposed as a measure of variability in case of scalar values.

To this regard we will focus on questions such as: how much *true* is the ground truth? To this respect, we will introduce the concept of *fineness* of the Gold Standard. How much *reliable* is the ground truth? We relate ground truth reliability to the extent the single "measuring instruments", often human annotators, are *accurate* in their measure (i.e., in mapping a property of the object of interest to a value) and *precise* with respect to each other, i.e., how much their measures/annotations vary (or agree upon each other) for a single object of observation. How much *informative* (or *representative*) is the ground truth with respect to the reference population? This is also related to its *conformity*, that is the degree of resemblance between the available data and the reference population from which they have been drawn. How much *uncertain* is the set (which we call Diamond Standard) of all of the observations from which the ground truth is derived? To try to address the above questions, we propose a general framework to circumscribe the main concepts regarding the quality of data feeding the learning process of Machine Learning. With reference to Fig. 1, we call *Gold Standard* the training set, that is the data set where each case is annotated with a unique "true" value for the target feature. We distinguish it from the *Diamond Standard*, that is the data set where multiple (m) annotators (also called raters or observers), have associated the description of the cases to the target class. We call *reductions*, the data transformations that produce the Gold Standard from the Diamond Standard: reductions necessarily entail some information loss, because they allow to pass from a multi-rater labelling to "the one best" labelling by a "collective" rater. Obviously, if $m = 1$, the Gold Standard and the Diamond Standard coincide. On the basis of the number and interpretative skills of the annotators the Diamond Standard represents a more or less approximate representation of the truth (still yet, a symbolic and *datafied* expression of the truth), that is, of an *unknown* (and *unknowable*) data set that we call the *UR-SET* (Ultimately Realistic Symbolic Expression of Truth[1]).

In the next sections, we will consider methods to assess the quality of the input of the learning process, that is the data with which it has been trained to produce an accurate output when applied to new instances of data: we will illustrate the common cases of *reliability* and *representativeness*, and will introduce three original dimensions, by distinguishing between the *Fineness* and *Dryness* (of the Gold Standard), and the *Trueness* of the Diamond Set.

3   Reliability

The intuitive notion of reliability is straightforward: how much can we *rely upon* our ground truth to make decisions? How much can a ML model rely on its training set to make realistic (beside accurate[2]) predictions? More technically, the reliability of a dataset regards the *precision* of the measures it contains, for each case that it represents. This allows us to speak of reliability of a Gold Standard only in terms of the reliability of the Diamond Standard from which

---

[1] Notably, the UR-SET could be annotated with a different alphabet than the Gold Standard and the Diamond Standard. For instance, while the Gold Standard uses a binary symbol set (e.g., positive/negative) the UR-SET could be annotated with a set encompassing a symbol expressing that a case is 25% positive and 75% negative.

[2] Accuracy is historically defined in terms of *closeness* between the prediction and the Gold Standard, not with respect to the reality.

it has been derived. The reliability of a Diamond Standard regards the extent this set expresses a *unitary* interpretation of the single cases observed, despite the multiplicity of views entailed by the different raters involved in interpreting each case. If all of the raters agree upon each and every case (or the single raters agree with themselves, as in the case of the PRO meaasures mentioned above), that is if no disagreement among the case's annotations has been observed, both the reliability (and the trueness, as we will see) are maximum[3].

Over time, many measures of inter-rater agreement, and hence reliability, have been proposed, like the Fleiss's Kappa, the Cohen's Kappa, or the Krippendorff's Alpha. These indices aim to go beyond the simple proportion of matched pairs (a score called Proportion of Agreement, and usually denoted as P_o). This aim is motivated for the important, and often neglected, limitation of the P_o: it includes the amount of agreement that could be due to chance, and hence it produces an overly optimistic measure of the real agreement. All of the proposed metrics present some limitations, for instance in regard to the presence of missing values, or to the nature of ratings (e.g., categorical or ordinal), and all of them are subject to a number of paradoxes, e.g., when the cases to be rated are not well-distributed across the rating categories [34].

Unfortunately, scholars interested in assessing the reliability of annotated data still often rely on one of the indices presenting the most severe methodological problems [15], i.e., the Kappa; and, what is worst, they still usually adopt the range divisions proposed by Landis and Koch in 1977 [30] to interpret the scores, i.e., a scale that is obsolete, related to the first formulations of the Kappa, is "clearly arbitrary" (as frankly admitted by the first proponents), and

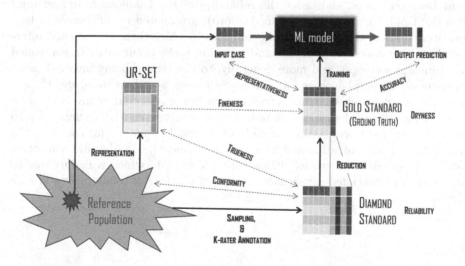

Fig. 1. The general framework and the main concepts illustrated in this contribution.

---

[3] That notwithstanding, it would be inaccurate to say that the Diamond Standard coincides with the UR-SET, which is unknowable.

manifestly inflating the degree of agreement (e.g., for the agreement to be considered "fair" it is sufficient that only 20% of times raters agree beyond the effect due to chance), likely one of the reasons for its fortune[4]. For these reasons we propose the adoption of more robust reliability measures, like the Krippendorf's Alpha, and to follow the indications for its interpretation given by Krippendorf [28]: he proposed to consider as sufficiently reliable for critical applications, like in the case of medical interpretation and prediction, collective annotations that would be associated with an Alpha of .8, or above, only. Krippendorff also considers two more robust criteria for acceptable reliability, both considering the distribution of the α (computed via bootstrapping): the first considers computing the confidence interval $[\alpha_{min}, \alpha_{max}]$ and then establishing acceptable reliability if the established threshold $\alpha_{required}$ (at least 0.8, as previously specified) is lower than α_{min}; the second approach, on the other hand, consists of computing the probability q that $\alpha \leq \alpha_{required}$ and then confronting this probability q with an a priori confidence threshold. These demanding requirements are seldom verified in the ML literature and, when they are, even less frequently met. We raised awareness on the issue of low reliability of the ground truth used to train medical AI in [5] and [7]. In this latter study we reported the low agreement between multiple raters in two settings from different medical specialties: cardiology and spine surgery. It is important to notice, yet, that disagreements do not occur only because some rater is less skilled than the others, and hence commits an interpretation error (due to what is called *label bias* [25]); in fact, this is seldom the case. More often, it is the *intrinsic ambiguity* of the *interpretand* phenomenon that brings raters to different, yet equally plausible, interpretations [5]. Other factors that could undermine the potential for agreement between raters, and hence the reliability of the Diamond Standard (and then the Gold Standard as mentioned above), are related to differences in how the raters react to the experimental conditions in which their opinions and interpretations are collected (since ground truthing tasks occur often in controlled experimental settings), and more generally, to the fact of being involved in an experiment. These phenomena are generally known as "Hawthorne effect" [36], but it is not clear whether the "awareness of being observed or involved in an experiment" affects the ratings more in terms of increasing the accuracy (up to levels that in real-world settings would not be tenable, mainly for conditions of uninterrupted concentration and focused commitment), or rather in terms of its reduction (an effect known as "laboratory effect" [20], which is mainly due to lack of real motivations, engagement or just of the fear of consequences in case of errors).

---

[4] To date, this single contribution has been cited almost 50,000 times, but likely more often by habit and imitation, than by the deliberate adoption of the assumptions therein discussed.

4   Representativeness and Conformity

"Representative" is a term that equally applies to individuals, with respect to a group from which they are ideally drawn; and to groups, with respect to wider groups, or populations, from which these groups are drawn as samples. To consider both these kinds of representativeness and, at the same time, avoid potential ambiguities, we distinguish between the *representativeness* of the Gold Standard, with respect to the single new case to predict; and the *conformity* of the Diamond Standard, with respect to the reference population. This analysis requires to focus on the *moments* of the probability distribution of our data: what we call *representativeness* regards the first moment, i.e., the centroid of the distribution, while *conformity* regards other higher-order moments, like variance, skewness, and kurtosis (the "shape" of the multi-dimensional distribution).

The simplest way to assess the *conformity* of the Diamond Standard is to consider, when available, the reference distributions of the single features, considering them separately: we call this basic type of population representativeness $conformity_u$, and this is based on the strong assumptions to know how the multivariate distribution of the population really is (e.g., from census information or other random sampling surveys), and that its change rate (or time constant) is negligible with respect to the sampling procedure.

Suppose that f is a categorical feature with k possible values, then we can test the *conformity* using the χ^2 goodness–of–fit test or the *G-test*; if f is an ordinal or continuous feature instead, we can apply the *Kolmogorov–Smirnov test*. In both cases, the obtained statistic (or the related p–value) represents a degree of the extent the Diamond Standard is similarly shaped with respect to the reference population.

When having access to the full joint distribution for the reference population we can extend the approach above described to define a multivariate definition of conformity, that we denote as $conformity_m$, using the multivariate versions of the respective statistical tests (see, as an example, [26] for a multivariate extension of the Kolmogorov–Smirnov test).

If we also have access to an analytic or model–based representation $\mathcal{M}$ of the reference population distribution we can give a third measure of conformity, that we term $conformity_p$, by directly computing the probability of the Diamond Standard D given $\mathcal{M}$, $P(D|\mathcal{M})$ and then sample (e.g., using Markov Chain Monte Carlo simulation techniques) the model in order to compute the probability q to obtain a probability $P \leq P(D|\mathcal{M})$ which can be taken as a measure of *conformity* (i.e., the greater q the greater our belief that D is indeed a fair representation of the reference population), because large values of q would imply that the Diamond Standard D is indeed "more probable" than most datasets generated according to the reference population distribution.

On the other hand, *representativeness* is defined between a given *input case* for the ML model, drawn from the reference population, and the Gold Standard: the Gold Standard is said to be representative of the input case if the input case resembles a "typical member" of the Gold Standard. This concept, while not usually evaluated, is important in checking whether the prediction that we would obtain from our model is meaningful; indeed, one of the major assumption of ML methodologies is that all the cases (the ones given as training examples as well as those which we are interested in making predictions on) come from the same distribution, that is are independent and identically distributed (IID). The most basic approach is to consider a case x representative of the Gold Standard G if it is "close" to its center, as described by the following algorithm:

Algorithm 1: Centroid–based Representativeness

Data: Gold Standard G, input case x
Result: Representativeness $r_c(G, x)$ of x

1 $c = \frac{1}{|G|} \sum_{p \in G} p$;

2 $dist(x, c) = \sqrt{\sum_{f \in F} (\frac{v_f^x - v_f^c}{v_f^{max} - v_f^{min}})^2}$;

3 $r_c(G, x) = 1 - \frac{dist(x, c)}{max\{dist(p, c) | p \text{ is not an outlier}\}}$;

---

where $\sum_{p \in G} p$, assuming that the instances belong to a vector space, is simply defined as the vector sum, F is the set of all features and the outlierness of a case is established via any outlier–detection algorithm.

The centroid–based representativeness r_c assumes values in $(-\infty, 1]$, with maximum value when x is exactly equal to the centroid of the Gold Standard. This basic technique, while simple also from a computational point of view, has various limitations: the most relevant one is that the whole distribution of G is not taken into account: the centroid in itself could be a non–representative point of G; x, while being quite distant from the center, could be in a region of the feature space which is actually homogeneous with respect to the distance and so on. A more valid approach would be to consider locality–based outlier–detection algorithm, such as the *Local Outlier Factor* [4], as described in the algorithm 2 which is based on the statistical transformation defined in [27].

Algorithm 2: Locality–based Representativeness

Data: Gold Standard G, input case x, number of neighbors k
Result: Representativeness $r_l(G, x, k)$ of x

1 $k - distance(x) = d(x, p_k)$ where p_k is the k-th nearest neighbor of x;

2 $\mathcal{N}_k(x) = \{p \in G | dist(x, p) \leq k - distance(x)\}$;

3 $S(x) = \text{Locality-Based-Outlier-Scoring}(x, \mathcal{N}_k(x))$;

4 $R(x) = max\{0, S(x) - 1\}$;

5 $r_l(G, x, k) = max\{0, erf(\frac{R(x) - \mu_R}{\sigma_R * \sqrt{2}})\}$

In the algorithm erf is the *Gaussian error function*, $S(x) \in [0, +\infty)$ and *Locality-Based-Outlier-Scoring* refers to any locality–based outlier detection algorithm. The locality–based representativeness can be understood as the probability of obtaining, from the Gold Standard G, a point similar to x, considering its connectivity degree (how much is it near to its nearest points) with respect to that of its neighbors. We also notice that Algorithm 2 can be used also in case of nominal attributes and missing values, by means of a suitable distance [40].

A last approach to define a measure of representativeness, which we denote as $r_p(\mathcal{M}_G, x)$, can be given when we have access to a generative ML model $\mathcal{M}_G$ for G. In this case, using a procedure analogous to the one used for defining $conformity_p$, we can compute the probability of x given the model $P(x|\mathcal{M}_G)$ and then sample the model to evaluate the probability q of getting a probability value as extreme as $P(x|\mathcal{M}_G)$. In the case of representativeness we could also refine this approach in order to define a local version of $r_p(\mathcal{M}_G, x)$ by limiting the sampled cases to ones belonging to a neighborhood of x.

An open question that these reflections invite to consider regards the feedback loop that could be established between the model's predictions (which affect the human decision making) and the reference population. If the decisions affected by the AI's advice can have an impact on the population from which new cases are to be extracted (like in case of prognostic models, where the model suggests how much an intervention could improve the health conditions of a patient, and hence also suggest who should receive a treatment, or an intensive one, and who should not), then it should be considered that the representativeness of the Gold Standard could change accordingly, usually for the worse. This would urge us for a continuous update of both the Diamond and the Gold Standard, or for the need to stratify the past interventions by distinguishing those who were likely impacted by the decision aid (directly or indirectly) and those who were not, and extract new cases for the ground truthing process from this latter portion of the reference population, using techniques akin to active learning [32].

5 Fineness of the Gold Standard

The $fineness(G, O)$ is the *probability* that the Gold Standard G, obtained from the Diamond Standard D by means of a *reduction* – e.g., taking the majority vote over a set O of observers (i.e., the mode for each case) – is equal to the true (unknowable) annotation (i.e., interpretation) of the portion of the reality of interest, what we call the $UR-SET$. For this reason, we consider $fineness(G, O)$ as a first measure of quality of the dataset which is fed into the ML model as a training set.

Let $O = \{o_1,...,o_m\}$ be m raters independently labeling the cases in dataset D; let also assume that each o_i has a constant error rate η_i. Assume that, in order to obtain the Gold Standard (G), for each case x we select the mode (i.e. the label who received the vote of the majority among the o_is) $\bar{o}$: thus, what is the probability that $\bar{o}(x)$ is a false label for x? This amounts to the probability that at least $\frac{m+1}{2}$ raters made an error, this probability can be computed via the *Poisson binomial distribution*:

$$P(error) = \sum_{k=\frac{m+1}{2}}^{m} \sum_{A \in F_k} \Pi_{i \in A} \eta_i \Pi_{j \notin A}(1 - \eta_j) \tag{1}$$

where F_k is the family of sets in which exactly k observers gave the wrong labeling.

Then, the probability to obtain a Gold Standard without errors is:

$$fineness(G, O) = (1 - P(error))^{|G|} \tag{2}$$

An interesting aspect of this is that the fineness of a Gold Standard (and thus the probability of no errors) is exponentially decreasing with the size of the Gold Standard itself. Via the Chernoff bound, and omitting some terms, we can upper bound $P(error)$ as:

$$P(error) \le e^{-\frac{m+1}{2} log \frac{m+1}{2\mu}} \tag{3}$$

where $\mu = \sum_i \eta_i$, thus the probability of an error decreases exponentially with both *increasing number of raters* and *decreasing expected errors*. By directly inserting this estimate into the bound for PAC learnability given in [1] we obtain that the true (but unknown) target is learnable, with probability $1 - \delta$ over samples and maximum error ϵ, when given at least:

$$\mathcal{O}\left(\frac{d \cdot log\frac{1}{\delta}}{\epsilon(1 - 2e^{-\frac{m+1}{2} log \frac{m+1}{2\mu}})^2} \right) \tag{4}$$

samples whose target is obtained by taking the majority vote as previously specified, where d is the Vapnik-Chervonenkis dimension [38] of the class of models adopted.

The inverse problem of determining the minimum number of raters needed to obtain a certain level of fineness can be solved via the method proposed in [23]. Then, to obtain a desired level of $fineness = 1 - \delta$ for each case $x \in D$ we should involve

$$\mathcal{O}\left(\frac{log\frac{|D|}{\delta}}{(1 - 2\eta_O)^2} \right) \tag{5}$$

raters, where η_O is the average error rate among O.

6   Trueness of the Diamond Standard

Where *fineness* is a propriety of the Gold Standard (with respect to the UR-SET), *trueness* is a propriety of the original Diamond Standard (always with respect to the UR-SET). The total trueness of the Diamond Standard is defined on the basis of the case-wise trueness. Basically, the trueness of a labeling $\langle o_1(x), ..., o_m(x) \rangle$ for a given case x is a measure of how much this labelling could be taken as a representation of the underlying (and unknown) true labeling, that is the corresponding case in the $UR-SET$. In other words, the trueness is the probability that this *diamond* (i.e., multi-facet, multi-rater) labeling actually corresponds to the true one. Basically we would assume that this probability is maximum when $o_1(x) = ... = o_m(x)$, that is, all of the raters agree with each other upon the labeling, while it is minimum when all the possible outcomes are equi–frequent. The trueness of the Diamond Standard as a whole can be computed in various ways starting from the the trueness of its units/cases, among which the simplest way is to take the average trueness for all its cases, with its %95 Confidence Interval (CI).

To quantify the *trueness* of a case x with its diamond labeling $o(x)$ we propose two approaches. Without loss of generalization, we will focus on the binary case (i.e. where the target can assume only values $\{0, 1\}$). Let $k \in [0, 1]$ be a threshold value, above which we get vote proportions that can be denoted as an *overwhelming majority* (usually proportions higher than 0.9 or even 0.95 and above according to the application domain); and let p be the observed probability of the majority labeling, taken as a rough estimate of the *trueness*.

Then the 95% confidence interval of p can be computed as

$$trueness'_c(o(x)) = p \pm 1.96\sqrt{\frac{p(1-p)}{m}} \tag{6}$$

where m is the number of observers, and we say that $o(x)$ has *acceptable trueness* if $inf(trueness_c(o(x))) \geq k$.

In the second approach, we know that the maximum number of *disagreements* is $M_d = \frac{m^2-1}{4}$ and we expect the $trueness(o(x))$ to decrease as the number of observed disagreements approaches M_d.

Thus if O_d is the number of observed disagreements, then

$$trueness''_c(o(x)) = 1 - \frac{O_d + \epsilon}{M_d + \epsilon} \tag{7}$$

where the ϵ acts as a smoothing factor (avoiding a value of 1 when $O_d = 0$ that could be misleading, since even in that the case there is a non–zero probability that the, unique, Diamond labeling is distinct from the true one). The two approaches have the following properties:

1. With fixed m, $trueness''_c$ has minimum value when $p = \frac{m+1}{2m}$ and maximum value when $p = 1$;
2. With fixed m, $trueness'_c$ has maximum width when $p = \frac{m+1}{2m}$ and minimum width when $p = 1$;
3. Increasing m the width of $trueness'_c$ decreases monotonically, this means that, fixing k and p, it is easier to obtain *acceptable trueness*;
4. If $p \in o(m^2)$ then $lim_{m \to +\infty} trueness''_c(o(x)) = 1$.

As suggested above, in order to extend this two case-wise definitions of trueness to the Diamond Standard trueness $trueness_D$, we can take different approaches. The most simple approach to extend the $trueness'_c$ definition is to say that the Diamond Standard D has *strong acceptable trueness* if $\forall_{x \in D}$ x has *acceptable trueness*. However, since this criterion of trueness is very restrictive, we can define other Diamond Standard–level measures of trueness, by making two assumptions: for the $trueness''_c$ definition we can assume that the trueness of the cases are distributed as independent Bernoullis; for both $trueness'_c$ and $trueness''_c$ we can assume an underlying distribution of the values of p (resp.

$trueness_c(\mathbf{o}(x))''$), which could be seen as a distribution of the *difficulty degrees* of assigning the correct labeling to the cases. Under the first approach we obtain the following expression of $trueness_D^{ind} = \Pi_{x \in D} trueness_c''(\mathbf{o}(x))$.

Under the second approach we first compute the average proportion (resp. trueness) and we can thus provide an interval estimate, about the average, of the value of trueness in two ways: assuming an underlying model distribution (with expected value equal to the computed average) and then compute (analytically or numerically) the 95% confidence interval; in a non–parametric way via a bootstrap–based estimate of the confidence interval (i.e. drawing a large number of samples with replacements of the original Diamond Standard and then computing the average proportion of trueness for each of these samples). In both cases we obtain an interval estimate $trueness_D^{int} = [trueness_D^{inf}, trueness_D^{sup}]$ and we say that the Diamond Standard D has *weak acceptable trueness* if $trueness_D^{inf} \geq k$.

It is noteworthy that the concepts of trueness and fineness are, obviously, related with each other: in particular, the greater the trueness of the Diamond Standard, the greater the fineness of the resulting Gold Standard. Most significantly, we could take $trueness_c'$ or $trueness_c''$ as estimates of $1 - P(error)$ to have an approximation of the degree of fineness of the resulting Gold Standard. If we assume that the error rates η_i (i.e., the probability of the annotation of the observer to be in perfect disagreement with the true symbolic representation, cf. accuracy in metrology) are constant for all the cases x, then given that $P(error)$ is also constant, we could simply average p (respectively, $trueness_c''$) over the whole dataset to obtain an estimate $\widetilde{P}(error)$ that we can connect to the fineness bounds obtained in Sect. 5. Moreover we could also assume that for each case x each observer o_i has a distinct error rate $\eta_i(x)$ (approximated by p or $trueness_c''(x)$), this setting is known as *Constant Partition Classification Noise* (CPCN) which, as shown in [35], is equivalent (in terms of learning complexity) to the setting described in Sect. 5.

7   Dryness of the Gold Standard

Dryness regards how much the information content of the Diamond Standard has dried off, or "shrinked", in the *reduction* of this latter into the Gold Standard. The reference is an homage to the seminal idea by Goguen of dry and wet information [17]: the more multiple, collaborative, social, and even ambiguous, the information, the "wetter" (that is "impregnated" with information) it is. Therefore, the higher the *information loss* implied by the reduction, the higher the dryness. Since the reduction implies that the information contained in m columns is reduced in the content of a single column, assessing the dryness of the resulting set can be useful to understand if some reduction is more information-preserving than others, and hence preferable.

In the following we will assume a nominal valued target, thus the target of the Diamond Standard is expressed in terms of a m–dimensional vector over a set Y (i.e. $\mathbf{o}(x) \in Y^m$) and we suppose that the target of the Gold Standard is generated from $\mathbf{o}(x)$ via a reduction $T: Y^m \mapsto \mathcal{C}(Y)$ where $\mathcal{C}(Y)$ is a set of structures, in a general sense, over Y (e.g. the set of probability distributions over Y). In general, the reduction T involves an information loss (or an increase in dryness) given by the fact that only observing $T(\mathbf{o}(x))$ it is impossible to (perfectly) recover $\mathbf{o}(x)$ (assuming that $\mathcal{C}(Y) \neq Y^m$ and $T \neq id_{Y^m}$); this means that T implicitly defines an inverse set–valued map $L: \mathcal{C}(Y) \mapsto \mathcal{P}(Y^m)$ allowing us to define a measure of dryness in both *quantitative* and *qualitative* terms.

In quantitative terms we will define the dryness of $T(\mathbf{o}(x))$ as:

$$dryness(\mathbf{o}(x), T) = \frac{|L(T(\mathbf{o}(x)))| - 1}{|Y|^m - 1} \tag{8}$$

which can be understood as the ratio of the *information contents* of $\mathbf{o}(x)$, in the denumerator, and $T(\mathbf{o}(x))$, in the numerator: in particular, the numerator is the number of objects in $|Y|^m$ satisfying the constraints imposed by L. From the values of $dryness(\mathbf{o}(x))$, for each case x, we can obtain the value of the dryness, under reduction T, for the whole Gold Standard as:

$$dryness(G, T) = \frac{1}{|G|} \sum_{x \in G} dryness(\mathbf{o}(x)) \tag{9}$$

Usually, in the nominal case, the reduction T is taken as the mode, that is $T(\mathbf{o}(x)) = mode(\mathbf{o}(x))$; in this case the numerator is given by all possible diamond labelings in which $mode(\mathbf{o}(x))$ is in fact the most frequent label, which can be approximated via the following bound:

$$dryness(\mathbf{o}(x), mode) = O\left(\frac{\sum_\pi \sum_{\pi_{|Y|} \leq \ldots \leq m^*} \binom{m}{m^*, \ldots, \pi_{|Y|}}}{|Y|^m - 1} \right) \tag{10}$$

where π is any assignment of $m - 1$ least frequent classes, π_i is the i-th least frequeny class in the assignment π and m^* is the frequency of the mode. However, other reductions could be defined, a first such example is the transformation $freq$ defined as:

$$freq(\mathbf{o}(x)) = \langle \frac{m_1}{m}, \ldots, \frac{m_{|Y|}}{m} \rangle \tag{11}$$

where m_i is the frequency of class $c_i \in Y$ in $\mathbf{o}(x)$. The dryness of $freq$ is defined as:

$$dryness(\mathbf{o}(x), freq) = \frac{\binom{m}{m_1, \ldots, m_{|Y|}} - 1}{|Y|^m - 1} \tag{12}$$

in which the numerator is given exactly by the number of diamond labelings in which the labels occur with exactly the frequency given by $freq(\mathbf{o}(x))$. Evidently, $dryness(\mathbf{o}(x), freq) \leq dryness(\mathbf{o}(x), mode)$ and, $freq$ is the reduction with minimal dryness among the ones that are order–irrelevant. However, besides the quantitative part of the dryness, there is also a qualitative part: each reduction defines which information is deemed relevant (and thus conserved), and which information is instead discarded. The *mode* reduction maintains only the most frequent label and discards every other information; on the other hand, the *freq* reduction keeps the proportions of each possible alternative and only "forgets" the order–part of the vector (i.e. which option each observer selected).

Another qualitative aspect of the dryness is given by the fact that we can provide two different interpretations of each reduction T:

1. The *epistemic* view, according to which we suppose that the true labeling of x in the UR–SET is a single label in Y and $T(\mathbf{o}(x))$ represents our degree of belief assigned to the alternatives for that label (e.g. *freq* represents our subjective posterior probability of which it is the real labeling);
2. The *ontic* view, according to which we suppose that in fact the true labeling of x in the UR-SET is not a label from Y but one from $im(T) = \mathcal{C}(Y)$ and reduction T allows us to estimate this label from the information given by $\mathbf{o}(x)$ (e.g. the ontic view associated with the *freq* reduction is that our phenomenon is indeed a non–deterministic one and $freq(\mathbf{o}(x))$ is an estimation of the propensities of the system to be in one of the alternative states).

If we look at the quantitative component of the dryness, the *freq* reduction is manifestly the optimal choice to construct the Gold Standard. However, the qualitative approach suggests that it may retain "too much" information: the exact proportions may be observed only "by accident" or they could be irrelevant. In the following, we will suggest two alternative reductions that are mid–way between *mode* and *freq* in terms of dryness.

7.1 Fuzzy–Possibilistic Reductions

Let $m^* = max_i(m_i)$ be the index of the most frequent labels in $\mathbf{o}(x)$, then we define the *possibilistic reduction* as:

$$poss(\mathbf{o}(x)) = \langle \frac{m_1}{m^*}, ..., \frac{m_{|Y|}}{m^*} \rangle \qquad (13)$$

Under the qualitative point of view, the *poss* reduction preserves the preference ordering among the possible alternatives and also a "relative" indication of degrees of preference of an alternative compared to the others: thus, the numerator of the dryness is the number of diamond labelings in which the proportions between the most frequent label and the other ones are determined by m and the values of $poss(\mathbf{o}(x))$. If we denote by σ the ordering of the labels in Y in order of decreasing value of $poss$, then we can bound the dryness as:

$$dryness(\mathbf{o}(x), poss) = o\left(\frac{\sum_{m_{\sigma_1}=\frac{1}{\rho_{|Y|}}}^{m} \binom{m}{m_{\sigma_1}, m_{\sigma_1} \cdot \rho_2, ..., m_{\sigma_1} \cdot \rho_{|Y|}} - 1}{|Y|^m - 1} \right) \qquad (14)$$

Under the *epistemic* interpretation, the *poss* reduction models our degree of belief in terms of a *possibility distribution* [42], which could be taken as representing an *imprecise probability distribution* [12] representing our belief in the relative preferences and their proportions but not the exact counts.

Under the *ontic* interpretation, on the other hand, the *poss* reduction represents a *fuzzy set*, that is, we assume that the different labelings given by the observers are not due to errors but due to the fact that the phenomenon itself is multi–faceted and, in some sense, *vaguely defined* and the labelings reported more frequently are *prototypical* for the observed instance of the phenomenon.

7.2   Three–Way Reduction

Three–way decision theory [41] refers to an extension of standard decision–theory in which the "decision maker" (in a general sense, including also an algorithm) has the ability to abstain (totally or partially) instead of expressing a decision.

We will describe two approaches to perform a three-way transformation. Let $\epsilon \in [0,1]$, $freq(\mathbf{o}(x))$ be the frequencies of the labels in Y and σ the ordering of the labels in decreasing frequency order. Then we say that $\mathbf{o}(x)$ is $(m, \epsilon) -$ *ambiguous* if

$$\forall i \in \{1, ..., m\}.|\sigma_1 - \sigma_i| \leq \epsilon \qquad (15)$$

Let m^* be the greatest m such that $\mathbf{o}(x)$ is $(m, \epsilon) - ambiguous$, then we define the tw_a reduction as:

$$tw_a(\mathbf{o}(x), \epsilon) = \{\sigma_1, ..., \sigma_{m^*}\} \qquad (16)$$

In this case the numerator of Eq. (8) is given by the number of diamond labelings for which the labels in $tw_a(\mathbf{o}(x), \epsilon)$ are the most frequent ones and their distance is at most ϵ.

The second approach, that we term *decision–cost theoretic*, descends from our previous work on three–way classification [9,10]. Let ϵ be an error cost, α be an abstention cost and $freq(\mathbf{o}(x))$, σ defined as above. Then we define the tw_d reduction as:

$$tw_d(\mathbf{o}(x), \epsilon, \alpha) = \begin{cases} \{\sigma_1, ..., \sigma_j\} & \alpha \cdot \sum_{i=1}^{j} \sigma_i + \epsilon \cdot \sum_{i=j+1}^{k} \sigma_i < \epsilon * (1 - \sigma_1) \\ \sigma_1 & \text{the inequality has no solution} \end{cases}$$

$$(17)$$

where j is the optimal index satisfying the inequality.

Future work will be devoted to understand how knowledge about the raters' skills, and confidence (even self-perceived) in the raters' interpretation, can be integrated in the reduction to make the Gold Standard finer (and reduce the information loss in the transformation from the Diamond Standard).

Example 1. *Let D be a Diamond Standard of 3 cases and*

$$\mathbf{o}(D) = \begin{bmatrix} 0 & 1 & 0 & 1 & 0 \\ 1 & 0 & 1 & 1 & 1 \\ 1 & 0 & 0 & 0 & 0 \end{bmatrix}$$

the respective labeling given by 5 observers.

Applying the mode reduction we obtain, $mode(\mathbf{o}(D)) = \begin{bmatrix} 0 & 1 & 1 \end{bmatrix}$ for which the dryness dryness$(\mathbf{o}(D), mode) = \begin{bmatrix} 15/31 & 15/31 & 15/31 \end{bmatrix}$. The total dryness of G (reduced from D in this way) is then the average, 0.48.

On the other hand, for transformation freq we obtain

$$freq(\mathbf{o}(D)) = \left[(0:3/5, 1:2/5) \, (0:1/5, 1:4/5) \, (0:4/5, 1:1/5) \right]$$

for which the dryness is $dryness(\mathbf{o}(D), freq) = \left[10/31 \; 5/31 \; 5/31 \right]$. *The total dryness of G (reduced from D in this second way) is then the average, 0.22.*

For the tw_a *reduction, setting* $\epsilon = 0.4$ *we have that* $tw_a(\mathbf{o}(D), \epsilon) = \left[\{0, 1\} \; 1 \; 0 \right]$ *for which the dryness is* $dryness(\mathbf{o}(D), tw_a) = \left[19/31 \; 6/31 \; 6/31 \right]$. *The total dryness of G (reduced from D in this third way) is then the average, 0.33.*

Finally for the poss reduction we have that

$$poss(\mathbf{o}(D)) = \left[(0:1, 1:2/3) \, (0:1/4, 1:1) \, (0:1, 1:1/4) \right]$$

for which the dryness is $dryness(\mathbf{o}(D), poss) = \left[10/31 \; 5/31 \; 5/31 \right]$. *Thus, the total dryness of G (reduced from D in this last way) is then the average, 0.22. Summarizing, when computing the values of* $dryness(G)$ *obtained with different reductions, we get that:*

$$dryness(D, poss) = dryness(D, freq) < dryness(D, tw_a) < dryness(D, mode)$$

We remind that the higher the dryness, the higher the information loss, and hence the informatively "poorer" the Gold Standard.

8   Some More Idle Reflections

Explainable AI (XAI) has recently been set forth as a necessary component of human agencies where decision making is supported by computational means[5].

Apart from a "XAI paradox", which we will mention at the end of this contribution, we agree that some form of XAI is necessary for human decision makers who use some kind of AI decision support to reach more informed decisions and be rightly held fully accountable for these decisions.

In the context of the XAI discourse, human decision makers must be able *to interpret* the AI system output, that is make sense of it in terms of *why* the system proposed a specific output for the provided input [24] and, to some extent, of *how* the system yielded this output, so as to take its advice into due consideration in making their decision. In this line, *interpretability* is often tightly related to *explainability* (so much that these two terms are often used interchangeably) and both are usually articulated in terms of the capability of the AI system "to explain its reasoning" [11]. Thus, the lack of a formal, or at least unique and non-ambiguous definition of explanation (and hence explainability), which is lamented by many observers (e.g., [11,31]), should not make us overlook the fact that the ability to interpret the system behavior by the humans, so that they can make an informed use of the system output, is often translated

---

[5] To this respect, here we are covering different cases than those covered by the GDPR article no. 22, which regards decisions that are solely based on automated processing, without human intervention [39].

into *a property of the system*, that is its capability to provide human decision makers with resolving clues about its functioning and "reasons" for a prediction. However, while this property can be linked to the presence or absence of specific functions that make some information available to the decision makers (e.g., what aspects of the phenomenon at hand, i.e., predictor variables, were more important for proposing a specific advice), self-explanations tell nothing about their suitability of being understood and hence of their potential to contribute to the interpretation of the system. This allows us to relate the notion of interpretability/explainability to the notion of *usability* of the system. A focus on usability suggests to assess AI not only in regard to task efficiency (e.g., time to completion) and effectiveness (e.g., error rate) but also in terms of user satisfaction. In the context of human decision making this regards the extent decision makers are satisfied by their interaction with the system; feel to be in control of the situation; believe to have got a sufficient number of indications to formulate an *informed decision*; feel to be able to account for it; are confident that the system supported them in considering all of the aspects that were due; and that it did not misled them. However, usability, as widely known, is not a property of the system, but rather of the coupling between the system and the human users; in other words, usability emerges in the interaction between the AI and its users, in the fit between system functionalities and the user skills. So does the XAI. In the light of seeing interpretability as a kind of usability (or better yet, as a way in which the usability of AI-driven decision support is manifested), we also advocate an *interpretable* and *explainable* AI [18] as a necessary condition for the embedding of AI in human agencies that are called to make critical decisions significantly affecting other people's life. Even more than this, we emphasize the importance to design for an *interpretand AI*, that is an *AI that must be interpreted* by the decision makers, so that that they build a *local narrative* to convince themselves, as well as the others, of the soundness and *reasonability* of the resulting decision. Thus, in the human-AI interaction, it is important to distinguish between a *right to explanation*, that is for the users to receive indications by the AI system that satisfactorily bring them to believe to have understood why the decision support gave them a certain advice; and the *obligation to interpretation*, that is for the users to have to adopt an active attitude to collect and interpret these indications: advocacy for explainable AI should not diminish responsibility for decision makers. This duty to active interpretation can be promoted, and even afforded, by the decision support system itself: to this aim we are testing a decision support system that is currently adopted in a large teaching hospital specialized in musculoskeletal disorders and surgery and is endowed with *programmed inefficiencies*, that is features aimed at purposely increase the "decision friction" (cf. [13]), by requiring an active stance by the users so as to minimize the risk of automation bias and deskilling [8].

9   Conclusion

In this paper, we focused on the importance of letting the decision makers know and understand the quality of the data used to train the models by which an AI can provide its predictions and advice. In fact, no model can bring meaningful output if the input data are not reliable: the notorious phrase "Garbage In, Garbage Out" here applies, and is the central tenet of our contribution, as the tongue-in-cheek title suggests.

To make the AI system more transparent, we propose to focus on the ground truth by which the AI has been trained. To make the ground truth more interpretable, we proposed a framework that distinguishes between Gold Standards and Diamond Standards, and encompasses some common (but relevant) quality dimensions, like representativeness and reliability, and some novel quality dimensions, like *fineness*, *trueness* and *dryness*, which we discuss and for which provide a preliminary yet formal specification.

These metrics are given for a twofold aim. First, their definition and application invite AI researchers to devise alternative ways to produce the ground truth from the observations and interpretations available (what we call alternative *reductions*), other than the simple majority vote, so that the quality of the training set could improve along multiple dimensions. However, this is still a technicality, although of no little importance. More importantly: since we usually *assume* that our ground truth is perfect, reflecting on its quality necessarily entails growing an informed *prudence* in regard to its reliability and adequacy for the task of supporting decision making in delicate domains. Thus, our ultimate main aim is to contribute to raising awareness of the impact of our assumptions, models, and representations in intensive cognitive tasks. The dimensions we started to envision are aimed at facilitating people to reflect on these aspects, rather than focus on model details and misleading performance metrics, like accuracy, which only regards the match between the AI predictions and the Gold Standard (see Fig. 1). From the design point of view, we should ask what an actually *useful* support from AI looks like. We hold that a useful AI is a *usable* AI, but not necessarily an AI providing decision makers with simple and clear-cut predictions, nor the system that combines its output with a plenty of indications and explanations. In the light of the research on the use of *computers as persuasive technologies* [2] (evocatively called *captology* by Fogg [16]), we should be aware of a potential conundrum on effective XAI, what we could call a *captological XAI paradox*: "AI can give us a wrong advice, and yet also in that case accompany it with plausible reasons that *prime* our interpretation and convince us. The more imperscrutable AI is, the more likely we can doubt it, and make sense of the available data with less interference". Obviously, awareness of this paradox should not convince us to stop pursuing a better XAI. All the opposite, it urges us to consider new and more effective ways by which technology itself can promote a reflective stance in the decision makers and a stronger will and commitment to take full responsibility of the vigilant use of that technology.

References

1. Angluin, D., Laird, P.: Learning from noisy examples. Mach. Learn. **2**(4), 343–370 (1988)

2. Atkinson, B.M.C.: Captology: a critical review. In: IJsselsteijn, W.A., de Kort, Y.A.W., Midden, C., Eggen, B., van den Hoven, E. (eds.) PERSUASIVE 2006. LNCS, vol. 3962, pp. 171–182. Springer, Heidelberg (2006). https://doi.org/10.1007/11755494_25
3. Brennan, P., Silman, A.: Statistical methods for assessing observer variability in clinical measures. BMJ: Br. Med. J. **304**(6840), 1491 (1992)
4. Breunig, M.M., Kriegel, H.P., Ng, R.T., et al.: Identifying density-based local outliers. SIGMOD Rec. **29**(2), 93–104 (2000)
5. Cabitza, F., Ciucci, D., Rasoini, R.: A giant with feet of clay: on the validity of the data that feed machine learning in medicine. In: Cabitza, F., Batini, C., Magni, M. (eds.) Organizing for the Digital World. LNISO, vol. 28, pp. 121–136. Springer, Cham (2019). https://doi.org/10.1007/978-3-319-90503-7_10
6. Cabitza, F., Dui, L.G., Banfi, G.: PROs in the wild: assessing the validity of patient reported outcomes in an electronic registry. Comput. Methods Program. Biomed. (2019)
7. Cabitza, F., Locoro, A., Alderighi, C., Rasoini, R., Compagnone, D., Berjano, P.: The elephant in the record: on the multiplicity of data recording work. Health Inform. J. (2019)
8. Cabitza F., Campagner A., Ciucci D., Seveso A.: Programmed inefficiencies in DSS-supported human decision making. In: Proceedings of 16th MDAI International Conference (2019, to appear)
9. Campagner, A., Cabitza, F., Ciucci, D.: Exploring medical data classification with three-way decision trees. In: Proceedings of the 12th BIOSTEC International Joint Conference - Volume 5: HEALTHINF, pp. 147–158 (2019)
10. Campagner, A., Cabitza, F., Ciucci, D.: Three–way classification: ambiguity and abstention in machine learning. In: Mihálydeák, T., et al. (eds.) IJCRS 2019. LNCS (LNAI), vol. 11499, pp. 280–294. Springer, Cham (2019). https://doi.org/10.1007/978-3-030-22815-6_22
11. Doshi-Velez, F., Kim, B.: Towards a rigorous science of interpretable machine learning. arXiv preprint arXiv:1702.08608 (2017)
12. Dubois, D., Prade, H.: Possibility theory and its applications: where do we stand? In: Kacprzyk, J., Pedrycz, W. (eds.) Springer Handbook of Computational Intelligence, pp. 31–60. Springer, Heidelberg (2015). https://doi.org/10.1007/978-3-662-43505-2_3
13. Edwards, P.N., Mayernik, M.S., Batcheller, A.L., et al.: Science friction: data, metadata, and collaboration. Soc. Stud. Sci. **41**(5), 667–690 (2011)
14. Esteva, A., Kuprel, B., Novoa, R.A., et al.: Dermatologist-level classification of skin cancer with deep neural networks. Nature **542**(7639), 115 (2017)
15. Feinstein, A.R., Cicchetti, D.V.: High agreement but low kappa: I. the problems of two paradoxes. J. Clin. Epidemiol. **43**(6), 543–549 (1990)
16. Fogg, B.J.: Persuasive computers: perspectives and research directions. In: CHI 1998, pp. 225–232. ACM Press (1998)
17. Goguen, J.: The dry and the wet. In: Proceedings of the IFIP TC8/WG8.1 Working Conference on Information System Concepts: Improving the Understanding, pp. 1–17 (1992)
18. Goebel, R., et al.: Explainable AI: the new 42? In: Holzinger, A., Kieseberg, P., Tjoa, A.M., Weippl, E. (eds.) CD-MAKE 2018. LNCS, vol. 11015, pp. 295–303. Springer, Cham (2018). https://doi.org/10.1007/978-3-319-99740-7_21
19. Gulshan, V., Peng, L., Coram, M., et al.: Development and validation of a deep learning algorithm for detection of diabetic retinopathy in retinal fundus photographs. Jama **316**(22), 2402–2410 (2016)

20. Gur, D., Bandos, A.I., Cohen, C.S., et al.: The "laboratory" effect: comparing radiologists' performance and variability during prospective clinical and laboratory mammography interpretations. Radiology **249**(1), 47–53 (2008)
21. Haenssle, H., Fink, C., Schneiderbauer, R., et al.: Man against machine: diagnostic performance of a deep learning convolutional neural network for dermoscopic melanoma recognition in comparison to 58 dermatologists. Ann. Oncol. **29**(8), 1836–1842 (2018)
22. Han, S.S., Park, G.H., Lim, W., et al.: Deep neural networks show an equivalent and often superior performance to dermatologists in onychomycosis diagnosis. PloS One **13**(1), e0191493 (2018)
23. Heinecke, S., Reyzin, L.: Crowdsourced PAC learning under classification noise. arXiv preprint arXiv:1902.04629 (2019)
24. Holzinger, A., Langs, G., Denk, H., Zatloukal, K., Mueller, H.: Causability and explainability of AI in medicine. Wiley Interdisc. Rev.: Data Min. Knowl. Discov. **9**(4), e1312 (2019)
25. Jiang, H., Nachum, O.: Identifying and correcting label bias in machine learning. arXiv preprint arXiv:1901.04966 (2019)
26. Justel, A., Peña, D., Zamar, R.: A multivariate Kolmogorov-Smirnov test of goodness of fit. Stat. Probab. Lett. **35**(3), 251–259 (1997)
27. Kriegel, H.P., Kröger, P., Schubert, E., Zimek, A.: Interpreting and unifying outlier scores, pp. 13–24 (2011)
28. Krippendorff, K.: Content Analysis: An Introduction to Its Methodology. Sage Publications, London (2018)
29. Lakkaraju, H., Kamar, E., Caruana, R., Leskovec, J.: Interpretable and explorable approximations of black box models. arXiv preprint arXiv:1707.01154 (2017)
30. Landis, J.R., Koch, G.G.: The measurement of observer agreement for categorical data. Biometrics. **33**(1), 159–174 (1977)
31. Lipton, Z.C.: The mythos of model interpretability. arXiv preprint arXiv:1606.03490 (2016)
32. MacKay, D.J.C.: Bayesian methods for adaptive models. Ph.D. thesis, California Institute of Technology (1992)
33. Popović, Z.B., Thomas, J.D.: Assessing observer variability: a user's guide. Cardiovasc. Diagn. Ther. **7**(3), 317 (2017)
34. Quarfoot, D., Levine, R.A.: How robust are multirater interrater reliability indices to changes in frequency distribution? Am. Stat. **70**(4), 373–384 (2016)
35. Ralaivola, L., Denis, F., Magnan, C.N.: CN = CPCN. In: ICML 2006. ACM (2006)
36. Stand, J.: The hawthorne effect - what did the original Hawthorne studies actually show. Scand J. Work Environ. Health **26**(4), 363–367 (2000)
37. Svensson, C.M., Krusekopf, S., Lücke, J., et al.: Automated detection of circulating tumor cells with naive Bayesian classifiers. Cytom. Part A **85**(6), 501–511 (2014)
38. Vapnik, V.N., Chervonenkis, A.Y.: On the uniform convergence of relative frequencies of events to their probabilities. Theor. Probab. Appl. 17, 264–280 (1971)
39. Wachter, S., Mittelstadt, B., Floridi, L.: Why a right to explanation of automated decision-making does not exist in the general data protection regulation. Int. Data Priv. Law **7**(2), 76–99 (2017)
40. Wishart, D.: k-means clustering with outlier detection, mixed variables and missing values. In: Schwaiger, M., Opitz, O. (eds.) Exploratory Data Analysis in Empirical Research, pp. 216–226. Springer, Heidelberg (2003). https://doi.org/10.1007/978-3-642-55721-7_23

41. Yao, Y.: An outline of a theory of three-way decisions. In: Yao, J.T., et al. (eds.) RSCTC 2012. LNCS (LNAI), vol. 7413, pp. 1–17. Springer, Heidelberg (2012). https://doi.org/10.1007/978-3-642-32115-3_1

42. Zadeh, L.: Fuzzy sets as a basis for a theory of possibility. Fuzzy Sets Syst. **100**, 9–34 (1999)

Automated Machine Learning for Studying the Trade-Off Between Predictive Accuracy and Interpretability

Alex A. Freitas[✉] ⓘ

School of Computing, University of Kent, Canterbury CT2 7NF, UK
A.A.Freitas@kent.ac.uk

Abstract. Automated Machine Learning (Auto-ML) methods search for the best classification algorithm and its best hyper-parameter settings for each input dataset. Auto-ML methods normally maximize only predictive accuracy, ignoring the classification model's interpretability – an important criterion in many applications. Hence, we propose a novel approach, based on Auto-ML, to investigate the trade-off between the predictive accuracy and the interpretability of classification-model representations. The experiments used the Auto-WEKA tool to investigate this trade-off. We distinguish between white box (interpretable) model representations and two other types of model representations: black box (non-interpretable) and grey box (partly interpretable). We consider as white box the models based on the following 6 interpretable knowledge representations: decision trees, If-Then classification rules, decision tables, Bayesian network classifiers, nearest neighbours and logistic regression. The experiments used 16 datasets and two runtime limits per Auto-WEKA run: 5 h and 20 h. Overall, the best white box model was more accurate than the best non-white box model in 4 of the 16 datasets in the 5-hour runs, and in 7 of the 16 datasets in the 20-hour runs. However, the predictive accuracy differences between the best white box and best non-white box models were often very small. If we accept a predictive accuracy loss of 1% in order to benefit from the interpretability of a white box model representation, we would prefer the best white box model in 8 of the 16 datasets in the 5-hour runs, and in 10 of the 16 datasets in the 20-hour runs.

Keywords: Automated Machine Learning (Auto-ML) ·
Classification algorithms · Interpretable models

1 Introduction

This work focuses on the classification task of machine learning, where each instance (example, or data point) consists of a set of predictive features and a class label. A classification algorithm learns a predictive model from a set of training data, where the algorithm has access to the values of both the features and the class labels of the instances, and then the learned model can be used to predict the class labels of instances in a separate set of testing data, which was not used during training.

© IFIP International Federation for Information Processing 2019
Published by Springer Nature Switzerland AG 2019
A. Holzinger et al. (Eds.): CD-MAKE 2019, LNCS 11713, pp. 48–66, 2019.
https://doi.org/10.1007/978-3-030-29726-8_4

Recently, classification algorithms have been used by an increasingly larger and more diverse set of users, including users with relatively little or no expertise in machine learning. In addition, a large amount of machine learning research has produced many different types of algorithms [5, 19] with increasingly greater complexity. Also, in general these algorithms have several hyper-parameters whose settings need to be carefully tuned to maximize predictive accuracy, for each input dataset.

As a result, recently there has been an increasing research interest in the area of Automated Machine Learning (Auto-ML). In the context of the classification task, Auto-ML methods usually try to solve the problem of finding the best classification algorithm and its best configuration (hyper-parameter settings) for any given dataset provided as input by the user. This is sometimes referred to as the CASH problem – Combined Algorithm Selection and Hyper-parameter optimization [16, 18].

There has also been a growing interest in learning interpretable classification models, motivated by several factors like the need to improve users' trust on the models' recommendations, legal requirements for explaining the model's recommendations in some domains, and the opportunity to provide users with new insight about the data and the underlying application domain [7]. Furthermore, several studies have discussed how to evaluate the interpretability of classification models – e.g., [7, 8].

Despite this increasing interest in the interpretability of classification models, the classification literature is still overwhelmingly dominated by the use of predictive accuracy as the main (and very often the only) evaluation criterion. As a result, the literature is currently dominated by black box classification models, produced by algorithms that were designed to maximize predictive accuracy only, without taking into account model interpretability.

This focus on predictive accuracy as the only criterion to evaluate a classification model is particularly strong in the area of Auto-ML, where the interpretability of classification models is normally ignored.

Hence, we propose a novel approach, based on Auto-ML, to investigate the trade-off between the predictive accuracy and the interpretability of classification-model representations. Note that the focus of this investigation is on the type of knowledge representation used by the learned classification models, rather than the contents of the models themselves. Broadly speaking, we consider as interpretable the following 6 types of model representation: decision trees, If-Then classification rules, decision tables, Bayesian network classifiers, nearest neighbours and logistic regression representations. Hence, in this work we distinguish mainly between learned models using these representations and learned models using other (non-interpretable or only partly interpretable) representations – as discussed in more details in Sect. 2.2.

Although using an interpretable knowledge representation is not a sufficient condition for a model to be really interpretable by a user, arguably an interpretable representation tends to be a necessary or at least highly desirable condition for obtaining model interpretability. In addition, the full notion of model interpretability involves very subjective, user-dependent issues, which are out of the scope of this work.

Hence, in this work we perform a number of experiments with Auto-WEKA, whose search space includes many classification algorithms for learning models with both interpretable and non-interpretable representations, and then analyze in detail the results to investigate to what extent (if any) the best interpretable-representation models produced by Auto-WEKA are sacrificing predictive accuracy by comparison with the best non-interpretable-representation models produced by Auto-WEKA. This is an interesting approach to analyze the trade-off between accuracy and interpretability because Auto-WEKA automatically selects the best algorithm and its best hyper-parameter settings in a way customized to each input dataset. Hence, the discussion on the trade-off between accuracy and interpretability is raised to a new, more challenging level than usual, where the question is how much accuracy (if any) an interpretable-representation model is sacrificing, not just by comparison with a strong algorithm, but rather by comparison with the strongest (most accurate) algorithm found by Auto-WEKA for each particular dataset at hand.

Note that, although there are several studies evaluating the performance of Auto-ML methods [6, 10, 14], in general these studies focus only on the predictive accuracy of the selected algorithms, ignoring the issue of the interpretability of their learned models. To the best of our knowledge, this current work is the first one to investigate the trade-off between the predictive accuracy and interpretability of classification models which were optimized to each input dataset by an Auto-ML method.

More precisely, this paper presents the following contributions. First, we investigate the influence of two different runtime limits (as 'computational budgets') given to Auto-WEKA on the predictive accuracy of the best algorithms selected by Auto-WEKA for each of the 16 datasets used in our experiments. Second, we investigate the frequencies with which different classification algorithms (using different knowledge representations for their learned models) are selected by Auto-WEKA for each dataset, across several runs with different random seeds used to initialize the Auto-WEKA's search. Third, as the main contribution of this work, we analyze the trade-off between the predictive accuracy and the interpretability of the model representations selected by Auto-WEKA for each dataset.

The remainder of this paper is organized as follows. Section 2 reviews background on Auto-ML and interpretable classification models. Section 3 describes the proposed experimental methodology. Section 4 reports the computational results of the experiments with 16 datasets. Section 5 summarizes the results, and Sect. 6 presents the conclusions and some future research directions.

2 Background

2.1 Background on Automated Machine Learning (Auto-ML)

With the increasing interest in the area of Auto-ML, several types of Auto-ML methods have been proposed in the literature [18], using a variety of search methods to perform a search in the space of candidate machine learning algorithms and their hyper-parameter settings. However, most Auto-ML methods use, as the search method, some variation of Bayesian Optimization (BO) [6, 16] or Evolutionary Algorithms (EAs) [3, 15]. Both BO and EAs are suitable for Auto-ML because they are derivative-free global search methods. That is, they do not require knowledge of the derivative of the objective function, which is suitable for the discrete search space of candidate solutions (involving choices of algorithms), and they perform a global search in the space of candidate solutions, coping with the trade-off between exploitation and exploration in a way that reduces the chances of getting trapped into local optima in the search space. There are also Auto-ML methods based on other types of search methods, like hierarchical planning [14].

Two popular Auto-ML tools, representing seminal work in this area, are Auto-WEKA [16] and Auto-sklearn [6], which search in the space of classification algorithms offered by the popular WEKA and scikit-learn machine learning libraries, respectively – both using BO as the search method. In this work we focus on the Auto-WEKA tool, mainly due to the wide range of classification algorithms considered by this tool – particularly because it includes classification algorithms learning 6 types of interpretable model representations, as discussed in Sect. 2.2. This is in contrast with e.g. Auto-sklearn, which has a considerably smaller diversity of algorithms learning such interpretable model representations.

The search space considered by Auto-WEKA also includes feature selection methods and their hyper-parameter settings. That is, for each input dataset, the output of Auto-WEKA will be at least a recommended classification algorithm and its hyper-parameter settings, and that output may or may not include also a feature selection algorithm and its hyper-parameter settings, applied in a data pre-processing step. In this work, however, we focus only on the classification algorithms selected by Auto-WEKA, since we focus on analyzing the trade-off between the predictive accuracy and interpretability of the models learned by the classification algorithms.

2.2 Background on Interpretable Classification Models

Classification algorithms can be categorized into groups based on the type of knowledge representation used by the classification models that they produce. We emphasize that this grouping is based on the knowledge representation used by the classification *model*, i.e., the output of a classification algorithm. This distinction is important because the same type of model can be learned by very different types of algorithms – e.g., decision trees can be learned by a conventional greedy search method or by a more global search method like evolutionary algorithms [1].

In this work we categorize classification models into 4 broad groups, based on two criteria: (a) whether or not the model is an ensemble (i.e. combining the predictions of a set of base classifiers), and (b) the model's type of knowledge representation – which can be broadly considered interpretable or non-interpretable. These two criteria are combined into the 2×2 matrix shown in Fig. 1.

The bottom-right quadrant of the matrix in Fig. 1 (non-ensemble, non-interpretable knowledge representation) contains models categorized as black boxes. That is, users cannot normally understand such black box models in their original form. Examples include, in general, artificial neural networks and support vector machines (SVMs). Note that it is possible to extract interpretable knowledge from a black box model [9], e.g. by extracting a set of rules from neural networks or from SVMs, but in this case of course it is the set of rules which would be interpreted, not the original black box model.

		Model's overall type	
		Ensemble	Non-Ensemble
Knowledge Representation	Interpretable	Grey box	White box
	Non-Interpretable	Black box	Black box

Fig. 1. Categorization of classification models into 'white box', 'black box' and 'grey box' models, based on whether or not the model is an ensemble (combining the outputs of multiple base classifiers) and whether or not the model's knowledge representation is interpretable.

The bottom-left quadrant of the matrix in Fig. 1 (ensemble, non-interpretable knowledge representation) contains models which are also categorized as black boxes. They are in general even harder to interpret than a non-ensemble black box model, due to the typically large number of non-interpretable models in the ensemble.

The top-right quadrant of the matrix in Fig. 1 (non-ensemble, interpretable knowledge representation) contains models categorized as white boxes (sometimes called glass boxes [12]). Such models are, at least in principle, directly interpretable by users. In practice, their degree of interpretability varies depending on several factors, including e.g. the user's understanding about the meaning of the features (attributes) occurring in the model and the user's understanding about the model's knowledge representation. In this work, we consider the following 6 types of knowledge representation as 'white box' models: decision trees, If-Then classification rules, decision tables, Bayesian network classifiers, nearest neighbours and logistic regression models. The interpretability of the former 5 types of model representations was discussed in detail in [7], whilst logistic regression is also usually recognized as an interpretable type of model in the literature.

Finally, the top-left quadrant of the matrix in Fig. 1 (ensemble, interpretable knowledge representation) contains models categorized as 'grey boxes'. This term is used here to refer to some kinds of ensemble models that are partly interpretable, although substantially less interpretable than white box models. Broadly speaking, we will refer to an ensemble as a grey box if its base classifiers are white box models, since in principle some approaches for interpreting such white box models can be applied to the ensemble's members and the results can then be combined to get some interpretability for the ensemble as a whole.

An example of how an ensemble can be partly interpreted involves random forests. In general random forest models are not directly interpretable by users, since they contain too many decision trees as base classifiers, and each tree by itself is also hardly interpretable – each tree tends to be large and to have its contents heavily influenced by random samplings of instances and features. Hence, a random forest model is not a white box model. However, random forest models can be partly interpreted by computing a measure of the importance of each feature across all trees in the forest, and ranking the features in decreasing order of importance. Several such feature importance measures have been proposed in the literature [4, 17]. By using feature importance measures, a random forest model can be considered a grey box model.

3   Experimental Methodology

3.1   Datasets Used in the Experiments

We report the results of experiments using 16 datasets, whose main characteristics are mentioned in Table 1. More precisely, this table shows, for each dataset, its number of training and testing instances, as well as its number of predictive features and class labels. The training and testing sets used here are in general the same as used in the first Auto-WEKA paper [16], and they were in general downloaded from: https://www.cs. ubc.ca/labs/beta/Projects/autoweka/datasets/. The exception is the Adult Census dataset, whose training and testing sets were downloaded from the well-known UCI dataset repository: http://mlr.cs.umass.edu/ml/datasets.html.

Table 1. Main characteristics of the datasets used in the experiments

Dataset	Training Inst.	Testing Inst.	Features	Class labels
Adult Census	32561	16281	14	2
Car	1209	519	6	4
CIFAR10-small	10000	10000	3,072	10
Convex	8000	50000	784	2
Dexter	420	180	2,000	2
GermanCredit	700	300	20	2
Gisette	4900	2100	5,000	2
KDD09-Appentency	35000	15000	230	2
Kr-vs-kp	2237	959	36	2
Madelon	1820	780	500	2
MNIST basic	12000	50000	784	10
Secom	1096	471	590	2
Semeion	1115	478	256	10
Shuttle	43500	14500	9	7
Waveform	3500	1500	40	3
Yeast	1038	446	8	10

3.2   Auto-WEKA's Parameters and Experimental Set up

The output of Auto-WEKA depends on several user-specified parameters. We specified the values of three of such parameters, as discussed next, and kept the other parameters at their default values.

First, the output of Auto-WEKA naturally depends on the runtime limit ('computational budget') specified by the user, i.e. how much time the system is allowed to spend in the search for the best classification/feature selection algorithm and its/their best hyper-parameter settings for the input dataset. We report results for Auto-WEKA running with 5 h and 20 h of runtime limit.

Second, like Auto-ML systems in general, Auto-WEKA is non-deterministic, i.e., its output (selected algorithm and hyper-parameter settings) depends on the random seed number used to initialize the search. We report results of Auto-WEKA with 5 different random seed numbers, for each of the two time limits, for each dataset.

Third, Auto-WEKA's evaluation function (used to guide the search) was modified from the default 'error rate' to the Area Under the ROC curve (AUROC) [13]. The rationale for this modification was that the error rate does not cope well with very imbalanced class distributions, which is the case for several datasets used in the experiments. In addition, the AUROC is one of the most used measures of predictive accuracy in practice. The AUROC measure takes values in the range [0..1], with the value 0.5 indicating a predictive accuracy equivalent to that of a random classifier, and 1 indicating the maximal predictive accuracy.

Auto-WEKA was run 10 times (5 seeds × 2 runtime limits) for each dataset. The total time taken by the experiments for each dataset was 125 h: 25 h for the 5 runs taking 5 h each, and 100 h for the 5 runs taking 20 h each. So, the total time taken by the experiments for all 16 datasets was 2,000 h. All experiments were run on a desktop computer with an Intel® Core(TM) i7-7700 CPU with 3.6 GHz and 16.0 GB of RAM memory.

3.3   The Type of Auto-WEKA's Output Analyzed in This Work

Recall that the output of Auto-WEKA consists of the best classification algorithm (with its best hyper-parameter settings) selected for the input dataset, and possibly also the best feature selection algorithm (with its best hyper-parameter settings) to be applied in a data pre-processing step. In this work we analyze only on the types of classification algorithms selected by Auto-WEKA, i.e., the analysis of the feature selection algorithms output by Auto-WEKA is out of the scope of this work. In addition, we focus on analyzing the output algorithms by themselves, i.e., an analysis of the selected hyper-parameter settings for each algorithm selected by Auto-WEKA is also out of the scope of this work.

Recall that the classification algorithms output by Auto-WEKA have been categorized into the three broad groups of white box, black box and grey box models, based on whether or not their learned model is an ensemble and on the broad interpretability of their model's knowledge representation, as discussed in Sect. 2.2.

A brief overview of the classification algorithms selected by Auto-WEKA in our experiments (reported in the next Section) is given next, first for ensembles and then for non-ensemble algorithms.

Ensemble Algorithms:

- AdaBoost-M1: It learns an ensemble of base classifiers by iteratively re-weighting instances – increasing the weights of instances misclassified in previous iterations.
- Bagging (Bag): It learns an ensemble of base classifiers, each of them is learned from randomly sampling instances.
- Random Committee (RandCom): It learns an ensemble of randomized base classifiers; each is learned from the same data, but using a different random seed number.
- Random Forest (RF): It learns a forest (set) of decision trees, each of them is learned by randomly sampling instances and features.
- Random SubSpace (RandSS): It learns an ensemble of randomized base classifiers; each is learned by randomly sampling features (creating different feature subspaces).
- Vote: An ensemble combining the outputs of different types of base classifiers.

Non-ensemble Algorithms:

- BayesNet: It learns a Bayesian network classifier, it can cope with dependences among features (unlike Naïve Bayes).
- Decision Table (DecTable): It learns a decision table model, finding a good set of features to be used in the table.
- Decision Stump (DecStump): It learns a decision stump, which is a decision tree with just one internal (non-leaf) node.
- IBk: A k-nearest neighbour (instance-based learning) classifier.
- JRip: It implements the RIPPER algorithm for learning a list of IF-THEN rules.
- KStar (K*): A specific type of k-nearest neighbour (instance-based learning) classifier that uses an entropy-based distance function.
- Logistic (Log): It learns a multinomial logistic regression model with a ridge estimator.
- LMT: It learns a Logistic Model Tree, i.e., a decision tree with logistic regression models at the leaf nodes.
- LWL: Locally Weighted Learning – It uses an instance-based learning algorithm to assign instance weights, which are then used by a suitable classifier.
- MLP: It learns a Multi-Layer Perceptron neural network using backpropagation.
- Naïve Bayes (NB): The simplest type of Bayesian network classifier; it assumes that features are independent from each other given the class variable.
- PART: Rule induction algorithm that iteratively learns a list of IF-THEN rules, by iteratively converting a learned partial decision tree into a rule.
- RepTree: A decision tree learning algorithm designed to be faster than other algorithms of this type – it sorts numeric attributes just once.
- SimpleLogistic (SimpLog): It learns linear logistic regression models.
- SMO: The Sequential Minimal Optimization algorithm for learning an SVM (Support Vector Machine) model.
- ZeroR: No learned model; it simply predicts the most frequently class in the data.

4   Computational Results

4.1   Analysis of the Influence of the Runtime Limit on Auto-WEKA's Predictive Accuracy

Table 2 shows the mean and standard deviation (over 5 runs varying the random seed) of the Area Under the ROC curve (AUROC) values obtained by the algorithms selected by Auto-WEKA, measured on the test sets, for the experiments with 5 h and 20 h of runtime limit. The last column of this table shows the difference between the mean AUROC with 20 h and the mean AUROC with 5 h. Hence, a positive (negative) value in that column indicates that increasing the runtime limit from 5 to 20 h had a positive (negative) effect on the AUROC. The AUROC difference in the last column tends to be larger for datasets with smaller AUROC values, which of course offer more opportunities for larger differences to arise.

In 12 of the 16 datasets included in Table 2, the difference of AUROC between the two runtime limits was very small, smaller than 1%. In the other 4 datasets, however, the runtime limit had a substantial effect: the longer run (20 h) led to a larger AUROC in two datasets (an increase of 11.6% for KDD09-Appentency and 2.8% for GermanCredit) but to a smaller AUROC in two other datasets (a decrease of 6.8% for Convex and 2% for Madelon). The AUROC values' standard deviations are in general small, except for the 5-hour runs in two datasets (Convex and KDD09-Appentency).

Table 2. Mean and (after the symbol±) standard deviation of the AUROC obtained by Auto-WEKA on the test set over 5 runs varying the random seed, with the runtime limit set to 5 h and 20 h, and the difference between the two AUROC values.

Dataset	AUROC (5 h)	AUROC (20 h)	AUROC difference (20h-AUROC–5h-AUROC)
Adult Census	0.9058 ± 0.003	0.9014 ± 0.010	-0.0044
Car	1.0 ± 0	1.0 ± 0	0
CIFAR10-small	0.7282 ± 0.020	0.7268 ± 0.013	-0.0014
Convex	0.6276 ± 0.121	0.5592 ± 0.024	-0.0684
Dexter	0.9588 ± 0.037	0.9578 ± 0.012	-0.001
GermanCredit	0.7182 ± 0.041	0.7464 ± 0.012	0.0282
Gisette	0.9878 ± 0.003	0.987 ± 0.003	-0.0008
KDD09-Appent.	0.663 ± 0.152	0.7794 ± 0.032	0.1164
Kr-vs-kp	0.9864 ± 0.019	0.9896 ± 0.013	0.0032
Madelon	0.836 ± 0.032	0.816 ± 0.042	-0.02
MNIST basic	0.989 ± 0.008	0.9878 ± 0.007	-0.0012
Secom	0.6978 ± 0.022	0.7018 ± 0.021	0.004
Semeion	0.9932 ± 0.004	0.9944 ± 0.002	0.0012
Shuttle	1.0 ± 0	1.0 ± 0	0
Waveform	0.972 ± 0.001	0.9704 ± 0.003	-0.0016
Yeast	0.828 ± 0.008	0.8292 ± 0.011	0.0012

We used the non-parametric Wilcoxon signed-rank test of statistical significance to compare the results for 5-hour and 20-hour runs shown in Table 2. Using a two-tailed test and significance level $\alpha = 0.05$ as usual, we obtained $p = 0.94$, so the difference of AUROC values among the 5-hour and 20-hour runs is clearly not significant.

Table 3. Distribution of classification algorithms selected by Auto-WEKA for each dataset across 5 runs, for the runtime limits of 5 h and 20 h. The number in brackets after an algorithm's name represents the selection frequency for that algorithm, out of the 5 runs. The absence of numbers in brackets means the algorithm was selected only once.

Dataset	Selected algorithms (5-hour runs)	Selected algorithms (20-hour runs)
Adult Census	BayesNet(4), RF	BayesNet (2), RF; SimpLog; NB
Car	MLP(2), Bag-SMO, SMO, AdaBoost-SMO	MLP (2), AdaBoost-SMO, AdaBoost-MLP, SMO
CIFAR10-small	RF(2), SMO, PART, NB	RF(3), NB(2)
Convex	RF(4), RandCom.-RepTree	RF(2), SMO, LMT, RandCom.-RepTree
Dexter	NB(2), SMO, MLP, RepTree	Logistic, Bag-J48, KStar, Vote-SimpLog, LMT
GermanCredit	SMO(4), Bag-RF	Vote-LMT, MLP, LWL-MLP, Bag-MLP, RandCom-MLP
Gisette	RF(2), Logistic, SimpLog, AdaBoost-RepTree	RF(3), Logistic, AdaBoost-RepTree
KDD09-Appentency	DecTable(2), Bag-PART, Bag-DecStump, ZeroR	DecTable(3), MLP, Bag-DecStump
Kr-vs-kp	AdaBoost-JRip, LMT, AdaBoost-RepTree, MLP, RandCom-MLP	AdaBoost-JRip, LMT, AdaBoost-PART, MLP, RandCom-MLP,
Madelon	RandSS-RepTree, RandCom-RepTree, RandCom-RF, IBK, RF	RF(2), Rand-SubSp-JRip, IBK, RandCom-RepTree
MNIST basic	IBk(2), RF(2), NBmultidim	IBK(2), RF(2), BayesNet
Secom	BayesNet(3), NB, Bag-BayesNet	BayesNet(4), Bag-BayesNet
Semeion	KStar(3), MLP, RandSS-KStar	KStar(3), RF, RandSS-KStar
Shuttle	RF(5)	RF(5)
Waveform	MLP(5)	MLP(3), SimpLog, Bag-MLP
Yeast	RF(2), MLP, Bag-LMT, RandCom-RF	RF(2), MLP, Bag-JRip Bag-MLP

We investigated in more detail the results for the KDD09-Appetency dataset, with the largest difference of AUROC between the two runtimes. The large increase in the AUROC value associated with the longer runs of 20 h is due mainly to the fact that, in the experiments with 5-hour runs, two of the 5 runs achieved a very low AUROC of 0.5 (equivalent to random predictions). In both these runs, the classifier selected by Auto-WEKA was a trivial classifier that simply predicted the most frequent class label to all instances, ignoring the features.

4.2 Analysis of the Distribution of the Classification Algorithms Selected by Auto-WEKA for Each Dataset, Varying Runtime Limit and Random Seed

Table 3 shows the distribution of classification algorithms selected by Auto-WEKA for each dataset, separately for the experiments with runtime limits of 5 h and 20 h. Recall that, for each runtime limit, Auto-WEKA was run 5 times for each dataset, varying the random seed across runs. For information about the algorithms' acronyms used in this table, the reader is referred to Sect. 3.3.

Table 3 shows that there is a wide variety of classification algorithms selected by Auto-WEKA across all datasets. This reinforces the motivation to use an Auto-ML system to try find the best algorithm for each dataset, supporting the results in [16].

There is also substantial variation among the algorithms selected for each dataset, confirming that the output of Auto-WEKA is sensitive to the random seed number used to initialize its search. However, for some datasets the selection of the best algorithm was reasonably stable across the runs varying the random seed. More precisely, there were 4 algorithms that were selected in the majority (i.e. at least 3) of the 5 runs for each of the two runtime limits (5 h and 20 h) for some dataset, as follows. First, Random Forest (RF) was chosen in all 10 runs (5 runs × 2 runtime limits) for the Shuttle dataset. Second, MLP was selected in 8 runs for the Waveform dataset: 5 times with the runtime limit of 5 h and 3 times with the limit of 20 h. Third, BayesNet was selected in 7 runs for the Secom dataset: 3 times with the runtime limit of 5 h, and 4 times with the runtime limit of 20 h. Fourth, KStar was selected in 6 runs for the Semeion dataset: 3 times for each of the two runtime limits. In addition, when the runtime limit was 5 h, RF was selected 4 times for the Convex dataset; and when the runtime limit was 20 h, RF was selected 3 times for the Gisette dataset and 3 times for the CIFAR10-small dataset.

One can also observe in Table 3 that, for the large majority of the datasets, the set of selected algorithms is broadly similar in the two scenarios of 5-hour and 20-hour runs. More precisely, for 13 of the 16 datasets, the intersection between the sets of algorithms selected by Auto-WEKA in the two scenarios has at least 3 (out of 5) algorithms. In one dataset (Shuttle) all 5 selected algorithms were the same (RF) in the two scenarios. However, in two datasets (Dexter and GermanCredit) there was no intersection between the sets of algorithms selected in the two scenarios. As mentioned earlier, for the GermanCredit dataset the longer runs led to a somewhat higher AUROC, but for the Dexter dataset the change of selected algorithms between 5-hour and 20-hour runs did not have any substantial effect on the AUROC.

Table 4 shows the selection frequency of each algorithm for all datasets as a whole. In Table 4 the algorithms are divided into the three previously discussed broad groups of algorithms that learn: (a) white box models, (b) black box models, and (c) ensembles, some of which can be considered as 'grey box' models if they use white box models as their base classifiers, as discussed in Sect. 2.2.

Table 4. Selection frequency for each type of model (black box, white box or ensemble model) selected by Auto-WEKA for all datasets as a whole, for each runtime limit (5 h or 20 h), and total frequency. In the rows for ensembles, the numbers in brackets are the numbers of ensemble models that can be categorized as 'grey boxes', in the sense of consisting of base classifiers that are a type of white box model.

Model type	Algorithm	Sel. Freq (5-hours runs)	Sel. Freq (20-hour runs)	Total Sel. frequency
Non-ensemble white box	BayesNet	7	7	14
	Naïve Bayes	4	3	7
	Naïve Bayes multinomial	1	0	1
	KStar	3	4	7
	IBK	3	3	6
	Decision table	2	3	5
	LMT	1	3	4
	SimpleLogistic	1	2	3
	Logistic	1	2	3
	PART	1	0	1
	RepTree	1	0	1
	Totals for white boxes:	25	27	52
Non-ensemble black box	MLP	11	9	20
	MLP-LWL	0	1	1
	SMO	7	2	9
	Totals for black boxes:	18	12	30
Ensemble (number of grey boxes)	AdaBoost	4 (3)	5 (3)	9 (6)
	Bagging	6 (4)	7 (4)	13 (8)
	Random Committee	5 (2)	4 (2)	9 (4)
	Random SubSpace	2 (2)	2 (2)	4 (4)
	Random forest	20 (20)	21 (21)	41 (41)
	Vote	0 (0)	2 (2)	2 (2)
	Totals for ensembles:	37 (31)	41 (34)	78 (65)
No model	ZeroR	1	0	1

Let us first discuss in more detail the results for white box and black box models. As a whole, white box models were selected more often than black box models in both the experiments with 5 h of runtime limit and the experiments with 20 h. The difference of selection frequency in favour of white box models is considerably larger for the 20-hour runs (27 white box models vs. 12 black box models) than for the 5-hour runs (25 vs. 18).

The most frequently selected type of white box model was BayesNet, which was selected 14 times in total (adding the selection frequencies for both runtime limits). In addition, Naïve Bayes was the second most frequently selected white box classifier, with a total frequency of 8 (including one selection of its variant Naïve Bayes multinomial); and since both BayesNet and Naïve Bayes are instantiations of a Bayesian network classifier, this broad type of model was selected in total 22 times.

The second most frequently selected broad type of white box model was nearest neighbours, with the KStar and IBk algorithms selected 7 and 6 times, respectively – i.e., 13 times in total.

Other types of white box models had smaller but still substantial selection frequencies, as follows. DecisionTable was selected 5 times. LMT (Logistic Model Trees) was selected 4 times. Note that a LMT model is a hybrid decision tree/logistic regression model (it is a decision tree with logistic regression models at the leaf nodes). A stand-alone logistic regression model was selected 6 times (3 times with the Logistic algorithm and 3 times with the SimpleLogistic algorithm).

Decision trees by themselves (i.e., not counting their use in ensembles) had a surprising low selection frequency. Not counting the 4 times a LMT model was selected, a stand-alone decision tree model was selected just once, with the RepTree algorithm – which was designed to be fast (not just to maximize accuracy), i.e., it may sacrifice some accuracy to gain computational efficiency.

The black box models selected by Auto-WEKA were less diverse than the white box models; more precisely, MLP was selected 21 times (one of them using LWL – Local Weighted Learning – to assign weights to instances), whilst SMO (a type of SVM algorithm) was selected 9 times.

We now turn to ensembles. As a whole, ensembles were the type of algorithm most frequently selected by Auto-WEKA, for both runtime limits (5 and 20 h). In total, ensembles were selected in 78 out of the 160 cases (i.e., in about 49% of the cases). The overall success of ensembles is not surprising, due to their advantages stemming from combining diverse base models to achieve a more effective classifier [20].

By far the most selected type of ensemble model was Random Forest, which was selected 41 times in total (i.e., in about 23% of the 180 cases). Bagging, AdaBoost-M1 and Random Committee were also selected quite often by Auto-WEKA, in total 13, 9 and 9 times, respectively. Random SubSpace and Vote were selected only 4 and 2 times, respectively.

Recall that we considered as 'grey box' models the ensembles that can be partly interpreted, due to their base classifiers being interpretable (white box) models. Hence, in the rows for ensemble models in Table 4, the numbers in brackets are the numbers of models that can be categorized as 'grey boxes'.

Note that, since random forests consist of partly random decision tree models, and many feature importance measures for random forests are available in the literature as mentioned earlier, all 41 random forest models mentioned in Table 4 can be considered grey box models. The other types of ensemble models in Table 4 also have a high proportion of grey box models in general. Actually, considering all types of ensemble models in Table 4 for the two runtime limits of 5 h and 20 h, 65 out of the 78 ensemble models (i.e., about 83%) can be considered grey box models. It should be emphasized, however, that a grey box model is still considerably less interpretable than a white box model, and it requires substantial post-processing for interpretability. That is, after the ensemble is constructed, typically we still need to run some post-processing procedure (e.g. the aforementioned feature importance measures). By contrast, such post-processing is not usually required in the case of white box models, which can be more directly interpreted. A detailed investigation of to what extent such grey box, partly interpretable models can be really (subjectively) interpreted by users in practice is beyond the scope of this work.

4.3 Analysis of the Trade-Off Between the Predictive Accuracy and the Interpretability of the Selected Classification Models

Recall that Auto-WEKA's search is guided by an evaluation function that is based on estimating only the predictive accuracy of the candidate algorithms, without considering the interpretability of their learned models. Despite this, for any given dataset, it is possible that the best algorithm selected by Auto-WEKA for a given input dataset is an algorithm that learns a white box model, in which case we would get the benefit of a model with an interpretable knowledge representation without sacrificing accuracy.

As mentioned in the Introduction, there is a growing importance of interpretability in the classification task of machine learning, due to the increasingly large number of applications of classification algorithms across many domains. Despite this, the literature is still overwhelmingly dominated by the goal of maximizing predictive accuracy, with relatively little emphasis on learning interpretable models. That is, most researchers and practitioners focus on using only black box or ensemble models, without even trying algorithms that learn at least potentially interpretable models. It is not clear how often this leads to missing the opportunity of learning an interpretable model that is almost as accurate as a black box or ensemble model. Hence, it is important to investigate this trade-off between predictive accuracy and interpretability by considering a wide range of algorithms.

Auto-ML systems like Auto-WEKA provide an interesting novel perspective for this investigation, because Auto-WEKA automatically searches for the best algorithm for the input dataset, in a search space that includes both many algorithms learning white box models and many algorithms learning black box or ensemble models.

In this context, the important research question addressed in this section is: to what extent does the best white box model recommended by Auto-WEKA (for the input dataset) sacrifice predictive accuracy, by comparison with the best non-white box (i.e. black box or ensemble) model recommended by Auto-WEKA?

To investigate this issue, for each dataset, and for each of the two runtime limits (5 h and 20 h), Table 5 reports two types of AUROC values, both measured on the test set: (a) the highest AUROC among the non-white box (i.e., black box and ensemble) models produced by the algorithms selected by Auto-WEKA in its 5 runs varying the random seed; and (b) the highest AUROC among the white box models produced by the algorithms selected by Auto-WEKA in its 5 runs. Each cell of Table 5 also indicates, below the AUROC value, the name of the algorithm(s) which obtained that result. If none of the 5 algorithms selected by Auto-WEKA for a given pair of dataset and runtime limit learns the type of model associated with the corresponding table column, the corresponding cell in Table 5 has the keyword 'none'.

Hence, in order to determine to what extent the selected white box models are sacrificing predictive accuracy by comparison with the best non-white box model found by Auto-WEKA, for each dataset and runtime limit, one can compare two pairs of columns in Table 5: the second and third columns (5-hour runs), and the fourth and fifth columns (20-hour runs). The best result for each dataset and each run time limit is shown in boldface font.

For the 5-hour runs, the best white box model achieved a higher AUROC than the best non-white box model in only 4 of the 16 datasets. In those 4 datasets, the gain in predictive accuracy associated with the best white box model (versus the best non-white box model) was: 0.5% for Adult Census, 0.9% for Semeion, 1.8% for Dexter, and 7.3% for KDD09-Appetency. However, no white box model was selected in the 5 Auto-WEKA runs for 6 datasets. Regarding the remaining 6 datasets, it is interesting to note that the loss of predictive accuracy associated with the best white box model (versus the best non-white box model) was very small (less than 0.5%) in 4 of those datasets. More precisely, these AUROC losses were: 0.1% for kr-vs-kp and MNIST Basic, 0.2% for Gisette, 0.4% for CIFAR10-small, 3.6% for Secom, 8% for Madelon.

For the 20-hour runs, the best white box model achieved a higher AUROC than the best non-white box model in 7 of the 16 datasets. In those 7 datasets, the gain in predictive accuracy (AUROC value) associated with the best white box model (versus the best non-white box model) was: 0.4% for Dexter and Semeion, 0.5% for Adult Census, 0.9% for MNIST Basic, 1.0% for KDD09-Appetency, 2.7% for Secom, and 2.9% for CIFAR10small. However, no white box model was selected in the 5 Auto-WEKA runs for 4 datasets. Regarding the remaining 5 datasets, it is interesting to note that the loss of predictive accuracy associated with the best white box model (versus the best non-white box model) was very small (less than 1%) in 3 of those datasets. More precisely, these AUROC losses were: 0.1% for kr-vs-kp, 0.2% for Gisette, 0.7% for Waveform, 2.8% for Convex, and 4.8% for Madelon.

We used the non-parametric Wilcoxon signed-rank test of statistical significance to compare the aforementioned two pairs of results in Table 5, i.e., to compare the results for the best non-white box vs. the results for the best white box model, for each runtime limit (5 h and 20 h). For this comparison, the cases where no white box model was selected were assigned an AUROC of 0. Using a two-tailed test and significance level $\alpha = 0.05$ as usual, we obtained $p = 0.0349$ and $p = 0.2846$ for the 5-hour and 20-hour runs, respectively. Hence, the difference of predictive accuracy between the best non-white box models and the best white box models is statistically significant (in favour of non-white box models) for the 5-hour runs, but not statistically significant for the 20-hour runs.

Table 5. AUROC (on the test set) of the best non-white box model (i.e. the best among black box and ensemble models) and the best white box models, separately for 5-hour and 20-hour runs. In each cell, the name of the algorithm(s) producing the corresponding best model is shown below the AUROC value. The best result for each pair of dataset and runtime limit is shown in boldface font.

Dataset	5-hour runs		20-hour runs	
	Best non-white box model	Best white box model	Best non-white box model	Best white box model
Adult Census	0.903 Rand. forest	**0.908 BayesNet**	0.903 Rand. forest	**0.908 BayesNet**
Car	**1.0 SMO, MLP, Bagg., AdaBo.**	None	**1.0 SMO, MLP, AdaBoost**	None
CIFAR10-small	**0.751 SMO**	0.747 Naïve Bayes	0.718 Rand. forest	**0.747 Naïve Bayes**
Convex	**0.844 Rand. forest**	None	**0.584 Rand. forest**	0.556 LMT
Dexter	0.973 MLP	**0.991 Naive Bayes**	0.965 Vote	**0.969 LMT**
GermanCredit	**0.753 SMO**	None	**0.762 Vote**	None
Gisette	**0.991 AdaBoost**	0.989 Log., SimpLog.	**0.991 AdaBoost**	0.989 Logistic
KDD09-Appentency	0.723 Bagging	**0.796 DecTable**	0.786 MLP	**0.796 DecTable**
Kr-vs-kp	**1.0 AdaBoost**	0.999 LMT	**1.0 AdaBoost**	0.999 LMT
Madelon	**0.891 Rand. Com.**	0.811 IBk	**0.859 Rand. forest**	0.811 IBk
MNIST basic	**0.996 Rand. forest**	0.995 IBk	0.986 Rand. forest	**0.995 IBk**
Secom	**0.735 Bagging**	0.699 Naive Bayes	0.708 Bagging	**0.735 BayesNet**
Semeion	0.987 MLP	**0.996 KStar**	0.993 Rand. SubSp.	**0.997 KStar**
Shuttle	**1.0 Rand. forest**	None	**1.0 Rand. forest**	None
Waveform	**0.973 MLP**	None	**0.973 MLP, Bagg.**	0.966 SimpleLogistic
Yeast	**0.835 Rand. Comm.**	None	**0.839 Rand. forest**	none
Num. of wins	12	4	9	7

5 Summary of Results and Discussion

Regarding the influence of the runtime limit on the predictive accuracy of Auto-WEKA, the difference between the mean AUROC values for the experiments with 5 h and 20 h of runtime limit was smaller than 1% in 12 of the 16 datasets; and overall (across all datasets) the difference was not statistically significant.

Regarding the frequencies with which different classification algorithms are selected by Auto-WEKA for each dataset, Auto-WEKA selected a wide variety of classification algorithms across the 16 datasets. This supports the motivation to use an Auto-ML system to try to find the best algorithm with its best hyper-parameter settings for each dataset.

For most datasets, the difference between the sets of classification algorithms selected by Auto-WEKA with 5-hour runs and 20-hour runs is not large, i.e., several selected algorithms tend to be the same for both runtime limits.

In any case, in practice it seems important to run Auto-WEKA several times for the same dataset by varying the random seed across the runs, since for most datasets there was a substantial diversity of selected algorithms across different runs – which was observed with both 5 h and 20 h of runtime limit.

Ensembles were selected by Auto-WEKA as the best algorithm in about 49% of the cases (in 78 out of 160 cases). The high prevalence of ensembles was consistently observed for both runtime limits (5 h and 20 h). In addition, the model type most frequently selected by Auto-WEKA was random forest, an ensemble considered a grey box model (see Sect. 2.2), which was selected 41 times in total – over both 5-hour and 20-hour runtime limits. Among non-ensembles, white box and black models were selected in 52 (32.5%) and 30 (18.75%) of the 160 cases, respectively.

The most frequently selected type of white box model was Bayesian network classifiers – more precisely, 14 selections of BayesNet and 8 selections of standard Naïve Bayes or its multinomial variant. Although a Naïve Bayes model can be easily interpreted due to its simplifying assumption that features are independent of each other given the class variable, the interpretation of BayesNet becomes more difficult as more and more feature dependencies are included in the network. In the general case of Bayesian networks, for instance, Heckerman et al. [11] have pointed out that users can get confused with the interpretation of (in)dependence relationships represented in Bayesian networks, and suggested an alternative knowledge representation of dependence networks that seems to have improved interpretability.

We also analyzed the difference of predictive accuracy (AUROC values) between the best white box model and the best non-white box model selected by Auto-WEKA for each dataset.

Overall, the best white box model achieved a higher AUROC than the best non-white box model in only 4 of the 16 datasets in the experiments with 5 h of runtime limit, and in 7 out of 16 datasets in the experiments with 20 h of runtime limit. However, the loss of predictive accuracy associated with the best white box model (versus the best non-white box model) was smaller than 0.5% for 4 datasets in the 5-hour experiments, and smaller than 1% for 3 datasets in the 20-hour experiments. The higher AUROC values associated with the best non-white box models was statistically significant in the 5-hour experiments, but not in the 20-hour experiments.

6   Conclusions and Future Work

We have proposed the use of Automated Machine Learning (Auto-ML) methods as a novel approach to investigate the trade-off between the predictive accuracy and interpretability of classification models. The experiments involved 160 runs of Auto-WEKA (a popular Auto-ML tool) – 10 runs for each dataset, varying the runtime limit (computational budget) and the random seed across the runs.

In this work classification algorithms were divided into the following groups (as summarized in Fig. 1): white box non-ensemble models (potentially fully interpretable), black box non-ensemble models (not interpretable) and ensembles – some of them considered partly interpretable grey box models; whilst other ensembles are black boxes.

Overall, the algorithm type most selected by Auto-WEKA were ensembles, with the random forest ensemble in particular being the most selected algorithm type. Among non-ensembles, algorithms producing white box models were selected more often than algorithms producing black box, and variations of Naïve Bayes and Bayesian network classification algorithms were the most selected type of algorithm producing white box models.

Finally, we used Auto-WEKA's automated search for the best algorithm for each dataset as an approach to address the following research question: "to what extent does the best white box model recommended by Auto-WEKA (for the input dataset) sacrifice predictive accuracy, by comparison with the best non-white box (i.e. black box or ensemble) model recommended by Auto-WEKA?"

The results have shown the loss of predictive accuracy (AUROC value) associated with the best white box model – by comparison with the best non-white box – is often small, in several cases being smaller than 1%.

In application domains where interpretability is very important, an accuracy loss of 1% seems an acceptable price to pay for the benefit of having a white box, interpretable model, instead of a non-interpretable model – see e.g. the discussion in [2], where interpretable logistic regression models were preferred by the user over substantially more accurate but non-interpretable neural network models in a medical domain.

If we consider an accuracy loss of 1% as acceptable in order to get the benefits of an interpretable model representation (which is an application domain-dependent decision in practice), the main conclusions are as follows. For the 5-hour experiments, we would prefer the best white box model over the best non-white box one in 8 out of the 16 datasets (with the best white box model being more accurate in 4 datasets). For the 20-hour experiments, we would prefer the best white box model in 10 of the 16 datasets (with the best white box model being more accurate in 7 datasets).

Note, however, that this work considered as white box all models using some interpretable knowledge representation, without analyzing the internal details of the models to check if they are really (subjectively) interpretable by users.

As future work, it would be interesting to perform experiments with other Auto-ML tools and more datasets. In addition, although we have to some extent discussed the potential interpretability of ensemble models where the base classifiers are white box models, this is a complex issue that deserves more investigation in future work.

References

1. Barros, R.C., Basgalupp, M.P., de Carvalho, A.C.P.L.F., Freitas, A.A.: A survey of evolutionary algorithms for decision tree induction. IEEE Trans. Syst. Man Cybern. Part C: Appl. Rev. **42**(3), 291–312 (2012)

2. Caruana, R., Lou, Y., Gehrke, J., Koch, P., Sturm, M., Elhadad, N.: Intelligible models for healthcare: predicting pneumonia risk and hospital 30-day readmission. In: Proceedings ACM SIGKDD International Conference on Knowledge Discovery and Data Mining (KDD 2015), pp. 1721–1730. ACM (2015)
3. de Sá, A.G.C., Freitas, A.A., Pappa, G.L.: Automated selection and configuration of multi-label classification algorithms with grammar-based genetic programming. In: Auger, A., Fonseca, C.M., Lourenço, N., Machado, P., Paquete, L., Whitley, D. (eds.) PPSN 2018. LNCS, vol. 11102, pp. 308–320. Springer, Cham (2018). https://doi.org/10.1007/978-3-319-99259-4_25
4. Epifanio, I.: Intervention in prediction measure: a new approach to assessing variable importance for random forests. BMC Bioinformatics **18**, 230 (2017)
5. Fernandez-Delgado, M., Cernadas, E., Barro, S., Amorin, D.: Do we need hundreds of classifiers to solve real world classification problems? J. Mach. Learn. Res. **15**, 3133–3181 (2014)
6. Feurer, M., Klein, A., Eggensperger, K., Springenberg, J., Blum, M., Hutter, F.: Efficient and robust automated machine learning. In: Proceedings Advances in Neural Information Processing Systems, pp. 2962–2970 (2015)
7. Freitas, A.A.: Comprehensible classification models. ACM SIGKDD Explor. **15**(1), 1–10 (2013)
8. Furnkranz, J., Kliegr, T., Paulheim, H.: On cognitive preferences and the interpretability of rule-based models. arXiv preprint: arXiv:1803.01316v2 [cs.LG], 10 March 2018
9. Guidotti, R., Monreale, A., Turini, F., Pedreschi, D., Giannotti, F.: A survey of methods for explaining black box models. arXiv:1802.01933v1 [cs.CY], 6 February 2018
10. Guyon, I., et al.: A brief review of the ChaLearn AutoML challenge: any-time any-dataset learning without human intervention. In: Proceedings ICML 2016 AutoML Workshop, vol. 64, pp. 21–30 (2016). Published as JMLR: Workshop and Conference Proceedings
11. Heckerman, D., Chickering, D.M., Meek, C., Rounthwaite, R., Kadie, C.: Dependency networks for inference, collaborative filtering and data visualization. J. Mach. Learn. Res. **1**, 49–75 (2000)
12. Holzinger, A.: Interactive machine learning for health informatics: when do we need the human-in-the-loop? Brain Inform. **3**(2), 119–131 (2016)
13. Japkowicz, N., Shah, M.: Evaluating Learning Algorithms: A Classification Perspective. Cambridge University Press, Cambridge (2011)
14. Mohr, F., Wever, M., Hüllermeier, E.: ML-Plan: automated machine learning via hierarchical planning. Mach. Learn. **107**(8–10), 1495–1515 (2018)
15. Olson, R.S., Bartley, N., Urbanowicz, R.J., Moore, J.H.: Evaluation of a tree-based pipeline tool for automating data science. In: Proceedings Genetic and Evolutionary Computation Conference (GECCO-2016), pp. 485–492 (2016)
16. Thornton, C., et al.: Auto-WEKA: combined selection and hyperparameter optimization of classification algorithms. In: Proceedings 19th ACM SIGKDD International Conference on Knowledge Discovery & Data Mining, pp. 847–855. ACM (2013)
17. Verikas, A., Gelzinis, A., Bacauskiene, M.: Mining data with random forests: a survey and results of new tests. Pattern Recogn. **44**, 330–349 (2011)
18. Yao, Q., et al.: Taking human out of learning applications: a survey on automated machine learning. arXiv preprint arXiv:1810.13306, 31 October 2018
19. Zhang, C., Liu, C., Zhang, X., Almpanidis, G.: An up-to-date comparison of state-of-the-art classification algorithms. Expert Syst. Appl. **82**, 128–150 (2017)
20. Zhou, Z.H.: Ensemble Methods: Foundations and Algorithms. CRC, Boca Raton (2012)

Estimating the Driver Status Using Long Short Term Memory

Shokoufeh Monjezi Kouchak[✉] and Ashraf Gaffar

Arizona State University, Tempe, AZ, USA
smonjezi@asu.edu

Abstract. Driver distraction is one of the leading causes of fatal car accidents. Driver distraction is any task that diverts the driver attention from the primary task of driving and increases the driver's cognitive load. Detecting potentially dangerous driving situations or automating some repetitive tasks, using Advanced Driver Assistance Systems (ADAS), and using autonomous vehicles to reduce human errors while driving are two suggested solutions to diminish driver distraction. These solutions have some advantages, but they suffer from their inherent inability to detect all potentially dangerous driving situations. Besides, autonomous vehicles and ADAS depend on sensors. As a result, their accuracy diminishes significantly in adverse conditions. Analyzing driver behavior using machine learning methods and estimating the distraction level of drivers can be used to detect potentially hazardous situations and warn the drivers. We conducted an experiment in eight different driving scenarios and collected a large dataset from driving data and driver related data. We chose Long Short Term Memory (LSTM) as our machine learning method. We built and trained a stacked LSTM network to estimate the driver status using a sequence of driving data vectors. Each driving data vector has 10 driving related features. We can accurately estimate the driver status with no external devices and only using cars Can-Bus data.

Keywords: Recurrent Neural Network · Driver distraction · Deep learning · Long Short Term Memory Network

1 Introduction

Everyday approximately nine people die and more than 1,000 are injured in car crashes that are caused by distracted drivers [1]. More than 90% of car accidents happen due to human error [2]. Using a cell phone or interacting with the car infotainment system significantly increases drivers' cognitive load and causes driver distraction. Car crash is the leading cause of teenage death [3]. [4] Mentions that 21% of teenagers involved in fatal car accidents were distracted by their cellphone. Teenagers are four times more likely to have car crashes while they are texting or talking on the phone [4]. Being an attentive driver can reduce the chance of car accidents significantly, but it is not the ultimate solution for driver distraction since people do distractive tasks such as texting although they know it is a serious issue for their safety and can lead to a car crash. AAA Traffic Safety 2016 report mentions that from 2,501 drivers that participated in

© IFIP International Federation for Information Processing 2019
Published by Springer Nature Switzerland AG 2019
A. Holzinger et al. (Eds.): CD-MAKE 2019, LNCS 11713, pp. 67–77, 2019.
https://doi.org/10.1007/978-3-030-29726-8_5

their research almost 90% said texting while driving is very dangerous, but 18% of them admitted to texting while driving in the past month [5].

Driver distraction is any task that diverts the driver attention from the primary task of driving and increases the cognitive load of the driver [6, 7]. There are four types of driver distraction including visual, manual, audio and cognitive distraction. Some distracting tasks such as texting cause a combination of these types of distraction, so they are more dangerous compared to tasks that only cause one type of distraction [8, 9]. For instance, texting causes manual, visual and cognitive distraction and it makes texting more dangerous than drunk driving. Based on NHTSA research texting while driving is six times more dangerous than driving intoxicated [10].

Enhancing the design of user interface in cars [11, 12] to make it more suitable for cars' environment can reduce the cognitive load of interacting with the car's info-tainment system. Autonomous vehicles and Advanced Driver Assistant Systems (ADAS) are two suggested solutions for enhancing driver safety [13–16]. ADAS can handle some specific dangerous situations but they can't detect all potentially menacing situations [17]. Besides, although autonomous vehicles can drive automatically in normal situations using complex sensors and sophisticated artificial intelligence methods, they can't handle some unexpected and extreme events. The human driver needs to be ready for taking control of the car in emergency situations [18, 19].

Driver behavior analysis methods have been used in [20–22] to detect the driver's abnormal behavior or the level of the driver's cognitive load to estimate and reduce driver distraction. A variety of machine learning methods such as Neural Network, Hidden Markov Model and Support Vector Machine have been used to analyze driver behavior using driver status data, car related data or combination of them [22]. Deep learning methods such as convolutional neural networks, deep neural networks and recurrent neural networks outperform traditional machine learning approaches in many safety-related applications such as pedestrian detection [13]. Driving is an intricated task, so deep learning methods are suitable choices to extract and learn complex patterns of driving.

Driving is a continuous task and the driving situation in each time step depends on several previous steps and influences several next steps. Markov Model is a machine learning approach that has memory making it a suitable choice for applications with dependent inputs. It has been used in many driving safety applications successfully [23, 24]. In Hidden Markov Model all possible actions and states need to be defined in advance, so it works accurately when we want to detect a specific condition or analyze few specific maneuvers. On the other hand, if we have a large number of states and possible actions or if all states can't be defined in advance Markov Model is not a suitable choice for our system. In these cases, Recurrent Neural Networks (RNN) can be used as a more appropriate method that can learn intricated patterns with no need to have previous information about the model [25–27]. In this paper, we use a Long Short Term Memory (LSTM) network which is a type of Recurrent Neural Network to estimate the driver behavior using both scaled and not scaled driving data. We collected a large dataset of driving and driver behavior data vectors by conducting an experiment in 8 different driving scenarios and used the dataset to train an LSTM network which estimates the driver behavior using driving data. In Sect. 2, we talk about RNN and related works. Section 3 is the experiment and Sect. 4 is data. In Sect. 5, we discuss the model that we built using an LSTM network. Section 6 is results, and Sect. 7 is the conclusion.

2 Recurrent Neural Network

Traditional Neural Networks assume that all input samples are independent of each other, so after training the network using each sample, all information about the sample removes and the next sample doesn't use any information from the previous ones. Besides, they use fixed size input and output data. These assumptions are not true in some real-life applications such as speech recognition, language translation, and autonomous driving. People consider their previous experiments and knowledge in each time to make a decision and unlike traditional neural network for many real-life applications, the feature vector in each moment depends on one or several previous samples. Moreover, in some applications such as language translation, we need to deal with variable length inputs and outputs [25].

A recurrent neural network (RNN) is a type of neural network that solved these shortages of traditional neural networks using a loop in the network that allows information to persist. A recurrent neural network can be considered as several copies of the same network that each network passes a message to a successor. The chain shape of unrolled RNN shows that they are a specific architecture of neural networks to use for learning a sequence of dependent data. In this type of network, each output influenced not only by the current input of the network but also all inputs that have been fed to the network until the current step. RNN networks outperform many machine learning methods in real-life applications such as speech recognition and language translation [26, 27]. RNNs have been used in many car-related applications such as autonomous driving, driver behavior analysis and driving safety. In these car applications, their performance was much better than other machine learning approaches that don't have memory.

Various psychological conditions like sleepiness, fatigue, and distraction have an adverse effect on driver performance and can lead to fatal car accidents. [28] Discussed a model that detects driver potentially dangerous psychological conditions such as fatigue using a brain-computer interface. It proposed a new recurrent neural network architecture called Recurrent Self-evolving Fuzzy Neural Network. This model finds the correlation between the driver's brain activity, that monitors using EEG, and the driver's fatigue level [28].

[29] Proposed a data collection and data analysis framework called "DarNet". It can detect and classify driver behaviors. The framework has two parts including a data collection system and data analyzing part. Images that are collected by a face camera and Internal Measurement Unit (IMU) data from a mobile device are the inputs of this framework. They used deep learning methods including Long Short Term Memory networks and Convolutional neural network to classify driver behavior and reached 87.02% accuracy.

In this paper, we discuss a model that predicts the driver status using a sequence of driving data vectors and a stacked Long Short Term Memory (LSTM) network. We only used cars Can-Bus data and we didn't use any external devices such as camera or external sensors to make our dataset. We try both scaled and not-scaled data then we compared the results of them. Driver status in this paper shows if the driver is interacting with car infotainment system or not. Besides, if the driver has interaction with car infotainment system, what are the features of this interaction. We defined four features for each interaction including the number of errors while interaction, the length of interaction or response time, the number of navigation steps that the driver needs to pass in order to complete the interaction and the driving mode.

3 Experiment

We conducted our experiment in the HCaI lab using a Drive Safety Research simulator DS-600 which is fully integrated driving simulation system that includes a minimum 180° wraparound display, multi-channel audio/visual system, full-width automobile cab (Ford Focus) including windshield, center console, driver and passenger seats, dash and instrumentation and real-time vehicle motion simulation. This simulator provides a variety of road types such as urban road and highway. Besides, the different driving mode such as night, rain, fog and snow can be chosen for each road. We designed an urban road with high traffic. Figure 1 shows the designed road.

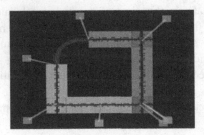

Fig. 1. Designed road

Four driving modes have been defined including Day, Night, Fog and Fog & night. The day mode is called the ideal mode, the Night mode and the Fog mode are adverse modes and the Fog & Night mode is the double adverse mode. We defined eight driving scenarios including four distracted and four non-distracted ones, so in each driving mode, we have distracted and non-distracted driving scenarios. In a distracted scenario, the driver interacts with the car interface continuously. An android application has been used as the car interface in this experiment. The application was hosted on the Android v4.4.2 based Samsung Galaxy Tab4 8.0 which was connected to the HyperDrive simulator. Figure 2(a) shows the place that we put the tablet on it to simulate the screen of car infotainment system. This application simulates car infotainment system interface in modern cars. We installed the tablet in the middle console next to the driver and ran an android application designed for the experiment. The main page of the application shows the main screen of car infotainment system and the driver can navigate in this application like car infotainment system.

In this experiment we wanted to detect if the driver had any interaction with our application. In non-distracted modes, the driver only focuses on the primary task of driving and we use the collected data in the non-distracted modes as the baseline of the model since they show no interaction. On the other hand, if the driver interacts with car infotainment system, we want to detect the features of this interaction since all interactions are not equally distracting.

We used the minimalist design discussed in [12] as the navigation model of the interface (Fig. 2(b)). In this navigation model, each command can be reached by 2, 3 or 4 steps of navigation, so we defined three types of tasks including 2-step, 3-step and 4-step tasks and use them to distract drivers in distracted scenarios. In distracted scenarios, we asked drivers to do some tasks on the interface considering this fact that driving is the primary task, and he/she shouldn't only focus on the requested task that has less priority compared to the primary task of driving. We define the task as reaching to a specific application on the interface that based on our interface design it needs 2 or 3 or 4 steps of navigation.

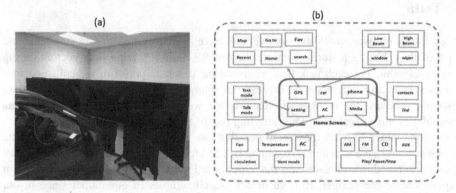

Fig. 2. (a) The blue circle shows the place that we put the tablet in the simulator (b) The navigation model that we used in our infotainment system (Color figure online)

We invited 35 volunteers to participate in this experiment by taking around 45 min of simulated drive. They were undergraduate and graduate students of Arizona State University in range 20 to 35 years old and they had at least two years of driving experience. Each volunteer was trained for 10 min before starting the experiment to become familiar with the simulator and the car's interface. After that, they drove in eight different scenarios. In non-distracted modes, they were asked to focus on the primary task of driving. In distracted scenarios, we chose some tasks from each type of tasks (2-step, 3-step and 4-step), and we tried to have equal number of tasks from each type. After that, we asked the driver to start the trip. After few seconds we started asking the driver to do some tasks on the interface, reach specific application on the interface, and we put the same gap between each two tasks since we want to have a border between the driving data related to each task. They were asked to give the highest priority to the primary task of driving and drive as realistic as possible. Besides, we added some events, such as pedestrians and bike drivers that jump into the road, randomly to each scenario to reduce the learning effect. We observed the driver behavior during each task and collected 4 data four each task:

1. The number of errors during a task: an error is defined as touch a wrong icon on the interface, not following traffic rules or have an accident.
2. Response time: the response time shows the length of interaction. We consider the length of interaction from the moment that we asked the driver to do the task and the moment that the task is done.
3. The number of navigation steps. We have three types of tasks including the 2-step, 3-step and 4-step task. The number of navigation steps is related to the task difficulty level since more navigation steps and longer tasks need more attention and cause higher level of distraction.
4. The driving mode which is the name of driving scenario. We use number 1 to 4 to shows non-distracted scenarios and 5 to 8 to display distracted scenarios.

4 Data

In each trip, on average 19000 data vectors have been collected by the simulator and each of them has 53 features, so we collected 5.3 million driving data vectors during this experiment. For each task that we asked the driver to perform, we collected four features including the mode of driving, the number of navigation steps, the driver response time and the number of errors. In all, we collected 2025 driver related data manually in this experiment. The sampling rate of the simulator was set on 60 samples per second which was the maximum possible rate. We wanted to collect as much data as possible and save it as a master data to be used for this and future experiments, so we decided to use this high sampling rate since it did not have any negative effect on our experiment or our budget. But we didn't use all samples in our model since in our application using 60 samples per second of driving is computationally very expensive for our machine. Therefore, we decided to compress the dataset. We replaced each 20 data vectors with their mean value. Future experiments will investigate if different methods of compressing this sampling rate (like median, mode, or any other function) would have an effect on the system training time or intelligence level.

For each driver feature vectors that we collected there are a large number of correlated driving data vectors that were collected by the simulator. The number of these vectors depends on the driver response time. To find the corresponding driving related vectors to each driver data vector we divided the collected driving vectors in each trip based on the length of the response time of all tasks that have been done during the trip. We calculated the sum of the response time for volunteer X in mode Y and calculate the portion of driving data vectors samples which are correlated to each driver data vector (1). In this equation, n shows the number of collected driver data vectors in a specific trip and the result shows the percentage of collected driving data vectors which is related to specific driver data vectors.

$$Percentage(i) = [response(i)/\sum_{k=0}^{n} response(k)] * 100 \qquad (1)$$

The driving data vectors have 53 features, and we chose 10 of them including speed limit, brake, velocity, steering wheel, longitude accelerating, lateral accelerating, lane position, accelerating, headway time and headway distance as inputs of our model. We used a paired t-test between distracted and non-distracted scenarios in each mode to detect the features that are significantly different between distracted and non-distracted modes (Table 1) and the combination of these 10 features shows significantly different between our scenarios.

Table 1. T-test results for distracted and non-distracted scenarios

	Day distracted vs. day non-distracted	Night distracted vs. night non-distracted	Fog distracted vs. fog non-distracted	Fog night distracted vs. fog night non-distracted
Velocity	Not significant	Extremely significant	Significant	Very significant
Lane position	Extremely significant	Extremely significant	Extremely significant	Extremely significant
Steering	Extremely significant	Very significant	Not significant	Extremely significant
Speed limit	Not significant	Not significant	Significant	Not significant
Accelerating	Extremely significant	Extremely significant	Extremely significant	Extremely significant
Brake	Very significant	Not significant	Not significant	Extremely significant
Longitude accelerating	Not significant	Extremely significant	Not significant	Not significant
Lateral accelerating	Not significant	Not significant	Not significant	Not significant
Headway time	Extremely significant	Not significant	Significant	Not significant
Headway distance	Extremely significant	Extremely significant	Extremely significant	Not significant

5 LSTM Model

We chose a multi-input single-output LSTM network for our model. The input of the network is a sequence of driving data and the output is a single driver data vector. We built three multi-layers LSTM networks with 2, 3 and 4 LSTM layers using Keras library. We put 50 neurons in each LSTM layer to check a different combination of batch sizes, learning rates and activation functions for this model. Finally, we chose 100 as the batch size, 0.0001 as the model's learning rate and ReLU as the activation function. We tried both scaled and not scaled data for training this network. We chose 80% of the dataset as training data and 20% as testing data. Besides, we used 20% of the training data as validating data during training process.

We trained the two-layer network with different numbers of LSTM neurons from 10 to 1000 and finally, we choose three numbers including 50, 150 and 300 as the number of LSTM neurons in each LSTM layer and analyze the effects of a low, medium and large number of LSTM neurons on the final result. Less than 50 neurons resulted in underfitting and increasing the number of LSTM neurons from 300 to 1000 didn't improve the accuracy of the network and only caused overfitting. After 300 LSTM neurons instead of increasing the number of neurons, we tried deeper networks to improve the accuracy of the model. After training fully connected LSTM networks, we tried 20%, 40%, and 50% dropout to reduce overfitting chance of the model but using drop-out didn't have a positive effect on the final accuracy.

6 Results

We tried pure data first and trained three different LSTM networks with two, three and four LSTM layers. The learning rate is 0.0001, the batch size is 100 and the activation function is ReLU for these networks. Table 2 shows the train, test and validation error of them using 50, 150 and 300 as the number of neurons in LSTM layers. The layer column shows the number of LSTM layers, the LSTM column shows the number of neurons in each LSTM layer, MAE is Mean Absolute Error and MSE is Mean Square Error of the model.

Table 2. Summary of multi-layers LSTM networks results with not-scaled data

Layer	LSTM	Train MAE	Train MSE	Val MAE	Val MSE	Test MAE	Test MSE
2-no scale	50	0.95	2.35	1.4	6.51	1.38	7.25
2-no scale	150	0.54	0.78	1.49	6.01	1.41	6.37
2-no scale	300	0.39	0.56	1.45	5.46	1.51	7.33
3-no scale	50	0.76	1.33	1.44	6.51	1.38	7.25
3-no scale	150	0.33	0.47	1.39	5.57	1.48	6.93
3-no scale	300	0.31	0.46	1.4	6.09	1.34	5.33
4-no scale	50	0.76	1.38	1.38	5.6	1.33	5.93
4-no scale	150	0.4	0.5	1.4	5.99	1.46	7.19
4-no scale	300	0.17	0.24	1.49	7.68	1.39	5.92

For the two-layer networks, using 50 neurons in LSTM layers resulted in the most accurate model since it has the minimum validation and test error. Besides, the gap between the train and test error is less than the rest of them. In the three-layer networks, although the network with 300 LSTM neurons has the minimum test error the performance of the network with 50 LSTM neurons is better since it has the minimum gap between the train and test error that shows the model trained well and it is not overfitted. Besides, the test error of this network is close to the network with 300 LSTM neurons. In the four-layer network, the model with 50 LSTM neurons resulted in the minimum validation and test error. Moreover, it has the minimum gap between the train and the test error, so it's trained better than the rest of the four-layer networks.

In sum, all three networks have their best performance using 50 neurons in their LSTM layers. Although adding more neurons enhanced their train error it didn't have a positive effect on the test error and the networks went toward overfitting. We scaled all input and output data then trained all networks again using scaled data. Table 3 shows the summary of three different networks with scaled data and 50,150 and 300 as the number of LSTM layer's neuron. In the two-layer LSTM network, we can see over-fitting in all networks that we trained and the minimum gap between train and test error is 1. When we increased the number of neurons from 50 to 300 the test error didn't change much but overfitting decreased. For the three-layer network, the network with 50 LSTM neurons shows the best performance and the least overfitting but the per-formance of two other networks is similar to the two-layer networks with the same number of neurons. Four-layer network has better performance and the least accuracy for all three number of neurons that we tried.

Table 3. Summary of multi-layers LSTM networks results with scaled data

Layer	LSTM	Train MAE	Train MSE	Val MAE	Val MSE	Test MAE	Test MSE
2-scaled	50	0.11	0.03	1.01	3.36	1	2.67
2-scaled	150	0.22	0.1	0.99	2.68	1.04	3.36
2-scaled	300	0.36	0.24	0.98	2.58	1.03	3.39
3-scaled	50	0.4	0.32	1.01	3.33	0.98	2.88
3-scaled	150	0.15	0.05	1.03	3.42	1.03	3
3-scaled	300	0.12	0.03	0.87	2.1	1	3.4
4-scaled	50	0.57	0.61	0.97	3.36	0.9	2.19
4-scaled	150	0.5	0.48	1	3.6	0.93	2.55
4-scaled	300	0.3	0.19	0.9	2.9	1.05	2.97

In sum, the four-layer network with 50 neurons resulted in the minimum test error which is 0.9 and it has the least gap between the train and the test error, so we can say that the network trained well and doesn't have overfitting. We tried more LSTM layers, but the final result was less accurate than the four-layer network. Increasing the number of neurons in shallower networks enhance the accuracy and in deeper networks, it just increased the overfitting chance.

7 Conclusion

Driver distraction is any task that diverts the driver attention away from the primary task of driving. Advances Driver Assistant Systems, Autonomous Vehicles and driver behavior analyzing are some suggested solutions to reduce driver distraction. Driver car infotainment system interaction is one of the main sources of driver distraction. Although interacting with car infotainment system, even for short time, causes dis-traction, the level of distraction and cognitive load which is caused by each task depends on many different factors such as the length of distraction, the context of driving and the complexity of the task. If the driver can do a specific task on the car infotainment system in a short time and find a specific application with few simple navigation steps, the distraction level that is caused by this task is much less than a long task such as texting or navigating in a complex interface to reach an application.

We defined four features for each interaction including number of errors, response time, number of navigation steps and the mode of driving. We used 10 driving data and a stacked LSTM network to detect the driver status that is defined by these four features. We reached 0.95 train MAE and 1.38 test MAE with not scaled data, two-layer LSTM network and 50 neurons in each LSTM layer. We trained the network with scaled data and reached 0.57 Train MAE and 0.9 test MAE with four-layer LSTM network and 50 neurons in each LSTM layer. In sum, we detect if the driver is distracted by interacting with car infotainment system or not and if he/she is interacting with the car infotainment system, what are the features of this interaction. As future work, we can extend this experiment using more driving scenario and new distracting task that cover a larger subset of the possible distracting task while driving. Besides, we can use the output of the model as the input of a decision system and give each feature a specific weight to calculate the driver distraction level more accurately.

Acknowledgement. We acknowledge the effort and participation of all ASU undergraduate and graduate students who enthusiastically participated in data collection and analysis of some of these experiments as part of their course load or as volunteers.

References

1. National Center for Statistics and Analysis, Distracted Driving: 2015, in Traffic Safety Research Notes. DOT HS 812 381. National Highway Traffic Safety Administration, Washington, D.C., March 2017
2. Injury Facts, Motor Vehicles Safety Issues. https://injuryfacts.nsc.org/motor-vehicle/motor-vehicle-safety-issues/. Accessed 20 Oct 2018
3. Ferguson, R., Green, A., Blau, E., Walker, L.: Teens in Cars. Safe Kids Worldwide, Washington, DC, May 2014
4. National Safety Council: Teens' Biggest Safety Threat is Sitting on the Driveway. https://www.nsc.org/road-safety/safety-topics
5. AAA Foundation For Traffic Safety, 2016 Traffic Safety Culture Index, February 2017. https://aaafoundation.org/wp-content/uploads/2017/11/2016TrafficSafetyCultureIndexReport.pdf. Accessed 10 Aug 2018
6. Gaffar, A., Monjezi Kouchak, S.: Using artificial intelligence to automatically customize modern car infotainment systems. In: Proceedings on the International Conference on Artificial Intelligence (ICAI), pp. 151–156 (2016)
7. National Center for Statistics and Analysis: Driver electronic device use in 2017, Traffic Safety Facts Research Note. Report No. DOT HS 812 665. National Highway Traffic Safety Administration, Washington DC, January 2019
8. Gaffar, A., Monjezi Kouchak, S.: Using simplified grammar for voice commands to decrease driver distraction. In: The 14th International Conference on Embedded System, pp. 23–28 (2016)
9. Vegega, M., Jones, B., Monk, C.: Understanding the effects of distracted driving and developing strategies to reduce resulting deaths and injuries: a report to congress. Report No. DOT HS 812 053. National Highway Traffic Safety Administration, Washington, DC, December 2013
10. Todd, W.: It is Time for a 'Parental Control, No Texting While Driving' Phone. Forbes Business, 18 September 2012

11. Gaffar, A., Monjezi Kouchak, S.: Quantitative driving safety assessment using interaction design benchmarking. In: IEEE Advanced and Trusted Computing (ATC 2017), San Francisco Bay Area, USA, 4–8 August 2017 (2017)
12. Gaffar, A., Monjezi Kouchak, S.: Minimalist design: an optimized solution for intelligent interactive infotainment systems. In: IEEE IntelliSys, the International Conference on Intelligent Systems and Artificial Intelligence, London, 7–8 September 2017
13. Campbell, M., Egerstedt, M., How, J., Murray, R.: Autonomous driving in urban environments: approaches, lessons and challenges. Philos. Trans. R. Soc. **368**(1928), 4649–4672 (2010). https://doi.org/10.1098/rsta.2010.0110
14. Gaffar, A., Monjezi Kouchak, S.: Undesign: future consideration on end-of-life of driver cars. In: IEEE Advanced and Trusted Computing (ATC 2017), San Francisco Bay Area, USA, 4–8 August 2017 (2017)
15. Lu, M., Werers, K., Heijden, R.: Technical feasibility of advanced driver assistance systems (ADAS) for road traffic safety. Transp. Plann. Technol. **28**(3), 167–187 (2005). https://doi.org/10.1080/03081060500120282
16. Monjezi Kouchak, S., Gaffar, A.: Determinism in future cars: why autonomous trucks are easier to design. In: IEEE Advanced and Trusted Computing (ATC 2017), San Francisco Bay Area, USA, 4–8 August 2017 (2017). https://doi.org/10.1109/uic-atc.2017.8397598
17. Brooks, C., Rakotonirainy, A.: In-vehicle technologies, advanced driver assistance systems and driver distraction: research challenges. In: International Conference on Driver Distraction, Sydney, Australia, 2–3 June 2005 (2005)
18. Finn, A., Scheding, S.: Developments and Challenges for Autonomous Unmanned Vehicles. Intelligent Systems Reference Library, vol. 3. Springer, Heidelberg (2010). https://doi.org/10.1007/978-3-642-10704-7. ISBN: 978-3-642-10703-0
19. Barabás, I., Todoruț, A., Cordoș, N., Molea, A.: Current challenges in autonomous driving. IOP Conf. Ser.: Mater. Sci. Eng. **252**, 012096 (2017)
20. Monjezi Kouchak, S., Gaffar, A.: Non-intrusive distraction pattern detection using behavior triangulation method. In: International Conference on Computational Science and Computational Intelligence (CSCI), 14–16 December 2017. https://doi.org/10.1109/csci.2017.140
21. Goodrich, M., Quigley, M.: Learning haptic feedback for guiding driver behavior. In: IEEE International Conference on Systems, Man and Cybernetics (IEEE Cat. No. 04CH37583) (2004)
22. Meiring, G., Myburgh, H.: A review of intelligent driving style analysis systems and related artificial intelligence algorithms. Sensors **15**(12), 30653–30682 (2015)
23. Gindele, T., Brechtel, S., Dillmann, R.: A probabilistic model for estimating driver behaviors and vehicle trajectories in traffic environments. In: 13th International IEEE Conference on Intelligent Transportation Systems (2010)
24. Hou, H., Jin, L., Niu, Q., Sun, Y., Lu, M.: Driver intention recognition method using continuous hidden Markov model. Int. J. Comput. Intell. Syst. **4**(3), 386–393 (2011)
25. Norvig, P., Russell, S.: Artificial Intelligence: A Modern Approach, 3rd edn. Pearson, London (2014). ISBN 0136042597
26. Gibson, A., Patterson, J.: Deep Learning. Oreilly, Sebastopol (2017). ISBN 978-1491914250
27. Mandic, D., Chambers, J.: Recurrent Neural Networks for Prediction: Learning Algorithms, Architectures and Stability. Wiley, Hoboken (2010). ISBN 0471495174
28. Liu, Y., Lin, Y., Wu, S., Chuang, C., Lin, C.: Brain dynamics in predicting driving fatigue using a recurrent self-evolving fuzzy neural network. IEEE Trans. Neural Netw. Learn. Syst. **27**, 347–359 (2017)
29. Streiffer, C., Raghavendra, R., Benson, T., Srivatsa, M.: Darnet: a deep learning solution for distracted driving detection. In: The 18th ACM/IFIP/USENIX Middleware Conference (2017). https://doi.org/10.1145/3154448.3154452

Using Relational Concept Networks
for Explainable Decision Support

Jeroen Voogd, Paolo de Heer$^{(\boxtimes)}$, Kim Veltman, Patrick Hanckmann,
and Jeroen van Lith

Netherlands Organisation for Applied Scientific Research (TNO) Defence, Safety
and Security, Oude Waalsdorperweg 63, 2597 AK The Hague, The Netherlands
{jeroen.voogd,paolo.deheer,kim.veltman}@tno.nl

Abstract. In decision support systems, information from many different sources must be integrated and interpreted to aid the process of gaining situational understanding. These systems assist users in making the right decisions, for example when under time pressure. In this work, we discuss a controlled automated support tool for gaining situational understanding, where multiple sources of information are integrated.

In the domain of operational safety and security, available data is often limited and insufficient for sub-symbolic approaches such as neural networks. Experts generally have high level (symbolic) knowledge but may lack the ability to adapt and apply that knowledge to the current situation. In this work, we combine sub-symbolic information and technologies (machine learning) with symbolic knowledge and technologies (from experts or ontologies). This combination offers the potential to steer the interpretation of the little data available with the knowledge of the expert.

We created a framework that consists of concepts and relations between those concepts, for which the exact relational importance is not necessarily specified. A machine-learning approach is used to determine the relations that fit the available data. The use of symbolic concepts allows for properties such as explainability and controllability. The framework was tested with expert rules on an attribute dataset of vehicles. The performance with incomplete inputs or smaller training sets was compared to a traditional fully-connected neural network. The results show it as a viable alternative when data is limited or incomplete, and that more semantic meaning can be extracted from the activations of concepts.

Keywords: Symbolic AI · Neural networks ·
Graph-based machine learning · Explainability · Decision support

1 Introduction

Military decision making often takes place in uncertain and complex situations, and can have a large impact on the opponent and also on civilians and cause

Published by Springer Nature Switzerland AG 2019
A. Holzinger et al. (Eds.): CD-MAKE 2019, LNCS 11713, pp. 78–93, 2019.
https://doi.org/10.1007/978-3-030-29726-8_6

collateral damage. Often the complete impact of a decision cannot be easily fore-seen. In order to make the right choice, situational understanding is needed. To obtain situational understanding, information from different sources needs to be integrated. For example, information from sensors (such as cameras, thermal sensors, but also social media and human observations) and expert knowledge e.g. obtained through experience, needs to be combined to build an understanding of the situation.

An example of a use-case where situational understanding is relevant is recognizing smuggling activities in a busy city. During a military mission in a hostile county, there is a lot of information that might be relevant to the mission, but not immediately linked to smuggling in an analyst's mind. Here, the types of vehicles in the area, the hours of active use of those vehicles, the locations and the people that suspected smugglers meet might play a large part in determining if there is an active smuggling operation going on. Even more general concepts might play a role as well, such as economy and law and order in the area, but also patterns of life, important dates, and the political situation. These factors can be described in flows of people, goods, and information, forming an intricate web of higher-level knowledge. Combining this knowledge in a semantic graph (see Fig. 1) can more easily provide new insights to commanders about which information might be relevant to the current situation. For example, a correlation might be found between the detected vehicle type and the time and area of usage, indicating smuggling activities. The values for connections between and correlations of these concepts need to be found or learned in order to make useful predictions.

There are, however, several factors that limit understanding. There may be too little information on some aspects, there may be unreliable or intentionally false information, and there may even be too much information. In those situations, human analysts and decision makers may not be well suited for sifting through the data to find trustworthy and useful information and being able to integrate it into the decision-making process, especially given the time pressure and the potential severity of the consequences.

A possible solution for finding the correlations in the available data is using Artificial Intelligence (AI) approaches. In recent years, developments in neural networks (NNs) have made Machine Learning (ML) a popular method for solving problems in the field of AI. However, for the use-case discussed above, such methods lack a critical property. When making decisions with the help of decision support systems, users require that these systems can explain themselves. They want to be able to ask why the system indicates that something is very likely present in the environment. Black box methods, such as neural networks, are unsatisfactory because they struggle to provide additional insights besides its output. What is needed is a so-called white box approach, which is able to accept and convey knowledge in a for humans understandable way, and can answer 'why' questions.

Another disadvantage of ML is the possibility that the algorithms learn to use prevalent, but unexpected, aspects of the data, which may lead to incorrect

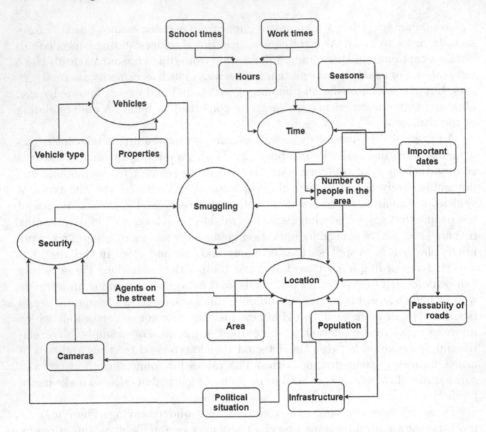

Fig. 1. A simplified semantic graph showing a number of the concepts that can play a role in a smuggling scenario.

decisions. A well-known example of the latter is an ML algorithm that learned to classify wolves versus huskies based on the presence of snow instead of learning the discerning characteristics of these animals [11].

Recently, research has partly shifted towards creating AI tools that are able to explain why and how a certain conclusion came to be [13]. Van Lent [8] called this type of artificial intelligence 'Explainable AI'. In a system with explainable AI users may be able to analyze the cause of (faulty) results. Furthermore, by improving the explainability of intelligent software, it has been argued that the users of these types of applications start to trust the systems more [17]. Having trustworthy, meaningful and understandable decision support is critical in the military domain.

In this work, we seek to support analysts with automated tooling. The goal is to be able to understand important aspects of the environment in which a military operation takes place. Often these operations take place in unfamiliar terrain and only a global understanding of the situation is available. The best possible understanding is based on all available knowledge and data.

The envisioned support tool integrates multiple sources of information, some of which consist of data streams, some are symbolic in nature such as the knowledge of human experts. We combine sub-symbolic information and approaches (sensor data, machine learning) with symbolic knowledge and technologies (expert knowledge, semantic networks) to work towards a decision support system that produces better results with fewer data required for learning compared to traditional neural network approaches, and supports explainability.

In this paper we first present a brief overview of related work (Sect. 2, and then discuss the RCN framework and its properties (Sect. 3). In Sect. 4 several experiments and their results are discussed. Lastly, in Sect. 5, our findings are discussed as well as some points of future work.

2   Related Work

One of the key challenges of neuro-symbolic model integration is to find suitable ways of merging symbolic and sub-symbolic models into a unified architecture. A good overview of the neuro-symbolic approach is provided by Besold et al. [1]. Our neuro-symbolic approach was partly inspired by Raaijmakers et al. [10], who presented a method for explaining sub-symbolic text mining with ontology-based information. However, this method does not allow a priori expert knowledge to be incorporated in the text mining itself; it is used to explain the mining results afterward. Gupta et al. [4] used an ontology to shape the architecture of the sub-symbolic model, which enables the model to become more explainable since high-level knowledge is embedded in the architecture. The idea to shape the architecture based on a symbolic model is used in the approach discussed in this paper.

Sabour et al. [12] describe a method that comes close to our approach: capsule networks. A capsule is a group of neurons whose activity vector represents the instantiation parameters of a specific type of entity such as an object or an object part. Active capsules at one level make predictions for the instantiation parameters of higher-level capsules. When multiple predictions agree, a higher level capsule becomes active. Sabour et al. show that a discriminatively trained, multi-layer capsule system achieves state-of-the-art performance on MNIST [7] and is considerably better than a convolutional network at recognizing highly overlapping digits. Although this approach was used for image processing and our approach is about concepts, the idea of clusters of neurons signifying one concept with relations to other clusters inspired us and is reflected in how the activation of concepts works in the methods discussed in this paper.

For this proposal, building on the work of Voogd et al. [15], an automated situational understanding support tool is created, which uses a Relational Concept Network (RCN), see Sect. 3.

Explainability in artificial intelligence is a topic that attracts much attention, see e.g Holzinger et al. [5] for an overview. A benefit of our approach is that it supports explainability. The expert knowledge that is incorporated in the network makes it possible to explain the output in a for humans understandable way, as it is the symbolic knowledge that humans can comprehend, given an appropriate user interface. The symbolic knowledge may contain relations that are causal in nature, but this is not guaranteed. Therefore our approach falls short of supporting explanations in terms of causality. But since it does refer to human understandable models, albeit mixed with link-strength based on data, it may be a way to obtain causability as defined in [5]. If in a future situation our approach is implemented as a decision support system, and the decisions together with observations of the results of acting on the decision are fed into the RCN, this could lead to detecting causality [9]. For our application domain, however, it is unfeasible to experiment freely with the decisions to actually obtain causality.

3   RCN Framework

The Relational Concept Network (RCN) consists of a directionally organized network of symbolic concepts (nodes) with relations (edges) connecting them based on expert knowledge. The concepts and their (logical) relations represent the domain knowledge provided by experts, ontologies, or other sources. The (activation) value of a concept is to be interpreted as the probability that this concept is present or relevant in the current situation.

Each concept has an internal predictor method to determine the concept's output activation based on its inputs (where an input comes from either external data or from another concept in the network model). This predictor method can range from a simple rule-based evaluation to a NN that has to be trained. The exact relation values between concepts are difficult to estimate for human. It is for example hard to determine how much the fact that a large car is moving fast would contribute to the chance that it is smuggling. Therefore, we use a sub-symbolic NN as each concept's internal method that, using training data, can learn the influence and strength of the relations, see Fig. 2 for a simple example.

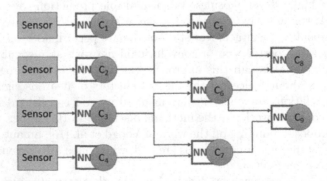

Fig. 2. A simple schematic representation of the RCN where symbolic concepts (C_x) have integrated neural networks.

Each concept is basically a small neural network with one neuron in its output layer: the concept. So, the whole RCN may be seen as one large neural network where some neurons have a concept name. Compared to traditional neural networks the expert knowledge effectively limits the number of links between neurons and therefore limits the state space of the NN, which should result in faster training and/or less need for training data.

A concept which represents a feature of the input data is called an *input concept*. The value is set using an external source, e.g. a sensor or human input (C_1 to C_4 in Fig. 2). *Intermediate concepts* receive their input from concepts and provide their output to other concepts (C_5 to C_7). *End concepts* are concepts without dependents, i.e. its output is not being used by other concepts in the network (C_8 and C_9).

The RCN is built as a framework to offer functionalities to create the situational understanding support system. Currently, network-related functionalities are implemented allowing searching for concepts, paths between concepts, dependency analysis, etc. The framework also enables defining and training concepts based on NNs, and querying the (trained) concept outputs. In the following subsections, these two functionalities will be explained further.

3.1   Training the RCN

When training the RCN, first domain knowledge of the expert is implemented in the model through defining concepts and making connections between them. This knowledge may come from different sources. The expert knowledge may come from different experts at different moments. It may also come from an existing RCN from a previous application. It is expected that especially conceptually higher level knowledge is more general and may be re-usable even when applied in different locations and times, such as a previous military mission.

Training the RCN is accomplished by training the integrated NNs separately. A training sample for a single concept consists of a vector of input data (the features) and the associated true value (the label). Because these NN predictors are small and typically have only a few inputs, they can be trained and evaluated quickly. This allows fast feedback on the performance of the proposed connection. A disadvantage of this strategy is that individual predictors can only be trained when there is ground-truth training data available for that concept. If a new connection is made to a concept, only that concept needs to be retrained. If the RCN is extended with a new concept, that concept needs to be trained, but also directly dependent concepts need to be retrained. If an expert changes links between concepts, they may 'break' knowledge that was inserted by another expert. This will sort itself out in the training phase because links that are not supported by the data will receive little weight and therefore have little influence on the outcome when using the RCN.

After training all concepts, the system can be used to provide indications of relevant concepts based on new data in the form of queries.

3.2   Querying the Concepts

After training, the RCN can be used by analysts to examine which concepts are found in the current situation. To do this, data from sensors and other sources are fed into the RCN at the appropriate concepts. These concepts are then used to calculate higher-level concepts. After these calculations, the probability values of all concepts are known. The analyst examines concepts with a high probability value to gain insight into what might be going on.

If only a subset of the input data is available, i.e. some input concepts do not receive data from external sources, thanks to the decoupled nature of the RCN, some intermediate concepts may still be (recursively) inferred, creating a chain of predictions. Furthermore, if an intermediate or end concept has several inputs, these inputs may differ in how much they contribute to the probability value of that concept. The contribution of an input to the output is its feature importance. Inputs with low feature importance may be ignored, making it possible for more high-level concepts to determine their probability value, be it with higher uncertainty (see also "Graceful degradation" in Sect. 4). Permutation Importance [6] is a method to determine the importance of individual features for a black box model. Since the method is model-agnostic it can be used with neural networks in concepts.

Users will - at least at first - be suspicious about the results produced by the RCN and want to be able to ask 'why' questions. The black box nature of a NN is not suited for these questions. The RCN supports explainability in the following way: if input is supplied to a trained RCN and then evaluated through the network of concepts, each concept obtains a probability value. Then, questions such as 'why does this concept have a high probability value?' can be answered by referring to which of the input concepts have a high probability and with what weight (i.e. the feature importance) it contributes to the concept under consideration, all the way back to the first layer of feature data, see Fig. 5.

Besides the probability value of the concepts, it is also useful for the analysts to query the performance of the knowledge that was supplied by the experts. After training, a set of validation data is used to determine the performance of the concepts. The analyst can check if there is expert knowledge that does not fit well with the data. If so, the analyst has to determine whether the data is of sufficient quality, or if the knowledge should be enhanced, i.e. remove connections to concepts with a low feature importance, and add connections to concepts that may prove to have a higher importance.

4   Experiments and Results

In this section, we discuss several tests and their results obtained by the RCN to verify the functionality as described in the previous chapter. It is examined how well the RCN can be trained, how modular the resulting model is thanks to its decoupled nature, how well it performs using smaller datasets and what the effect is on explainability.

For the tests described below, we have used the Large-scale Attribute dataset (LAD) [18], which is an annotated image dataset of six different super classes of objects. For these tests, we only used the 'Vehicles' super class, which includes land/air/water and civilian/military vehicles. Each item in the dataset is annotated on several properties ranging from used materials or components to the usage scenario, as well as the subclass it belongs to (e.g. whether it's a car, an ambulance, a train or an airplane). Figure 3 illustrates an instance of the dataset. Note that we only use the annotations as concepts, not the images themselves.

Van

Speed: moves fast:	True
Color: is white:	True
Safety: is dangerous:	False
Passenger capacity: medium:	True

Fig. 3. A sample from the LAD dataset, with some of the many annotated properties.

Thanks to the diversity of the annotated properties in the LAD, it is possible to structure the properties into a hierarchy of more basic concepts (e.g. materials and components) to concepts with a higher abstraction or complexity (e.g. usage scenario, safety).

Combining Expert Knowledge with Data
Current AI techniques are not capable of fully learning both sub-symbolic and symbolic data, therefore the symbolic concepts and their logical relations are defined by an expert. In order to allow the expert to define the RCN by specifying concepts and relations, a graphical user interface (GUI) has been made that allows an expert to pick and place concepts (which are derived based on the labels in the data) and connect them based on expert rules. Note that an expert rule is not an if-then relation, but a combination of concepts that the user/expert has specified as relevant.

In Fig. 4 an example is shown for the various vehicle types in the dataset, the resulting multi-layered hierarchy of connected concepts is a sparse network. In this example the user specified (among many other expert rules) that the relations needed for the concept "Function: can dive" are "Shape: is globular", "Shape: is ellipsoidal", "Parts: has a propeller". The RCN is then generated from this expert-made network and the user can start the learning phase. Afterwards, for a given input, the probabilities of higher level concepts and feature importance of the links can be visualized in the same GUI.

Trainability
An obvious requirement is that the system has to be able to learn, i.e. fit the structure in the data onto the structure defined by the relations between the concepts. This translates into the training of the sub-symbolic parts integrated into the concepts of the RCN.

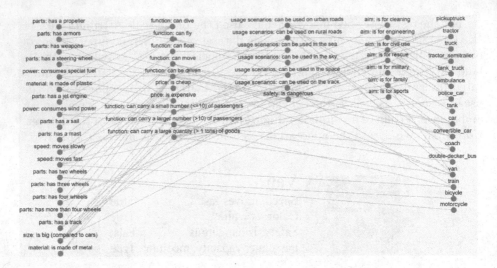

Fig. 4. The GUI to build the RCN and control the learning phase

As a performance measure, balanced accuracy [2] is used, which relates the average of combined precision and recall obtained on each class, corrected for any class imbalances. This provides a robust performance measure when dealing with varying amounts of data per class. If the network of concepts is close to what is present in the structure of the data, it should result in a well-trained RCN. i.e. a balanced accuracy score between 0.5 and 1.0.

To test this, the available data is split into a training set (80%) and a test set. As a test for this requirement we need to show that the RCN is capable of learning the desired concept relations. In addition, we will compare the performance of the RCN with a single traditional fully-connected NN (FCNN) that uses all concepts as input and provides predictions for all output concepts. The neural networks are trained using a standard multilayer perceptron [3] using 3 hidden layers of 8 hidden nodes each.

The results (see Fig. 6 showing the averaged balanced accuracy score over all concepts) indicate that most of the concepts can be trained to a perfect score and the RCN performs well. However, compared to a traditional NN the overall performance of the RCN is lower. There are some possible explanations for these observations. The relations between concepts were chosen based on the concept labels, without looking at the data. Therefore some relations that were defined may not be supported by the data. Another relevant aspect is the fact that the FCNN might learn patterns in the data that can give the correct answer in the current context, but in fact have no connection to the actual (symbolic) relation, similar to the previously mentioned example of the animal classifier that relies on the presence of snow in an image [11]. Lastly, the NN architecture (number and size of hidden layers) integrated in the concepts was chosen based on a rule of thumb and fixed for all concepts. However, some concepts may benefit from a larger or smaller architecture. Furthermore, when inspecting the dataset, some odd data points were observed. We do not know how the labels are assigned to the data, but there are some assignments that were inconsistent with what is semantically expected for that class. For example, some instances of the concept 'submarine' had no value for the concept 'function: can dive', which should be the case for all submarines.

Explainability

The RCN should allow a user to find out why a concept is indicated as strongly present in the data. To test this form of explainability, the RCN is first trained, then one single data point is applied to the input of the RCN and all probability values for the concepts are determined. Then a textual or graphical representation of how concepts are influenced by 'lower' concepts can be made. This can be examined by the user to check whether it makes sense.

In Fig. 5, a textual example of explainability is shown for a specific concept ("aim: is for military"), the dependencies on lower level concepts and their strength are shown.

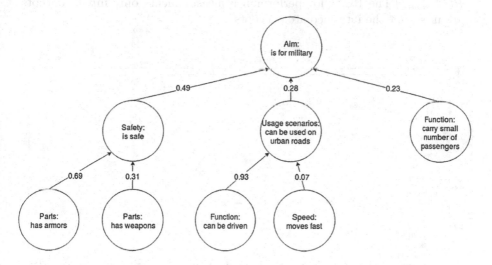

Fig. 5. The hierarchical structure of a concept, its sub-concepts and the feature importance values of those concepts that provide input for a higher-level concept.

Modularity

One of the advantages of the RCN compared to a NN is that it does not require all inputs and corresponding labels to be available at one time. The RCN can be neatly decomposed in concepts with their input concepts up to the point for which data is available. It can thus be trained piece by piece.

To make sure this works, our current implementation allows that concepts can be trained independently (or in sets). Only those concepts for which there is data in the training set are trained. When data for a concept becomes available at a later date, that concept can then trained, keeping the rest of the RCN (and its trained concepts) in tact. Note that this is not the same as true incremental learning of the RCN where the sub-symbolic parts of the concepts get to learn step-by-step in a more or less random order.

Improve Prediction Performance with Small Training Sets
The use of expert knowledge in the RCN effectively reduces the phase space available compared to the use of a traditional NN. For small training datasets, and if the expert knowledge fits this data well, this should result in a higher performance after training than for a traditional NN.

To test this, smaller and smaller training datasets are used to train the three different configurations have been tested:

- $FCNN_{full}$: A fully Connected NN that uses all input and intermediate concepts of the RCN as input features,
- FNN_{inputs}: A fully connected NN that uses the same input concepts as the RCN, but does not use the intermediate concepts,
- RCN_{inputs}: The RCN; for performance measurements only input concepts are used, not the intermediate concepts.

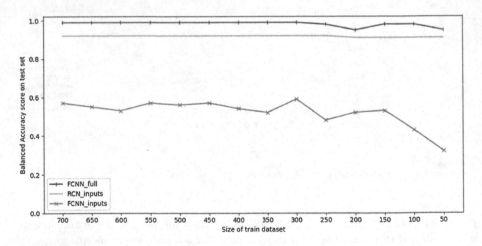

Fig. 6. The performance of different methods trained on the dataset, either using traditional fully connected NNs, or the RCN.

The results (see Fig. 6) show that $FCNN_{full}$ scores a near-perfect classification on most of the sizes. There does seem to be slight drop in performance for smaller datasets, this small drop might indicate a sensitivity to the data present. However, as this drop is not clearly observed in the performance of the other algorithms, it might be the case that it is not the data itself that causes this decrease in balanced accuracy, but an effect of the small size of the dataset.

The $FCNN_{inputs}$ has a much lower score with a strongly decreasing score for smaller datasets. The RCN_{inputs} shows a slightly lower performance than $FCNN_{full}$, but its performance remains more stable at smaller datasets.

The scores of the RCN_{inputs} compared to $FCNN_{full}$ is lower because the $FCNN_{full}$ are free to include subtle relations that are not included in the expert rules. Additionally some expert rules are poorly chosen compared to the data used. The instability of both the FCNNs performances at small dataset sizes indicate that the expert rules in the RCN_{inputs} are meaningful for stability.

Graceful Degradation

In the previous paragraph the size of the training dataset was decreased, but each entry in the dataset still had values for all properties, i.e. all input and intermediate concepts were still present in each data point. In practice it may often be the case that for only some of the input concepts data is available. Where traditional NNs can only function if all input data is available, the RCN allows for some concepts to still be evaluated even if only a subset of input data is available.

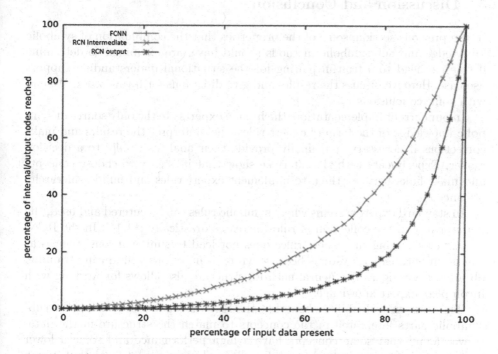

Fig. 7. The degradation of reachable concept outputs when decreasing the completeness of an input feature, shown for the traditional fully connected NNs, and the RCN. (Color figure online)

A function can be defined purely based on the connections between the concepts in the RCN that returns all concepts that can be evaluated for a given subset of input data. This reachability function is evaluated by averaging over the many possible configurations of available input for the RCN. Two values are calculated: the percentage of output concepts that are reachable and the percentage of intermediate concepts that are reachable.

In Fig. 7, the bottom graph (purple) shows that for traditional NN all input must be available. The percentage of output concepts that are reachable as a function of the percentage of available input concepts is the blue (middle) line, if the reachable intermediate concepts are included the red line (top) is observed. The higher the line the more concepts can evaluate their probability and become useful to an analyst.

After training it becomes clear that the feature importance can differ significantly, see Fig. 5. This opens the possibility to disregard connections with a low feature importance, making the network more sparse and resulting in an even higher reachability for a given percentage of available input. The drawback, however, is an increased error in the output.

5 Discussion and Conclusion

In the previous section, some of the properties that the combination of symbolic knowledge and sub-symbolic methods should have, are examined to determine if this can lead to a fruitful pairing for the situational understanding support use-case. Here we discuss the results and give directions for future work. We end with some conclusions.

In our current implementation, the human experts are the only source of symbolic knowledge in the form of expert rules, they interpret the results and make corrections if necessary, and finally provide their analysis results to a decision maker. Experiments with the interface show that it is easy to choose concepts and make links between them to implement expert rules and build a hierarchy of concepts.

In standard expert systems where symbolic rules can be entered and used, the maintenance of the collection of rules increases drastically [14,16]. In the RCN, adding or changing the expert rules does not lead to similar inconsistencies as can occur with expert systems because wrong connections will, during training, obtain a low weight. The neural network structure also allows for working with incomplete expert knowledge.

The tests on trainability show that the RCN is capable of training the subsymbolic parts integrated in the concepts to match the structure in the data. It was found that some concepts have a high performance and some a lower performance. The concepts that work well, implement expert rules that match well with the data. Low performing concepts are a signal for the user to examine why this is the case. There may be several reasons: they represent a wrong expert rule, the data is incomplete or has errors, or the structure in the data is too complex for the current rule given the size of the sub-symbolic part. If the last reason is the case, the sub-symbolic part can be increased, which would also require more data for training, or additional expert rules are required: more or different connections between concepts are necessary, and/or more concepts with accompanying expert rules need to be introduced.

Modularity is currently supported by collecting all relevant data at the concepts and then do a learning step for separate concepts. After training, the RCN can be used for inferring probabilities of higher level concepts in input data. During use, modularity is also supported, as the tests on graceful degradation show that it is possible to evaluate the probabilities for subsets of concepts, depending on the available input data and the connections from the expert rules. In real-world applications, it is more than likely that input data is only available for small sections of the RCN.

It is, however, desirable to have a continuous learning approach where incoming data with information on several hierarchical levels of the RCN can be used for learning. In our current implementation, we use the scikit-learn toolkit which supports incremental learning but this has not been tested yet.

The tests on performance as a function of training data set size indicate that the RCN's performance is less dependent on the amount of available training data compared to a NN. This is an important finding from the perspective of the use-case since it can be expected that at least for some concepts only little data will be available, making the use of a traditional NN infeasible. On the current dataset with imperfect expert rules, the RCN has a relatively high performance, close to that of a traditional FCNN trained on the same data, while providing the advantages of explainability and more flexibility in training data.

Although the basis for a decision support system that incorporates both symbolic and sub-symbolic data has been shown in the RCN approach, there are many things that can be enhanced. Hence, there are several plans for future implementations.

Currently, human experts are the only source of symbolic knowledge. This can be expanded by for example the use of ontologies that provide relevant domain knowledge. Often ontologies use well-defined relations between concepts. A way needs to be found to implement these in the RCN, and it needs to be studied how these can benefit a user by giving more insightful explainability. If implemented, the expert can also use this richer type of relations when specifying rules.

An additional source of symbolic knowledge may come from lessons learned from previous military operations that can be entered in the form of expert rules. It may even be possible to re-use the RCN as a whole. It is expected, however, that lower level concepts may require significant retraining. Higher level concepts may be more general in nature and therefore transferable from one situation to the next.

If an expert rule performs badly while using good data, it needs correcting. This can be done by a kind of hypothesis testing: changing the expert rule, checking its performance, changing it again, etc. until a good performance has been obtained. This can partly be automated by having the RCN check many different combinations of connections to a concept and suggesting the best performing set to the expert. Additionally, the predictor, here a NN with a number of hidden layers and nodes in each layer, can be changed to better suit the complexity of the mapping it needs to learn.

Another element that can be improved in the fact that the current network is a directed graph: the relation between nodes only works one way. This limits the rules that can be entered. For many concepts in a domain, it may not be clear what causes what, only that they are even related. This requires bidirectional connections. A potential problem arises when anything can be connected to anything: the output of a concept can indirectly become its own input. This will have to be solved, for example by freezing activations or repeatedly resampling.

One important aspect that is currently missing in the RCN is an indication of the uncertainty in the output. Uncertainty should be available integrally from lowest level input to the output of the highest level concepts. This will be important for analysts to consider when building their situational understanding. If implemented, this may be used as a means of graceful degradation: if a concept has inputs with low feature importance for which no data is available, they may be ignored. This will, however, increase the uncertainty of the output, which should be visible to the user.

All in all, the methods discussed in this paper show that combining symbolic knowledge (from experts and expert systems) with sub-symbolic methods (such as neural networks and sensor data), can lead to a fruitful pairing. The combination of the flexibility of neural networks with the domain knowledge of experts results in a form of hybrid AI, where symbolic insights from an expert can be made more precise by training sub-symbolic networks using data. The main advantages are the flexibility this approach offers, explainability of the results, and a reduction of data requirements.

References

1. Besold, T.R., et al.: Neural-symbolic learning and reasoning: a survey and interpretation. arXiv preprint arXiv:1711.03902 (2017)
2. Brodersen, K.H., Ong, C.S., Stephan, K.E., Buhmann, J.M.: The balanced accuracy and its posterior distribution. In: 2010 20th International Conference on Pattern Recognition, pp. 3121–3124. IEEE (2010)
3. Gardner, M.W., Dorling, S.: Artificial neural networks (the multilayer perceptron)– a review of applications in the atmospheric sciences. Atmos. Environ. **32**(14–15), 2627–2636 (1998)
4. Gupta, U., Chaudhury, S.: Deep transfer learning with ontology for image classification. In: 2015 Fifth National Conference on Computer Vision, Pattern Recognition, Image Processing and Graphics (NCVPRIPG). IEEE (2015)
5. Holzinger, A., Langs, G., Denk, H., Zatloukal, K., Mueller, H.: Causability and explainability of AI in medicine. WIREs Data Min. Knowl. Discov. e1312 (2019). https://doi.org/10.1002/widm.1312
6. Korobov, M., Lopuhin, K.: Permutation importance (2017). https://eli5.readthedocs.io/en/latest/blackbox/permutation_importance.html
7. LeCun, Y., Cortes, B.: The MNIST databasse of handwritten digits (2010). http://yann.lecun.com/exdb/mnist/
8. van Lent, M., Fisher, W., Mancuso, M.: An explainable artificial intelligence system for small-unit tactical behavior. In: IAAI Emerging Applications (2004)
9. Pearl, J., Mackenzie, D.: The Book of Why. Basic Books, New York (2018)

10. Raaijmakers, B.: Exploiting ontologies for deep learning: a case for sentiment mining. Proc. Comput. Sci. (2018). SEMANTiCS 2018–14th International Conference on Semantic Systems (2018)
11. Ribeiro, M.T., Singh, S., Guestrin, C.: Why should I trust you?: Explaining the predictions of any classifier. In: Proceedings of the 22nd ACM SIGKDD International Conference on Knowledge Discovery and Data Mining, pp. 1135–1144. ACM (2016)
12. Sabour, S., Frosst, N., Hinton, G.E.: Dynamic routing between capsules. In: 31st Conference on Neural Information Processing Systems (NIPS 2017), Long Beach, CA, USA (2017)
13. Samek, W., Wiegand, T., Müller, K.R.: Explainable artificial intelligence: understanding, visualizing and interpreting deep learning models. ITU J.: ICT Discov. (1) (2017)
14. Vargas, J.E., Raj, S.: Developing maintainable expert systems using case-based reasoning. Expert Syst. $10(4)$, 219–225 (1993)
15. Voogd, J., Hanckmann, P., de Heer, P., van Lith, J.: Neuro-symbolic modelling for operational decision support. NATO MSG-159 Symposium, STO-MP-MSG-159.3 (2018)
16. Wagner, W.P., Otto, J., Chung, Q.: Knowledge acquisition for expert systems in accounting and financial problem domains. Knowl.-Based Syst. $15(8)$, 439–447 (2002)
17. Wang, N., Pynadath, D.V., Hill, S.G.: Trust calibration within a human-robot team: comparing automatically generated explanations. In: The Eleventh ACM/IEEE International Conference on Human Robot Interaction, pp. 109–116. IEEE Press (2016)
18. Zhao, B., Fu, Y., Liang, R., Wu, J., Wang, Y., Wang, Y.: A large-scale attribute dataset for zero-shot learning. CoRR abs/1804.04314. arXiv:1804.04314 (2018)

Physiological Indicators for User Trust in Machine Learning with Influence Enhanced Fact-Checking

Jianlong Zhou[1]([⊠]), Huaiwen Hu[2], Zhidong Li[1], Kun Yu[1], and Fang Chen[1]

[1] University of Technology Sydney, Sydney, Australia
{jianlong.zhou,zhidong.li,kun.yu,fang.chen}@uts.edu.au
[2] Data61, CSIRO, Sydney, Australia
vivianhhw@gmail.com

Abstract. Trustworthy Machine Learning (ML) is one of significant challenges of "black-box" ML for its wide impact on practical applications. This paper investigates the effects of presentation of influence of training data points on machine learning predictions to boost user trust. A framework of fact-checking for boosting user trust is proposed in a predictive decision making scenario to allow users to interactively check the training data points with different influences on the prediction by using parallel coordinates based visualization. This work also investigates the feasibility of physiological signals such as Galvanic Skin Response (GSR) and Blood Volume Pulse (BVP) as indicators for user trust in predictive decision making. A user study found that the presentation of influences of training data points significantly increases the user trust in predictions, but only for training data points with higher influence values under the high model performance condition, where users can justify their actions with more similar facts to the testing data point. The physiological signal analysis showed that GSR and BVP features correlate to user trust under different influence and model performance conditions. These findings suggest that physiological indicators can be integrated into the user interface of AI applications to automatically communicate user trust variations in predictive decision making.

Keywords: Influence · Machine Learning · Trust · Physiological features

1 Introduction

We have witnessed a rapid increase in the availability of data sets in various fields, for example in infrastructure, transport, energy, health, education, telecommunications, and finance. Together with the dramatic advances in Machine Learning (ML), getting insights from these "big data" and data analytics-driven solutions are increasingly in demand for different purposes. While we continuously

A. Holzinger et al. (Eds.): CD-MAKE 2019, LNCS 11713, pp. 94–113, 2019.
https://doi.org/10.1007/978-3-030-29726-8_7

find ourselves coming across ML-based Artificial Intelligence (AI) systems that seem to work or have worked surprisingly well in practical scenarios (e.g. the self-driving cars, and the conversational agents for self-services), ML technologies still face prolonged challenges with low user acceptance of delivered solutions as well as seeing system misuse, disuse, or even failure. These fundamental challenges can be attributed to the nature of the "black-box" of ML methods for domain experts when offering ML-based solutions [36]. For example, for many non-ML users, they simply provide source data to an AI system, and after selecting some menu options, the system displays colorful viewgraphs and/or recommendations as output [37]. It is neither clear nor well understood *why ML algorithms made this prediction*, or *how trustworthy this output or decision based on the prediction was*. These questions demonstrate that both the explanation of and trust in ML play significant roles in affecting the user acceptance of ML in practical applications. The explanation is closely related to the concept of interpretability, which is referred to as the ability of an agent to explain or to present its decision to a human user, in understandable terms [5,29]. Trust is defined as "the attitude that an agent will help achieve an individual's goals in a situation characterized by uncertainty and vulnerability" [22].

As a result, recent research suggests model *interpretability/explanation* as a remedy for the "black-box" ML methods [24,26,36]. While there is much work in progress towards improving ML interpretability [15,18,30], the ideal state of having explainable, evidence driven ML-based decisions still remains a challenge [26]. To date, most of the work on ML interpretability has focused explicitly on ML model explanation itself, developing various explanation approaches to show why a prediction is made. However, the ML model explanation is just one component of the ML pipeline. Furthermore, what and how explanation information are presented to end users for the deployment to boost user trust plays significant roles in an ML-based intelligent (AI) system. Taking the influence of training data points on predictions [18] in supervised learning as an example, the explanation with influence allows to capture the weight/contribution of each training data point on the prediction of a testing data point. However, these explanations are highly biased towards ML experts' views, and are largely dependent on abstract statistical algorithms, which introduce further complexities to domain users. While domain users are more interested in what influence information affect and how these influence information are presented to them to boost their trust in predictions or decisions based on predictions.

Therefore, besides explanation, Mannarswamy et al. [26] proposed that the ability to provide justifiable and reliable evidences for ML-based decisions would increase the trust of users. Yin et al. [32] found that the stated model accuracy had a significant effect on the extent to which people trust the model, suggesting the importance of communication of ML model performance for user trust. Recently, *fact-checking*, which provides "evaluation of verifiable claims made in public statements through investigation of primary and secondary sources" [19], is increasingly used to check and debunk online information because of credibility challenges of the internet content [7]. Furthermore, previous research found the physiological correlations with decision making [39], it is possible that user trust in predictive decision making can be evaluated by monitoring specific physiological signals for intelligent user interface of AI applications.

Motivated by these investigations, this paper introduces *fact-checking* into ML explanation by referring training data points as facts to users to boost user trust. These training data points are selected based on their influence level of predictions. We aim to investigate what influence of training data points and how they affect user trust in order to enhance ML explanation and boost user trust. We tackle this question by allowing users check the training data points that have the higher influence on the prediction and the training data points that have the lower influence on the prediction. The model performance is also introduced into the pipeline to find how both the influence and model performance affect user trust. Physiological signals are also collected and analysed to find their correlations to trust under different influence and model performance conditions.

2   Related Work

2.1   Explanation for Machine Learning

In the early years, visualization is primarily used to explain simple ML algorithms. For example, different visualization methods are used to examine specific values and show probabilities of selected objects visually for Naïve-Bayes [4], decision trees [2], or SVMs [9]. Advanced visualization techniques are then used as an interaction interface for users in data analysis. Guo et al. [13] introduced a graphical interface named Nugget Browser allowing users to interactively submit subgroup mining queries for discovering interesting patterns dynamically. Zhou et al. [37] revealed states of key internal variables of ML models with interactive visualization to keep users aware what is going on inside a model. More recent work tries to use visualization as an interactive tool to facilitate ML diagnosis. ModelTracker [1] provides an intuitive visualization interface for ML performance analysis and debugging. Chen et al. [10] proposed an interactive visualization tool by combining ten state-of-the-art visualization methods in ML to help users interactively carry out multi-step diagnosis for ML models. Recently, visualization approaches are also proposed to explain complex deep neural networks [14]. For example, saliency maps are used to explain contributions of different points of a data to predictions [6].

Besides visualization, various mathematical approaches are proposed to explain ML models. Robnik-Sikonja et al. [30] explained classification models by evaluating contributions of features to classifications based on the idea that importance of a feature or a group of features in a specific model can be estimated by simulating the lack of knowledge about the values of the feature(s). Besides feature contributions, explanation of individual instance contributions to ML models was investigated to allow users to understand why a classification/prediction is made. For example, Landecker et al. [21] developed an approach of contribution propagation to give per-instance explanations of a network's classifications. Koh et al. [18] used influence functions to evaluate influence of each training data point on predictions.

These approaches explain ML models mostly from an ML expert's perspective, which introduce further complexities to domain users and make users more difficult to understand complex algorithms. Furthermore, these explanations mostly focus on the stage of ML models and pay less attention to the stage of deployment of ML models.

2.2 User Trust in Machine Learning

As the ultimate frontline users of ML-based systems, humans are the key stakeholders and human factors such as user trust are essential in extracting and delivering more sensible and effective insights from data science technologies [12]. From this perspective, Zhou et al. [35,39] argued that communicating user cognitive responses such as trust benefits the evaluation of effectiveness of ML approaches. Therefore, different approaches are investigated to reveal human cognition states such as trust in predictive decision making scenarios [35,37].

Moreover, various researches have been investigated to learn user trust variations in ML. Ye and Johnson [31] experimented with three types of explanations (trace, justification and strategy) for an expert system, and found that justification (defined as showing the rationale behind each step in the decision) was the most effective type of explanation in changing users' attitudes towards the system. Kizilcec [17] proposed that the transparency of algorithm interfaces can promote awareness and foster user trust. It was found that appropriate transparency of algorithms through explanation benefited the user trust. However, too much explanation information on algorithms eroded user trust. Ribeiro et al. [28] explained predictions of classifiers by learning an interpretable model locally around the prediction and visualizing importance of the most relevant features to improve user trust in classifications. Other studies that empirically tested the importance of explanation to users, in various fields such as the health informatics, consistently showed that explanations significantly increase users' confidence and trust [8].

2.3 Physiological Responses in Decision Making

In Human-Computer Interaction (HCI), physiological responses are used to understand an individual's decision making process [39]. For example, GSR refers to how well the skin conducts electricity when an external direct current of constant voltage is applied [11]. It yields continuous signals that are related to activity in the sympathetic branch of the anatomical neural system during tasks. It is well established that skin conductance covaries with the arousal dimension of affect, indexing its intensity. The Iowa Gambling Task (IGT) [3] demonstrated that GSR can be used as a process indicator of affective processes when making decisions. Zhou et al. [39] showed that decision making can be measured with GSR in order to allow users to perceive the quality of their decisions and the level of difficulty involved in making decisions. Therefore, GSR can serve as an objective, non-verbal, non-voluntary indicator and a physiological measure that is relatively free from demand characteristics and reporting biases in decision making. However, little work has been investigated on the variations of GSR in user trust in a predictive decision making scenario under various conditions such as uncertainty. Furthermore, sympathetic activation has been found to cause changes in heart rate, stroke volume and peripheral cardiovascular resistance [25]. These effects can be sensed by Blood Volume Pulse (BVP), which measures the blood volume in the skin capillary bed in the finger with photoplethysmography (PPG). BVP is often used as an indicator of affective processes and emotional arousal, which play an essential role in rational decision making, learning and cognitive tasks [33]. Zhou et al. [16] showed a set of BVP features for indexing cognitive load.

These previous work motivates us to consider both algorithmic explanations and model performance in the interpretability of ML, aiming to find what explanations and how these explanations affect user trust in ML. We also aim to investigate physiological indicators which may correlate with user trust in predictive decision making. This paper uses the influence of training data points as an example and investigates what influence and how they affect user trust in a predictive decision making scenario.

3 Hypotheses

The following hypotheses are posed in this study:

- H1: The presentation of influence of training data points on predictions will affect the user trust and result in the increase of user trust in predictions;
- H2: The training data points which have the higher influence on predictions will have the higher effect on user trust than those with the lower influence;
- H3: Higher model performance together with the presentation of influence of training data points will result in the higher user trust;
- H4: There are correlations between physiological indicators and user trust under different influence and model performance conditions.

4 Method

In this section, a framework of fact-checking for boosting user trust is firstly presented. A case study is then introduced. After that, the influence of training data points is formulated to understand contributions of training data points to test data predictions. Finally, the fact-checking visualization is proposed to present influence of training data points on test data predictions to users.

4.1 Framework of Fact-Checking for Boosting User Trust

We present a framework of fact-checking for boosting user trust in a predictive decision making scenario (see Fig. 1). In a typical conventional ML pipeline, a training data is used to train an ML model and predictions are made based on the trained model (as shown in the lower unshaded part in Fig. 1). There is no information on the ML explanation in order to promote the trustworthiness of the prediction. Motivated by the online fact-checking services for strengthening trust [7], an influence-enhanced fact-checking approach is added on the top of the conventional ML pipeline in the proposed framework (as shown in the upper shaded part in Fig. 1) to explain predictions and boost user trust in predictions. Firstly, the influence of all training data points for the prediction of a testing data point is calculated with influence functions as presented in the following subsection. All training data points are then ranked in descending order based on the calculated influence values. Training data points which have the higher influence values (e.g. the top 10 training data points in the ranking) and training data points which have the lower influence values (e.g. the bottom 10 training data points in the ranking) are obtained respectively based on the ranking. These training data points function as facts which are the most similar points to the testing data point and the least similar points to the testing data point respectively. The parallel coordinate based visualization as presented in the following subsection is used to visualize these selected ranked training data points allowing users to compare the facts with the testing data points to boost trust in predictions.

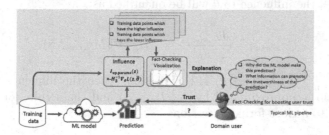

Fig. 1. A framework of fact-checking for boosting user trust.

4.2   Case Study

This paper used water pipe failure prediction as a case study for predictive decision making (replicated in lab environment). Water supply networks constitute one of the most crucial and valuable urban assets. The combination of growing populations and aging pipe networks requires water utilities to develop advanced risk management strategies in order to maintain their distribution systems in a financially viable way [23, 40]. Pipes are characterized by different attributes, referred as features, such as laid year, material, diameter size, etc. If pipe failure historical data is provided, future water pipe failure rate is predictable with respect to the inspected length of the water pipe network [23, 40]. Such models are used by utility companies for budget planning and pipe maintenance. However, different models with various presentation of influence of training data points and various prediction performance (accuracy) may be achievable resulting in different possible management decisions. The experiment is then set up to determine what influence and model performance may affect the user trust.

4.3   Influence of Training Data Points

Consider a machine learning based prediction problem from an input space $X \in \mathbb{R}^D$ (e.g. water pipe attributes of D dimensions) to an output space $Y \in \{0, 1\}$ (e.g. labels on failures of water pipes), with given training data points $z_1, ..., z_N$, where $z_i = (x_i, y_i) \in X \times Y$, each y_i is a failure observation of each pipe in one year. Based on the data points, a model parameter $\hat{\theta} \in \Theta$ can be learned by minimizing the loss function $\sum_{i=1}^{N} L(z_i, \theta)$. The influence is then defined as how important is a training data point z_{train} to the prediction of a testing data point z_{test}. The influence is then calculated for each pair of (z_{train}, z_{test}).

The intuitive way to get influence is to compare the difference of the prediction results, i.e. y_{test} with and without z_{train} used in the training. The method cannot be scaled up as it requires retraining the model for all training data points for each testing data, which means $N + 1$ retraining are needed. This is infeasible for large datasets usually with millions of data points. In this study, the influence function is used to avoid the retraining. It is used to trace a model's prediction through the learning algorithm and back to its training data, thereby identifying training data points most responsible for a given prediction [18]. For any training data point z_{train}, if its weight is to be upweighted by an infinitesimal amount $\epsilon > 0$ from $\frac{1}{N}$, the influence to $\hat{\theta}$ will be quantified as:

$$I(z_{train}) = \frac{d\hat{\theta}_{\epsilon, z_{train}}}{d\epsilon} = -H_{\hat{\theta}}^{-1} \nabla_\theta L(z_{train}, \hat{\theta}) \tag{1}$$

where $H_{\hat{\theta}} = \frac{1}{n} \sum_{i=1}^{n} \nabla_\theta^2 L(z_i, \hat{\theta})$ is the Hessian. Then the influence by removing z_{train} can be approximated by $-\frac{1}{N} I(z_{train})$ and chain rule, for z_{test}, it is proportional to a close-form expression:

$$I_{removing}(z_{train}, z_{test}) \propto \nabla_\theta L(z_{test}, \hat{\theta}) I(z_{train}) \tag{2}$$

which can be used in our influence evaluation as we only need the influence ranking of all training data points in the user trust evaluation. The details of these influence functions can be found in [18].

4.4   Fact-Checking Visualization

A training data point in ML usually has multiple features/attributes. The parallel-coordinates is one classical approach to visualize multi-attribute data points. One advantage of this technique is that it can provide an overview of data trend, where each attribute is represented with one axis in parallel coordinates. In this study, we present a visualization approach for presenting multiple data attributes based on parallel coordinates as shown in Fig. 2: each vertical axis represents one data attribute with the sorted descending order, and a polyline connecting points on each vertical axis represents a data point. In this study, each polyline represents one water pipe with various attributes. Various pipe attributes belonging to one pipe are encoded with the same color. Testing pipe is encoded with red color. The influence of each training pipe on the prediction of a test is encoded with the width of polylines, the wider the polyline, the higher the influence. Such color and line width encoding approach provides an overview of data trend of pipes and their associated attribute details, which can improve the information browsing efficiency. For example, Fig. 2 demonstrates how similar the training pipes are with the testing pipe in red color. If training pipes are considered as facts for predictions, this parallel coordinates based visualization is fact-checking visualization. The pipe attributes visualized in Fig. 2 include pipe size, length, pipe age, failure times during the observation period, and whether it was failed in the checked year (0 means no failure and 1 means failure occured).

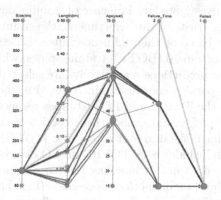

Fig. 2. Fact-checking visualization.

5   Experiment

This section sets up an experiment to examine our hypotheses with the case study of a decision making scenario on water pipe management.

5.1   Experimental Data

Water pipe failure prediction uses historical pipe failure data to predict future failure rate [23]. The historical data contain failure records of water pipes, and various attributes of water pipes, such as laid year, length, diameter size, surrounding soil type, etc. In this study, actual water pipe attributes (features) and the associated historical failure data from a region of a city were used in this experiment. The pipe features used in the experiment include the pipe age, pipe size (diameter), length, and failure times during the observation period. There are 108,745 failure records with 9,062 pipes. 80% of data was used to train the model and the rest was used for the testing. Convolutional neural network (CNN) [20,38] was trained to model the water pipe failures.

In this study, two CNN models (Model 1: 3 hidden layers with number of units of 64, 128, 256, and max iteration number of 1300 respectively; Model 2: 3 hidden layers with number of units of 32, 64, 128, and max iteration number of 700 respectively) were trained using different network settings, resulting in the model accuracy of 90% and 55% respectively. These two model performances were used as high model performance (90%) and low model performance (55%) respectively to find differences of user responses in the experiment. Furthermore, the influence of each training pipes on a prediction was calculated with the use of influence functions introduced in the previous section.

5.2   Task Design

Tasks are designed to investigate how influence of training data points and model performance affect user trust. In this experiment, the training pipes are ranked in the descending order based on their influence values. Based on the ranking, the top 10 (TOP10) and bottom 10 (BOT10) training pipes, which have the highest and lowest influence on predictions respectively, are selected. The fact-checking visualization based on parallel coordinates introduced in the previous section is then used to visualize the TOP10 and BOT10 pipes respectively. Based on the TOP10 and BOT10 pipes' visualization, this experiment divides fact-checking visualization settings for tasks into four categories: (1) *TOP10*, (2) *BOT10*, (3) *TOP10&BOT10* which includes both TOP10 and BOT10 visualizations in tasks, and (4) *Control* which does not include any influence visualization on training pipes. By considering both model performance cases (high and low performance) and fact-checking visualization conditions, we finally got 8 tasks as shown in Table 1. These 8 tasks were conducted two rounds with all same settings except testing pipes used. Two training tasks were also conducted by each participant before formal tasks. In summary, there were 18 tasks conducted (8 tasks × 2 rounds + 2 training tasks = 18 tasks) by each participant.

Table 1. Task setup in the experiment.

		Influence			
		TOP10	BOT10	TOP10&BOT10	Control
Model performance	High	T1	T2	T3	T4
	Low	T5	T6	T7	T8

The decision tasks investigated are: each participant was told that he/she would take the asset management responsibility of a water company. The water company plans to repair pipe failures in the next financial year. He/she was asked to make a decision on whether to replace a testing pipe, using water pipe failure prediction models learned from the historical water pipe failure records. Each task was divided into three stages: at the beginning of each task, participants were told that a pipe is predicted to fail next year with a prediction accuracy of 90% (High) or 55% (Low); then different fact-checking visualizations based on task settings (see Table 1) are displayed; lastly, participants were asked to make a decision of whether or not to replace the pipe based on the prediction. Participants were told that they were competing against other people to reach the best budget plan in a given time period in order to push them to make their efforts for tasks. The task orders were randomized during the experiment.

5.3   Participants and Data Collection

22 participants were recruited, who are mainly researchers and students with the range of ages from twenties to forties and an average age of 30 years. Of all participants, 5 were females. After each decision making task, participants were asked to rate the trust level of predictions on which decisions were made using a 9-point Likert scale (from 1: least trust, to 9: most trust). Participants were asked to rate how helpful the presentation of influence is for decision making. At the end of each round, participants were also asked to rate the usefulness of influence on helping them more confident in decision making. Besides subjective ratings, skin conductance responses of subjects with GSR sensors and blood volume pulse information with BVP sensors from ProComp Infiniti of Thought Technology Ltd were collected during task time.

6   Analysis of Subjective Ratings

In this study, we aim to understand: (1) the effects of influence on user trust under a given model performance, and (2) the effects of model performance on user trust under a given influence condition respectively. Therefore, for the evaluation of each aims, we first performed Friedman test and then followed it up with post-hoc analysis using Wilcoxon signed-rank tests (with a Bonferroni correction) to analyze differences in participant responses of trust under a fixed condition (e.g. trust changes with different influence types under the fixed high model performance). Trust values were normalized with respect to each subject to minimize individual differences in rating behavior (see Eq. 3):

$$T_i^N = \frac{T_i - T_i^{min}}{T_i^{max} - T_i^{min}} \tag{3}$$

where T_i and T_i^N are the original trust rating and the normalized trusting rating respectively from the user i, T_i^{min} and T_i^{max} are the minimum and maximum of trust ratings respectively from the user i in all of his/her tasks.

6.1   Influence and Trust

Figure 3(a) shows mean normalized trust values over different influence settings under high model performance (error bars represent the 95% confidence interval of a mean and it is same in other figures). Friedman's test gave statistically significant differences in trust among four influence conditions, $\chi^2(3) = 21.675$, $p = .000$. Then post-hoc Wilcoxon tests (with a Bonferroni correction under a significance level set at $p < .013$) was applied to find pair-wise differences between influence conditions. The adjusted significance alpha level of .013 was calculated by dividing the original alpha of .05 by 4, based on the fact that we had four influence conditions to test.

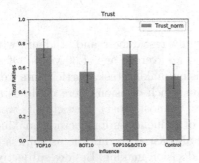 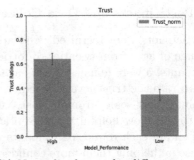

(a) Normalized trust by influence under high model performance.　(b) Normalized trust by different model performance.

Fig. 3. Normalized trust values.

The post-hoc tests found that participants had significantly higher trust in predictions when influences of TOP10 training pipes were presented than those without influence information presentation (Control condition) ($Z = 102.0, p < .001$). Participants also showed significantly higher trust in predictions when influences of both TOP10 and BOT10 were presented than that without influence information presentation (Control condition) ($Z = 120.5, p < .004$). The results suggest that the presentation of influence of training data points on predictions significantly increases the user trust in predictions as we hypothesized (H1). It was also found that participants had significantly higher trust in predictions when influences of TOP10 training pipes were presented than that when influences of BOT10 training pipes were presented ($Z = 61.5, p < .001$). This implies that the training data points having the higher influence on predictions have the higher effect on user trust than that having the lower influence (H2).

However under low model performance, statistically significant differences of trust among different influence conditions have not been found.

These results suggest that the presentation of influence of training data points on predictions significantly increases the user trust in predictions, but only for training data points with higher influence values under the high model performance condition.

6.2 Model Performance and Trust

Figure 3(b) shows mean normalized trust values under two model performance conditions (high and low). A Wilcoxon test found that participants had statistically higher trust in predictions under high model performance than that under low model performance ($Z = 854.5, p = .000$). This result confirms the findings in [32]. We then further drilled down to compare user trust differences over model performance under different influence conditions. Figure 4 shows mean normalized trust values over two model performance conditions (high and low) under different influence settings. It was found that participants showed significantly higher trust under high model performance than that under low model performance over all four influence settings (TOP10: $Z = 11.5, p < .000$; BOT10: $Z = 52.0, p < .000$; TOP10&BOT10: $Z = 77.5, p < .000$; Control: $Z = 77.5, p < .001$). The results suggest that high model performance together with influence information result in the higher user trust in predictions (H3). These findings go on to support the idea that people trust more in predictions with high model performance.

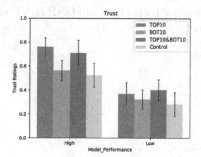

Fig. 4. Normalized trust by model performance under different influences.

7 Physiological Indicators

In this section, GSR and BVP signals are analyzed to investigate their variations under different conditions. The GSR and BVP data analysis process is divided into following steps: (1) signal smoothing, (2) data normalization, (3) feature extraction, and (4) feature significance test.

Similar to the analysis of subjective ratings in the previous section, various GSR and BVP features are also analysed to find: (1) the effects of influence on GSR and BVP features under a given model performance, and (2) the effects of model performance on GSR and BVP features under a given influence condition respectively. Therefore, for the evaluation of each aims, we first performed one-way ANOVA test and then followed it up with post-hoc analysis using t-test (with a Bonferroni correction) to analyze differences in physiological features under a fixed condition.

7.1   Signal Smoothing

The first step of physiological signal analysis is the signal smoothing. For GSR signals, we use convolution filter (similar to a low pass filter) to remove noise. All GSR signals are convoluted to a Hann window function to remove the noise [34]. For BVP signals, we use spectrogram to detect corrupt signals and abnormal outliers, then remove all corrupt data from the dataset [25].

7.2   Normalization

After signal smoothing, we normalize smoothed signal using subject-wise Z-Normalization to omit subjective differences between different signals [34]. The subject-wise normalization means that the mean and variance used in the normalization as in Eq. 4 are from signals of all tasks from each subject.

$$S_N = \frac{S - \mu}{\sigma} \tag{4}$$

where S is the original GSR/BVP value, S_N is the normalized GSR/BVP value, μ and σ are mean and variance respectively of GSR/BVP signals among all tasks conducted by each subject.

7.3   GSR

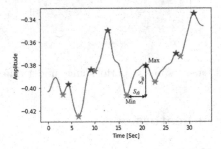

Fig. 5. An example of GSR signals after noise filtering. (Color figure online)

GSR Feature Extractions. In this paper, GSR features are defined based on signal extremas. The extremas are extracted from the normalized GSR signal. Figure 5 shows an example of extremas (red star as local maxima and yellow star as local minima) of GSR. We extracted and analysed both extreme-based and statistical features. All the features are listed in Table 2 [39]. The definition of duration S_{di} and magnitude S_{mi} are shown in Fig. 5. t is the time spent on each task. The estimated area can be regarded as the area of the triangle made by S_d and S_m, which is $S_a = \frac{1}{2}S_dS_m$ [39].

Table 2. GSR features

GSR features	Notes
Mean of GSR μ_g	Summation of all GSR values divided by task time
GSR variance σ_g	Variance of GSR values over task time
Number of peaks S_f	Number of peaks in a GSR signal
Sum of duration S_d	Sum of duration time of all tasks: $S_d = \sum S_{di}$
Sum of magnitude S_m	Sum of magnitude: $S_m = \sum S_{mi}$
Estimated area S_a	Sum of estimated area: $S_a = \sum S_{ai}$
Number of response per second S_{fs}	$S_{fs} = S_f/t$
Duration per second S_{ds}	$S_{ds} = S_d/t$
Magnitude per second S_{ms}	$S_{ms} = S_m/t$
Maximum of duration S_d^{max}	$S_d^{max} = max(S_d)$
Maximum of magnitude S_m^{max}	$S_m^{max} = max(S_m)$
Maximum of estimated area S_a^{max}	$S_a^{max} = max(S_a)$
Average gradient G_a	$G_a = \frac{1}{S_f} \sum S_{mi}/S_{di}$
Maximum gradient G^{max}	$G^{max} = max(S_{mi}/S_{di})$

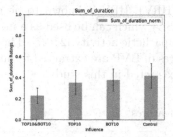 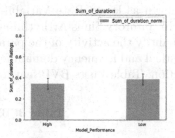

(a) Sum of duration over different influence conditions. (b) Sum of duration over different model performance conditions.

Fig. 6. GSR feature of sum of duration.

GSR Features and Influence. Under the high model performance, one-way ANOVA tests found that there are significant differences in GSR values among different influence conditions for GSR sum of duration S_d ($F(3, 112) = 2.874, p = .039$) (see Fig. 6(a)). The post-hoc t-tests (with a Bonferroni correction under $\alpha = .05/4 = .013$, based on the fact that 4 levels were tested) was used to examine the pair difference between influence conditions in GSR feature of S_d. The post-hoc tests found that participants had significantly lower GSR S_d values when influences of TOP10&BOT10 training pipes were presented than that without influence information presentation (Control condition) ($t = -3.039, p < .004$). Participants also showed relatively lower GSR S_d values when influences of TOP10&BOT10 training pipes were presented than that when BOT10 training pipes were presented. The results suggest that the presentation of influence of training data points on predictions especially TOP10 training points significantly decreases the GSR S_d values under high model performance.

GSR Features and Model Performance. Figure 6(b) shows mean GSR S_d values over two model performance conditions (high and low). Although there is no significant difference found in GSR S_d values over two model performance conditions, a trend shows that GSR S_d values under high model performance condition is relatively lower than that under low model performance condition. The findings suggest that the high model performance condition has a trend to decrease GSR values.

7.4 BVP

BVP Features. BVP is a periodical signal and associated with three major frequency bands: Very Low Frequency (VLF) (0.00–0.04 Hz), Low Frequency (LF) (0.05–0.15 Hz), and High Frequency (HF) (0.16–0.40 Hz). The LF/HF ratio is calculated by finding the ratio of low frequency energy to high frequency energy in the spectrum. Furthermore, the BVP sensor measures one of physiological changes known as Heart Rate Variability (HRV). HRV is known to be closely related to Respiratory Sinus Arrhythmia (RSA) which can be used as a measurement to quantify the activity of the parasympathetic activity [25,27]. Therefore, both statistical and frequency domain features of BVP are extracted for analysis in this section. Table 3 lists BVP features extracted in this study.

Table 3. BVP features

BVP features	Notes
Mean of BVP μ_b	Summation of all BVP values divided by task time
BVP variance σ_b	Variance of BVP values over task time
Number of peaks S_p	Number of peaks in a BVP signal
BVP peak mean μ_{bp}	Summation of all BVP values divided by number of peaks
BVP peak variance σ_{bp}	Variance of BVP peak values
LF	Power Spectral Density (PSD) for low frequency
HF	Power Spectral Density (PSD) for high frequency
LF HF ratio S_r	$S_r = LF/HF$

BVP Features and Influence. Under the high model performance, one-way ANOVA tests found significant differences in BVP peak mean μ_{bp} values ($F(3, 120) = 4.705$, $p = .004$) and BVP peak variance σ_{bp} values ($F(3, 120) = 4.961, p = .003$) among different influence conditions respectively. Figure 7 shows BVP peak mean values over four influence conditions. The post-hoc t-tests(with a Bonferroni correction as mentioned previously) were conducted to examine the pairwise difference between influence conditions in BVP features. For the BVP peak mean μ_{bp}, it was found that there were statistically significant lower values in TOP10&BOT10 than that in other three influence conditions TOP10 ($t = -2.921, p = .005$), BOT10 ($t = -3.45, p = .001$), and Control ($t = -3.644, p = .001$) respectively. Similarly, The BVP peak variance σ_{bp} showed significantly lower values in TOP10&BOT10 than that in other three influence conditions TOP10 ($t = -2.947, p = .005$), BOT10 ($t = -3.57, p = .001$), and Control ($t = -3.788, p = .000$) respectively.

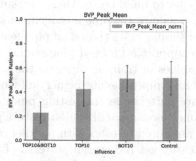

Fig. 7. BVP feature of BVP peak mean over four influence conditions.

These results show that the presentation of influence of training data points on predictions especially both TOP10 and BOT10 training points at the same time significantly decreases BVP values such as μ_{bp} under high model performance. However, all extracted BVP features did not show significant differences over model performance conditions, despite the trend with relative lower GSR values under high model performance related to influence presentations of TOP10 training points.

In summary, we found that the presentation of influence of training data points on predictions especially TOP10 training points significantly decreases both GSR and BVP values such as μ_{bp}, but only under high model performance. Furthermore, a trend shows that both GSR and BVP values are relatively lower under high model performance than that under low model performance. By considering the relations between trust and influence/model performance concluded in the previous section, the findings in this section on GSR and BVP features can be used as indicators of user trust in predictive decision making under different influence and model performance conditions as we expected in H4.

8 Discussions

As discussed in earlier sections, trust is a challenging concept to investigate in machine learning based solutions. This paper intends to study human-machine trust in a specialized predictive decision making scenario. As machines are becoming more intelligent, however in many scenarios, instead of full autonomy, Human-Machine Teaming (HMT) is required, where humans interact with the intelligent (AI) system to understand why the AI system is suggesting something that the human should do or not do. Therefore, both interaction with and transparency of the system help humans make effective uses of the AI system for trusting decisions.

In the water pipe failure prediction example as mentioned, when pipe management staff want to use an AI tool to make decisions on pipe replacement, they need to be confident that there is a clear rationale for the ML to predict a pipe's future failure, in order to build trust. Therefore, similar to precedent that humans justify actions by analogy, the pipe management staff interact with such kind of AI system to find similar cases (based on pipe features such as material, age, size, and length) to support a planned pipe management protocol. In our approach, the influence values of training pipes on the prediction were used to help users to locate/identify pipes having higher influence values (which may show more similar feature patterns to the testing pipe) or pipes having lower influence values (which may show more dissimilar feature patterns to the testing pipe). These pipes were presented to users with parallel coordinates based visualizations to help users easily get the overall patterns of features of pipes. The interaction with the visualization of pipes functioned as the fact-checking for the prediction to help users understand why a similar or different decision was made, thereby increases the transparency of the system and boosts user trust.

As we have seen that participants showed significantly higher trust when TOP10 visualization was presented. It was also found that participants showed significantly higher trust under high model performance. GSR and BVP features showed correlation to both influence and model performance conditions, suggesting that GSR and BVP features can be used as indicators for trust variations in predictive decision making.

In order to make ML-driven AI applications not only intelligent but also intelligible, the user interface of AI applications needs to allow users to access the most influential facts to predictions by visualizations. Such influence-enhanced fact-checking allows users find similar facts to the testing data point to get the rational behind for the justification of their actions, therefore boosting user trust.

A weakness of this training data based influence interpretation approach is the privacy issue of training data. The proposed approach is not applicable if the training data is sensitive and/or needs to be made private. However, there are still many applications where it is not an issue (such as the water pipe failures).

9   Conclusions and Future Work

This paper investigates the influence enhanced fact-checking for the ML explanation to boost user trust in a predictive decision making scenario. Both influence of training data points on predictions and model performance were examined to find their effects on trust. Physiological features were analysed and showed their correlations to influence and model performance conditions. A user study found that the presentation of influence of training data points on predictions significantly increased the user trust in predictions, but only for training data points with higher influence values under the high model performance condition, where users were expected to be able to justify their actions with more similar facts to the testing data point. These findings suggested that the access of the most influential facts to predictions by users in the user interface of AI applications would help users get the rational behind their actions and therefore benefit the user trust in predictions.

Our future work will focus on the setup of ML models to automatically predict user trust in decision making based on physiological features, which contributes to the ultimate goal of intelligent user interface of AI applications.

Acknowledgements. This work is partly supported by the Asian Office of Aerospace Research & Development (AOARD) under grant No. AOARD 216624.

References

1. Amershi, S., Chickering, M., Drucker, S.M., Lee, B., Simard, P., Suh, J.: ModelTracker: redesigning performance analysis tools for machine learning. In: Proceedings of the 33rd Annual ACM Conference on Human Factors in Computing Systems, pp. 337–346 (2015)
2. Ankerst, M., Elsen, C., Ester, M., Kriegel, H.P.: Visual classification: an interactive approach to decision tree construction. In: Proceedings of KDD 1999, pp. 392–396 (1999)
3. Bechara, A., Damasio, H., Damasio, A.R., Lee, G.P.: Different contributions of the human amygdala and ventromedial prefrontal cortex to decision-making. J. Neurosci. **19**, 5473–5481 (1999)
4. Becker, B., Kohavi, R., Sommerfield, D.: Visualizing the simple Bayesian classifier. In: Fayyad, U., Grinstein, G.G., Wierse, A. (eds.) Information visualization in data mining and knowledge discovery, pp. 237–249 (2002)
5. Biran, O., Cotton, C.: Explanation and justification in machine learning: a survey. In: Proceedings of the 2017 IJCAI Explainable AI Workshop, pp. 8–13 (2017)
6. Brahimi, M., Arsenovic, M., Laraba, S., Sladojevic, S., Boukhalfa, K., Moussaoui, A.: Deep learning for plant diseases: detection and saliency map visualisation. In: Zhou, J., Chen, F. (eds.) Human and Machine Learning. HIS, pp. 93–117. Springer, Cham (2018). https://doi.org/10.1007/978-3-319-90403-0_6
7. Brandtzaeg, P.B., Følstad, A.: Trust and distrust in online fact-checking services. Commun. ACM **60**(9), 65–71 (2017)
8. Calero Valdez, A., Ziefle, M., Verbert, K., Felfernig, A., Holzinger, A.: Recommender systems for health informatics: state-of-the-art and future perspectives. In: Holzinger, A. (ed.) Machine Learning for Health Informatics. LNCS (LNAI), vol. 9605, pp. 391–414. Springer, Cham (2016). https://doi.org/10.1007/978-3-319-50478-0_20
9. Caragea, D., Cook, D., Honavar, V.G.: Gaining insights into support vector machine pattern classifiers using projection-based tour methods. In: Proceedings of KDD 2001, pp. 251–256 (2001)
10. Chen, D., Bellamy, R.K.E., Malkin, P.K., Erickson, T.: Diagnostic visualization for non-expert machine learning practitioners: a design study. In: 2016 IEEE Symposium on Visual Languages and Human-Centric Computing (VL/HCC), pp. 87–95, September 2016

11. Figner, B., Murphy, R.O.: Using skin conductance in judgment and decision making research. In: A Handbook of Process Tracing Methods for Decision Research: A Critical Review And User's Guide, pp. 163–184 (2010)
12. Fisher, D., DeLine, R., Czerwinski, M., Drucker, S.: Interactions with big data analytics. Interactions 19(3), 50–59 (2012)
13. Guo, Z., Ward, M.O., Rundensteiner, E.A.: Nugget browser: visual subgroup mining and statistical significance discovery in multivariate datasets. In: Proceedings of the 15th International Conference on Information Visualisation, pp. 267–275 (2011)
14. Hartono, P.: A transparent cancer classifier. Health Inform. J. (2018)
15. Ilyas, A., Engstrom, L., Athalye, A., Lin, J.: Black-box adversarial attacks with limited queries and information. In: Dy, J., Krause, A. (eds.) Machine Learning. PMLR, vol. 80, pp. 2142–2151. Stockholmsmässan, Stockholm (2018)
16. Zhou, J., Arshad, S.Z., Luo, S., Yu, K., Berkovsky, S., Chen, F.: Indexing cognitive load using blood volume pulse features. In: Proceedings of the 2017 CHI Conference on Human Factors in Computing Systems, CHI EA 2017, May 2017
17. Kizilcec, R.F.: How much information?: Effects of transparency on trust in an algorithmic interface. In: Proceedings of the 2016 CHI Conference on Human Factors in Computing Systems, pp. 2390–2395 (2016)
18. Koh, P.W., Liang, P.: Understanding black-box predictions via influence functions. In: Proceedings of the 34th International Conference on Machine Learning, ICML 2017, Sydney, NSW, Australia, pp. 1885–1894, 6–11 August 2017
19. Kriplean, T., Bonnar, C., Borning, A., Kinney, B., Gill, B.: Integrating on-demand fact-checking with public dialogue. In: Proceedings of the 17th ACM Conference on Computer Supported Cooperative Work and & Social Computing, CSCW 2014, pp. 1188–1199 (2014)
20. Krizhevsky, A., Sutskever, I., Hinton, G.E.: ImageNet classification with deep convolutional neural networks. In: Advances in Neural Information Processing Systems, vol. 25, pp. 1097–1105 (2012)
21. Landecker, W., Thomure, M.D., Bettencourt, L.M.A., Mitchell, M., Kenyon, G.T., Brumby, S.P.: Interpreting individual classifications of hierarchical networks. In: 2013 IEEE Symposium on Computational Intelligence and Data Mining (CIDM), pp. 32–38, April 2013
22. Lee, J.D., See, K.A.: Trust in automation: designing for appropriate reliance. Hum. Factors 46(1), 50–80 (2004)
23. Li, Z., et al.: Water pipe condition assessment: a hierarchical beta process approach for sparse incident data. Mach. Learn. 95(1), 11–26 (2014)
24. Lipton, Z.C.: The mythos of model interpretability. In: Proceedings of the 2016 ICML Workshop on Human Interpretability in Machine Learning (WHI 2016), New York, NY, USA (2016)
25. Luo, S., Zhou, J., Duh, H.B.L., Chen, F.: BVP feature signal analysis for intelligent user interface. In: Proceedings of the 2017 CHI Conference Extended Abstracts on Human Factors in Computing Systems, CHI EA 2017, pp. 1861–1868 (2017)
26. Mannarswamy, S., Roy, S.: Evolving AI from research to real life - some challenges and suggestions. In: Proceedings of the Twenty-Seventh International Joint Conference on Artificial Intelligence, IJCAI-18, pp. 5172–5179. International Joint Conferences on Artificial Intelligence Organization, July 2018
27. Nilsson, M., Funk, P.: A case-based classification of respiratory sinus arrhythmia. In: Funk, P., González Calero, P.A. (eds.) ECCBR 2004. LNCS (LNAI), vol. 3155, pp. 673–685. Springer, Heidelberg (2004). https://doi.org/10.1007/978-3-540-28631-8_49

28. Ribeiro, M.T., Singh, S., Guestrin, C.: "Why should I trust you?": Explaining the predictions of any classifier. arXiv:1602.04938 [cs, stat], February 2016
29. Richardson, A., Rosenfeld, A.: A survey of interpretability and explainability in human-agent systems. In: Proceedings of IJCAI/ECAI 2018 Workshop on Explainable Artificial Intelligence (XAI), pp. 137–143 (2018)
30. Robnik-Sikonja, M., Kononenko, I., Strumbelj, E.: Quality of classification explanations with PRBF. Neurocomputing **96**, 37–46 (2012)
31. Ye, L.R., Johnson, P.E.: The impact of explanation facilities on user acceptance of expert systems advice. MIS Q. **19**(2), 157–172 (1995)
32. Yin, M., Vaughan, J.W., Wallach, H.: Does stated accuracy affect trust in machine learning algorithms? In: Proceedings of ICML2018 Workshop on Human Interpretability in Machine Learning (WHI 2018), July 2018
33. Zhai, J., Barreto, A., Chin, C., Li, C.: Realization of stress detection using psychophysiological signals for improvement of human-computer interactions. In: Proceedings of IEEE SoutheastCon 2005, pp. 415–420 (2005)
34. Zhou, J., Arshad, S.Z., Wang, X., Li, Z., Feng, D., Chen, F.: End-user development for interactive data analytics: uncertainty, correlation and user confidence. IEEE Trans. Affect. Comput. **9**(3), 383–395 (2018)
35. Zhou, J., Bridon, C., Chen, F., Khawaji, A., Wang, Y.: Be informed and be involved: effects of uncertainty and correlation on user confidence in decision making. In: Proceedings of ACM SIGCHI Conference on Human Factors in Computing Systems (CHI2015) Works-in-Progress, Korea (2015)
36. Zhou, J., Chen, F. (eds.): Human and Machine Learning: Visible, Explainable, Trustworthy and Transparent. Springer, Cham (2018). https://doi.org/10.1007/978-3-319-90403-0
37. Zhou, J., Khawaja, M.A., Li, Z., Sun, J., Wang, Y., Chen, F.: Making machine learning useable by revealing internal states update - a transparent approach. Int. J. Comput. Sci. Eng. **13**(4), 378–389 (2016)
38. Zhou, J., Li, Z., Zhi, W., Liang, B., Moses, D., Dawes, L.: Using convolutional neural networks and transfer learning for bone age classification. In: 2017 International Conference on Digital Image Computing: Techniques and Applications (DICTA 2017), pp. 1–6 (2017)
39. Zhou, J., et al.: Measurable decision making with GSR and pupillary analysis for intelligent user interface. ACM Trans. Comput.-Hum. Interact. **21**(6), 33 (2015)
40. Zhou, J., Sun, J., Wang, Y., Chen, F.: Wrapping practical problems into a machine learning framework: using water pipe failure prediction as a case study. Int. J. Intell. Syst. Technol. Appl. **16**(3), 191–207 (2017)

Detection of Diabetic Retinopathy and Maculopathy in Eye Fundus Images Using Deep Learning and Image Augmentation

Sarni Suhaila Rahim[1,2(✉)], Vasile Palade[1], Ibrahim Almakky[1], and Andreas Holzinger[3]

[1] Faculty of Engineering, Environment and Computing, Coventry University,
Priory Street, Coventry CV1 5FB, UK
{ad0490, ab5839, ab8961}@coventry.ac.uk
[2] Faculty of Information and Communication Technology,
Universiti Teknikal Malaysia Melaka, Hang Tuah Jaya,
76100 Durian Tunggal, Melaka, Malaysia
sarni@utem.edu.my
[3] Institute for Medical Informatics, Statistics and Documentation,
Medical University Graz, Graz, Austria
andreas.holzinger@medunigraz.at

Abstract. Diabetic retinopathy is a significant complication of diabetes, produced by high blood sugar level, which causes damage to the retina. Effective diabetic retinopathy screening is required because diabetic retinopathy does not show any symptoms in the initial stages, and can cause blindness if it is not diagnosed and treated promptly. This paper presents a novel diabetic retinopathy automatic detection in retinal images by implementing efficient image processing and deep learning techniques. Besides diabetic retinopathy detection, the developed system integrates a novel detection of maculopathy into one detection system. Maculopathy is the damage to the macula, the eye part that is responsible for central vision. Therefore, the combined detection of diabetic retinopathy and maculopathy is essential for an effective screening of diabetic retinopathy. The paper investigates the capability of image pre-processing techniques based on data augmentation as well as deep learning for diabetic retinopathy and maculopathy detection. Computer-assisted clinical decision-making is inevitably transforming the diabetic retinopathy detection and management today, which is crucial for clinicians and patients alike. Therefore, a high degree of accuracy, with which computer algorithms can detect the diabetic retinopathy and maculopathy, is absolutely necessary.

Keywords: Diabetic Retinopathy · Maculopathy · Eye screening ·
Colour fundus images · Image processing · Data augmentation · Deep learning

© IFIP International Federation for Information Processing 2019
Published by Springer Nature Switzerland AG 2019
A. Holzinger et al. (Eds.): CD-MAKE 2019, LNCS 11713, pp. 114–127, 2019.
https://doi.org/10.1007/978-3-030-29726-8_8

1 Introduction

The occurrence of diabetes is increasing globally at an accelerating rate. Diabetic Retinopathy (DR) contributes significantly as some of the main causes of vision loss, if it is not diagnosed and managed properly. In order to minimize the risk of blindness happening caused by the diabetic retinopathy, diabetes patients should control the blood sugar levels, blood pressure and cholesterol, in addition to undergoing regular eye screening.

Diabetic retinopathy and maculopathy screening helps identify high-risk individuals of having sight impairment. Therefore, an effective screening of diabetic retinopathy is essential for early action, as well as in the preventive management of diabetic complications. Moreover, screening is able to detect eye problems before starting to interfere with our vision, and the treatment can help prevent or reduce vision loss if the problems are caught early. The retinal screening helps give information about the condition progression, and determine the treatment type if the signs of diabetic retinopathy or maculopathy are detected.

A timely and complete eye examination that comprises dilated ophthalmoscopy or high quality fundus images assessment in patients without previous treatment of DR or other eye disease are the accepted methods in screening [1]. Eye fundus photography is most frequently used in clinical studies, and widely used for telemedicine and patient education as well. Moreover, fundus photography offers a colour or red-free image, and provides many advantages compared with the predecessor, colour photographic film [1]. Nowadays, digital retinal imaging is widely used as it provides high-resolution, faster images and easily image enhancement amenability. In diabetic retinopathy screening, the features of DR are characterized in order to demonstrate more precise details than clinical examination. Modifications of conventional imaging techniques and new developments in the area, technology innovations, such as automated image interpretation, large data sets usage and mobile applications, will improve the pathogenesis of DR [1]. In addition, Baumal et al. [1] suggest that task management and longitudinal treatment help prevent vision loss and recover a significant amount of vision in the patients.

Diabetic retinopathy is a complex disease with diverse clinical findings. Among the diabetic retinopathy signs are microaneurysms, haemorrhages, exudates and neovascularisation. This paper is focusing on the diabetic retinopathy and maculopathy detection. The yellow lesions found near the macula (also a disease of the macula) is termed as maculopathy. The macula, which is the centre of the retina, functions as a central mechanism that provides our vision. The macula area is considered as a very sensitive area, where the centre of the macula, called fovea, is a tiny area that is accountable for both detailed and colour vision [2]. Thus, the detection of maculopathy is vital as the loss of vision happening at the fovea part causes blindness. The presence or absence of the maculopathy condition will decide the requirement of appropriate treatment or referral. The referral to the ophthalmologist is assigned if maculopathy is detected. On the other hand, if maculopathy is not present, referral is not necessary and the screening will be repeated in a one-year period. The combination of diabetic retinopathy and maculopathy detection, therefore, is important in order to assist the diabetic retinopathy screening management. Figure 1 represents the eye fundus image with diabetic retinopathy, showing maculopathy in colour image.

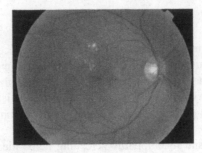

Fig. 1. Maculopathy representation in colour image [1] (Color figure online)

A variety of ways and solutions have been proposed by researchers working and focusing on the maculopathy detection, in order to detect and classify the fundus images into different stages of maculopathy, such as mild, moderate and severe maculopathy [34–39]. However, in this paper, the research work proposed other incorporation mechanisms of diabetic retinopathy and maculopathy, where the detection of both are based on the diabetic retinopathy signs discovery and also following ophthalmologists' practice. The severity level reported in the literature is based on the diabetic retinopathy features detection, rather than the maculopathy severity. The proposed classification refers to whether or not maculopathy is present, i.e., with maculopathy and without maculopathy. As a result, the new cases are: No Diabetic Retinopathy (DR), Mild DR with/without maculopathy, Moderate DR with/without maculopathy, Severe DR with/without maculopathy, Proliferative DR with/without maculopathy and Advanced Diabetic Eye Disease (ADED). This categorization is beneficial because two important detection can be identified in only one process of screening. Furthermore, the urgency of the referral, which should happen within four weeks, as proposed by the National Institute for Clinical Excellence [3], is applied to those who have any form of maculopathy, regardless of mild, moderate or severe levels. In this case, the severity of the maculopathy is therefore not significant, provided that its presence or absence has been determined. Therefore, this paper presents a novel development of diabetic retinopathy alongside maculopathy detection, by introducing effective image pre-processing techniques in conjunction with deep learning for classification. The new system has been tested on a new developed database collected from Melaka Hospital, Malaysia.

The paper is organized as follows. Section 2 presents some previous related work on automated methods for the detection of diabetic retinopathy, comprising of developed diabetic retinopathy detection systems with deep learning as well as developed maculopathy detection systems. Section 3 explains the proposed system for the diabetic retinopathy detection alongside maculopathy in eye fundus images, by implementing effective image pre-processing and deep learning techniques. Finally, Sect. 4 presents some conclusions and future work.

2   Related Previous Work

There are some developed automated systems reported in literature to detect and diagnose diabetic retinopathy. In addition, some researchers proposed the detection of maculopathy to support the management of diabetic retinopathy. It can be summarized that various techniques and methods were proposed for the image pre-processing, feature extraction and classification phases in order to produce reliable diabetic retinopathy detection systems.

2.1   Diabetic Retinopathy Detection

For the detection of diabetic retinopathy purpose, various machine learning techniques and methods were proposed, used and reported in the literature [4–39]. Meanwhile, deep learning has been used in diabetic retinopathy detection in [10–25]. However, the reported detection systems were concentrating on the diabetic retinopathy as general detection, and also on the diabetic retinopathy signs detection, using various machine learning methods, and deep learning among them.

In our earlier work, a basic automated system for general diabetic retinopathy detection, employing a combination of non-fuzzy techniques, has been proposed initially [26]. Following this, we investigated the capability of different fuzzy image processing techniques for the detection of diabetic retinopathy and maculopathy in retinal images in [27], which was enhanced in [28] with retinal structures' segmentation. Different machine learning techniques were used for the classification part to categorize the images into more detailed classes of the disease. The results show that employing fuzzy image processing in addition to the retinal structure localization and extraction can help produce a more reliable diabetic retinopathy screening system. Therefore, the proposed system in this paper, introducing a combination of techniques for the image pre-processing part as well as the deep learning, for better classification of diabetic retinopathy and maculopathy detection.

Data augmentation is implemented to artificially enlarge the datasets to overcome the shortcomings of using small image datasets, reducing overfitting on the image data and increasing the algorithm performance. Data augmentation has been used in the automated detection of diabetic retinopathy systems using deep learning. Lam et al. [10] implemented random augmentation of images for an automated detection of diabetic retinopathy in order to improve the capability of network localization and also reduce overfitting. Among the data augmentation techniques implemented were random zeros padding, zoom, rolling and rotation. Meanwhile, five different transformation types, which are rotation, flipping, shearing, rescaling and translation have been proposed by Xu et al. in [12] for the automatic classification of diabetic retinopathy using deep convolutional neural networks. Other augmentation methods, which are duplication and rotation with several degree angles, were implemented in [15]. Rakhlin [13] implemented some pre-processing techniques on the retinal images for the diabetic retinopathy detection with the integration of deep learning classification, including normalization, scaling, centering and cropping. Pratt et al. [16] implemented data augmentation, where each image was randomly rotated within the range of 0 to 90°, randomly horizontally and vertically flipped and also randomly horizontally and vertically shifted on the pre-processed images, for diabetic retinopathy detection and classification. Besides the basic transformations, other pre-processing based augmentations for color enhancements could also be used for the data augmentation. Ghosh et al. [17] proposed the adjustment of image brightness, followed by rotations of 90 and 180°, which were eventually able to increase the class size six times and adapt to different orientations and lighting conditions.

2.2 Maculopathy Detection

The localisation and the detection of both the macula and the fovea are essential in identifying maculopathy. The lesions in the macula region generate maculopathy, while the fovea is located at the centre of the macula. Some automatic localization and detection of the macula in digital eye fundus images have been proposed in [30–34]. Meanwhile, the detection of diabetic maculopathy in retinal images have been investigated in [34–39].

Tariq et al. [36] developed an automated detection and grading system of diabetic maculopathy using digital eye fundus images. The proposed system involves pre-processing, exudates and macula detection, some feature extraction, and finally the classification stage using a Gaussian Mixture Model classifier, where the input image was graded into three categories: healthy, non-clinically significant macular edema, and clinically significant macular edema. The same diabetic maculopathy classification (normal, non-clinically significant macular edema, and clinically significant macular edema), as proposed by Tariq et al. [36], was also used by Chowriappa et al. in [39]. The proposed system extracted the textural features and classified the eye fundus images into their classes of disease severity using four ensemble classifiers, employing the tree-based J48, naïve Bayes, sequential minimal optimization and also the hidden naive Bayes classifiers.

Meanwhile, a computer system for the purpose of the detection of diabetic maculopathy in human eye fundus images, employing morphological operations, was proposed by Vimala and Kajamohideen in [34]. The green component from the colour input image was extracted, followed by median filtering and also contrast limited adaptive histogram equalization techniques. The macula detection was obtained by employing top-hat transform and bottom hat transform techniques. Some colour and also texture features were extracted to grade the pre-processed image into two classes: exudates present or exudates absent, using a Support Vector Machine as classifier. Punnolil [35] presents the diagnosis system of diabetic maculopathy severity by employing image pre-processing techniques (colour normalization), the detection of optic disc, both macula and fovea localization, and then detecting exudates and hemorrhages. After these, several features were extracted and classified into the maculopathy severity grading (normal, mild, moderate and severe) using Support Vector Machine as classifier.

Siddalingaswamy and Prabhu [37] proposed a system of automatic grading of diabetic maculopathy severity level. The developed system initially performed the green component extraction, optic disc detection, fovea and macular region detection, then the detection of hard exudate lesions using mathematical morphological and clustering techniques. The level of maculopathy severity is classified as normal, mild, moderate and also severe, based on the exudates location in marked macular region. Another automated computer-based system for maculopathy diagnosis in diabetic retinopathy screening was presented by Hunter et al. in [38], where the detection and filtering of candidate lesions, extraction of features and classification by a multilayer perceptron were implemented.

In summary, the detection of maculopathy is really important because the untreated affected macula will eventually contribute to the loss of vision. Therefore, for this challenging problem, some researchers are currently contributing and proposing solutions for the detection of maculopathy in retinal images. However, in the previously reported maculopathy detection systems, image augmentation based techniques have not been implemented during the pre-processing stage in conjunction with deep learning for the classification stage. Therefore, this paper presents a novel development of diabetic retinopathy and maculopathy detection system based on such a combination of techniques.

3   Proposed Approach

In this paper, a deep learning approach that utilizes "on-the-fly" data augmentation techniques is proposed. The combination of normal and diabetic retinopathy eye fundus images from a novel dataset, which was collected from the Eye Clinic, Department of Ophthalmology, Melaka Hospital, Malaysia, is used to evaluate the model. The new dataset, with a total of 600 colour eye fundus images, contains images of size 3872 × 2592 pixels saved in JPEG format. The dataset is presented in detail in [29].

Through the proposed approach, input retina images are first pre-processed to facilitate the classification process. After that, the processed images are fed into a deep convolutional neural network (DCNN) for classification. This DCNN is trained using augmented images of the retina with varying levels of retinopathy and maculopathy.

3.1   Image Pre-processing

All input images, from both the training and testing sets, are reduced in size from their original 3872 × 2592 pixels to 242 × 162 pixels. This reduction was to maximize the performance of the model, while preserving as many features as possible from the original image. The aspect ratio of the images was also preserved to maintain the original shapes and spatial features contained in the original images.

3.2   Deep Convolutional Neural Network

A DCNN model was designed to classify the input images into their respective classes. The model starts with an image input layer, where the image size is specified, which in this case is 3 × 162 × 242. Following that, four convolutional layers are implemented, where each is followed by a 2-dimensional batch normalization layer, a rectified linear unit (ReLU) and a max-pooling layer. The filter sizes are gradually reduced through the four layers to reduce the inputs (3 × 162 × 242) into (9 × 14 × 128). The specific parameters used for each layer can be found in Fig. 2. Following the last convolutional layer, the features are flattened into a fully connected layer, which also implements a ReLU activation function:

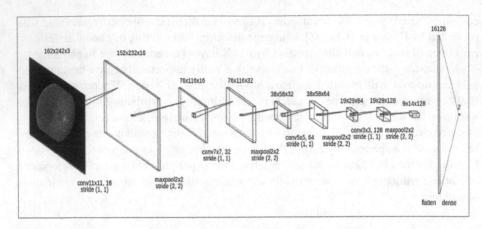

Fig. 2. Convolutional neural network structure used

$$f(x) = \max(0, x) \tag{1}$$

This layer is then connected to an output layer that has a varying number of neurons, depending on the categorization (shown in Table 1). The last classification layer implemented a SoftMax activation function that is tasked with producing the probabilities of each image belonging to a specific class.

3.3 Training

Following the image pre-processing stage, the model was trained using the resized images. During the training stage, on-the-fly data augmentation was implemented to enhance the number of training examples in the dataset. This stage helps with preventing the model from overfitting, while also helping with "calibrating" the high number of parameters in deep models, such as the one used in this work. The implemented data augmentation techniques are exactly aimed at the nature of retina images, where unlike natural scene images, the images are more standardized in terms of contrast and angles. Therefore, random rotation was the only data augmentation technique that was implemented with the resized images rotated randomly by an angle between −20 and 20°.

The model was trained separately for the three different class taxonomy categories shown in Table 1. The first category splits into two main cases: "no diabetic retinopathy" and "diabetic retinopathy". The categorization based on maculopathy detection is the second one, classifying into two other cases: "maculopathy detected" and "maculopathy not detected". The third categorization, representing the experts' original classification, provides more details and involves ten stages of retinopathy.

All three models were trained using the Cross Entropy Criterion to calculate the error at the last classification layer. The final error for the model is calculated using the cross entropy function (C):

$$C = \sum_{i=1}^{N} \sum_{j=1}^{K} t_{ij} \ln y_{ij} \qquad (2)$$

where N is the total number of images, K is the number of classes, t_{ij} is the indicator that sample i belongs to class j, and y_{ij} is the model's output for sample i for class j. Stochastic gradient descent with momentum (SGDM) was used for optimization with an initial learning rate of 0.0001 and 0.9 momentum. Each model was trained for a total of 100 epochs using the 70% random split of the dataset that form the training set. For each of the three models, two different variants were trained, one with and another without data augmentation.

3.4 Results

All trained models were tested using the same test set without any alteration other than the resizing. A summary of the classification accuracies for the different categories is shown in Table 2. The effect of the proposed data augmentation techniques are apparent with the enhanced performance in all categories. This means that even considering the small number of training samples in the dataset, our proposed classification models were able to achieve reasonable accuracies. It is also clear that the classifier trained for two classes was able to achieve the highest accuracy because each class has comparably higher number of samples than the other category.

Table. 1. Different categorizations

Categorization I		Categorization II	
Retinopathy stage	No. of images	Retinopathy stage	No. of images
1. No DR	276	1. Maculopathy detected	131
2. DR	324	2. Maculopathy not detected	469
Total	600		600

Categorization III	
Retinopathy stage	No. of images
1. No DR	276
2. Mild DR without maculopathy	72
3. Mild DR with maculopathy	27
4. Moderate DR without maculopathy	85
5. Moderate DR with maculopathy	83
6. Severe DR without maculopathy	23
7. Severe DR with maculopathy	11
8. PDR without maculopathy	6
9. PDR with maculopathy	10
10. ADED	7
Total	600

Meanwhile, for more clarity, the generated confusion matrix is presented in Fig. 3(a) and (b) to show the relative performance of the classifier. The confusion matrix for both variant models of categorization I and categorization II show that the sensitivity value (the percentage of abnormal images which have been classified as abnormal) is higher than the specificity (the percentage of normal images classified as normal). The sensitivity and the specificity values for both variants of the second categorization are similar. The models for the first and second categorizations show that the classification accuracy model with data augmentation is higher or similar with that of the model without data augmentation. However, for the third categorization, the classification accuracy for the model without data augmentation is higher than that of the model with data augmentation. This happened due the fact that the categorization III provides a hugely imbalanced number of images for some cases, particularly for the severe cases of DR. Although categorization II (maculopathy detection) provides an imbalanced classification between the two main cases, the classification accuracy for categorization II is the highest among the three categorization. The model was able to detect the maculopathy presence well, as the maculopathy can be seen clearly from the quality images provided, and, therefore, the model was capable to differentiate the severity of maculopathy lesions in the eye fundus images.

Table 2. Summary of results

Network model	Validation accuracy		
	Categorization I	Categorization II	Categorization III
CNN without data augmentation	61.11%	77.22%	44.20%
CNN with data augmentation	63.33%	77.22%	42.54%

It can be concluded that using balanced or near balanced datasets, and suitable data augmentation otherwise, help increase the classification accuracy. These two factors should be considered in the development of better detection and classification models.

4   Conclusions and Future Work

An approach for the detection of diabetic retinopathy and maculopathy in colour eye fundus images implementing data augmentation techniques and deep learning has been proposed in this paper. In summary, it is challenging to detect the diabetic retinopathy and maculopathy, particularly using a small and imbalanced dataset for classification. The use of image pre-processing techniques for the data augmentation helped improve the classification performance. The classification models can be further enhanced by employing different image augmentation techniques or different combinations of pre-processing techniques, including fuzzy techniques, as in our previous work [27–29], such as fuzzy transform, fuzzy histogram equalization, fuzzy filtering, etc. In addition, the retinal structures segmentation, such as the extraction of blood vessels and the localization of the optic disc can be implemented in order to increase the maculopathy detection performance.

Deep learning models have been deployed for the classification of diabetic retinopathy and maculopathy classification. Problem specific data augmentation was implemented to overcome the different challenges presented by the classification task. The proposed models classify the input images into three different taxonomies of classes, for the purpose of generating a diversity of results and performance analysis. The three types of classification consist of two types of 2-class classification and one type of 10-class classification. The classification can be enhanced by using another categorization involving four cases: no retinopathy class, non-proliferative diabetic retinopathy (mild, moderate and severe cases) class, proliferative diabetic retinopathy class and finally the advanced diabetic eye disease class. Pre-trained image classification networks also should be considered, as they have already been trained to identify specific visual features that could be useful when generalized on different tasks such as this. Additionally, different parameters and further exploration of deep learning architectures should be performed in order to generate better classification and

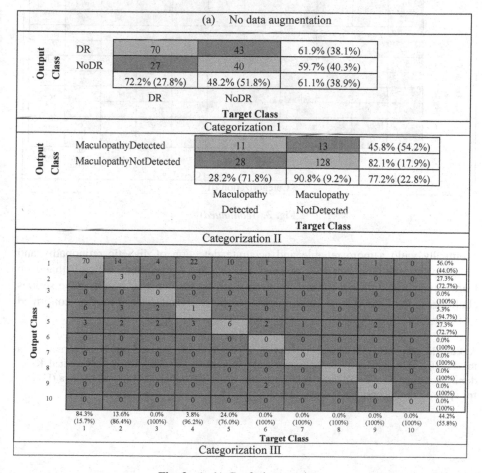

Fig. 3. (a, b) Confusion matrices

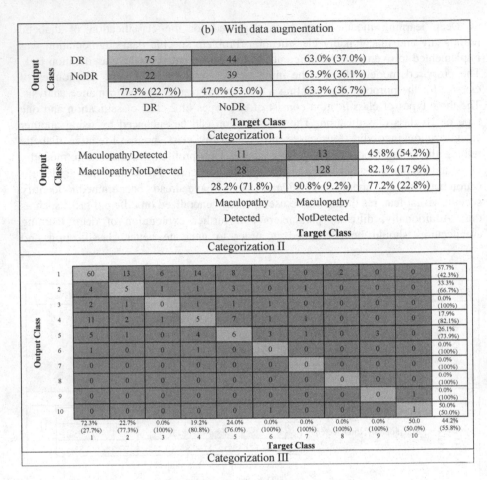

Fig. 3. (*continued*)

eventually yield a more reliable and accurate detection of diabetic retinopathy and maculopathy. A future aspect to investigate is due to a big problem, which is that deep learning methods turn out to be difficult to interpret for humans, which create serious challenges, including that of interpreting a predictive result when it may be confirmed as incorrect [40].

Acknowledgements. This project is part of a postdoctoral research currently being carried out at the Faculty of Engineering, Environment and Computing, Coventry University, United Kingdom. The deepest gratitude and thanks go to the Universiti Teknikal Malaysia Melaka (UTeM) for sponsoring this postdoctoral research.

References

1. Baumal, C.R., Duker, J.S.: Current Management of Diabetic Retinopathy. Elsevier, St. Loius (2018)
2. Taylor, R., Batey, D.: Handbook of Retinal Screening in Diabetes: Diagnosis and Management. Wiley, Chichester (2012)
3. National Institute for Clinical Excellence.: Management of type 2 diabetes. Retinopathy-screening and early management. NICE, London (2002)
4. Mookiah, M.R.K., et al.: Evolutionary algorithm based classifier parameter tuning for automatic diabetic retinopathy grading: a hybrid feature extraction approach. Knowl.-Based Syst. **39**, 9–22 (2013)
5. Priya, R., Aruna, P.: Review of automated diagnosis of diabetic retinopathy using the support vector machine. Int. J. Appl. Eng. Res. **1**(4), 844–863 (2011)
6. Priya, R., Aruna, P.: SVM and neural network based diagnosis of diabetic retinopathy. Int. J. Comput. Appl. **41**(1), 6–12 (2012)
7. Priya, R., Aruna, P., Suriya, R.: Image analysis technique for detecting diabetic retinopathy. Int. J. Comput. Appl. **1**, 34–38 (2013)
8. Shome, S.K., Vadali, S.R.K.: Enhancement of diabetic retinopathy imagery using contrast limited adaptive histogram equalization. Int. J. Comput. Sci. Inf. Technol. **2**(6), 2694–2699 (2011)
9. Sreng, S., Maneerat, N., Hamamoto, K., Panjaphongse, R.: Automated diabetic retinopathy screening system using hybrid simulated annealing and ensemble bagging classifier. Appl. Sci. **8**(7), 1198 (2018)
10. Lam, C., Yi, D., Guo, M., Lindsey, T.: Automated detection of diabetic retinopathy using deep learning. In: AMIA Joint Summits on Translational Science Proceedings, AMIA Joint Summits on Translational Science 2017, pp. 147–155 (2018)
11. Voets, M., Mollersen, K., Bongo, L.A.: Replication study: development and validation of a deep learning algorithm for detection of diabetic retinopathy in retinal fundus photographs (2018). https://arxiv.org/pdf/1803.04337.pdf
12. Xu, K., Feng, D., Mi, H.: Deep convolutional neural network-based early automated detection of diabetic retinopathy using fundus image. Molecules **22**(12), 1–7 (2017)
13. Rakhlin, A.: Diabetic retinopathy detection through integration of deep learning classification framework (2017). https://www.biorxiv.org/content/biorxiv/early/2018/06/19/225508.full.pdf
14. Gulshan, V., Peng, L., Coram, M., et al.: Development and validation of a deep learning algorithm for detection of diabetic retinopathy in retinal fundus photographs. JAMA **316**(22), 2402–2410 (2016)
15. Rajanna, A.R., Aryafar, K., Ramchandran, R., Sisson, C., Shokoufandeh, A., Ptucha, R.: Neural networks with manifold learning for diabetic retinopathy detection. In: Proceedings of IEEE Western NY Image & Signal Processing Workshop (2016). https://arxiv.org/pdf/1612.03961.pdf
16. Pratt, H., Coenen, F., Broadbent, D.M., Harding, S.P., Zheng, Y.: Convolutional neural networks for diabetic retinopathy. Procedia Comput. Sci. **90**, 200–205 (2016)
17. Ghosh, R., Ghosh, K., Maitra, S.: Automatic detection and classification of diabetic retinopathy stages using CNN. In: 4th International Conference on Signal Processing and Integrated Networks (SPIN), Noida, India, pp. 550–554. IEEE (2017)
18. Chudzik, P., Majumdar, S., Caliva, F., Al-Diri, B., Hunter, A.: Microaneurysm detection using deep learning and interleaved freezing. In: Proceedings SPIE 10574, Medical Imaging 2018: Image Processing 1057411, pp. 1–9 (2018)

19. Lam, C., Yu, C., Huang, L., Rubin, D.: Retinal lesion detection with deep learning using image patches. Invest. Ophthalmol. Vis. Sci. **59**(1), 590–596 (2018)
20. Hatanaka, Y., Ogohara, K., Sunayama, W., Miyashita, M., Muramatsu, C., Fujita, H.: Automatic microaneurysms detection on retinal images using deep convolution neural network. In: International Workshop on Advanced Image Technology (IWAIT), pp. 1–2 (2018)
21. Dai, L., et al.: Clinical report guided retinal microaneurysm detection with multi-sieving deep learning. IEEE Trans. Med. Imaging **37**(5), 1149–1161 (2018)
22. Harangi, B., Toth, J., Hajdu, A.: Fusion of deep convolutional neural networks for microaneurysm detection in color fundus images. In: 2018 40th Annual International Conference of the IEEE Engineering in Medicine and Biology Society (EMBC), pp. 3705–3708 (2018)
23. Shan, J., Li, L.: A deep learning method for microaneurysm detection in fundus images. In: 2016 IEEE First International Conference on Connected Health: Applications, Systems and Engineering Technologies (CHASE), pp. 357–358 (2016)
24. Haloi, M.: Improved microaneurysm detection using deep neural network (2016). https://arxiv.org/pdf/1505.04424.pdf
25. Tan, J.H., et al.: Automated segmentation of exudates, haemorrhages, microaneurysms using single convolutional neural network. Inf. Sci. **420**, 66–76 (2017)
26. Rahim, S.S., Palade, V., Shuttleworth, J., Jayne, C.: Automatic screening and classification of diabetic retinopathy fundus images. In: Mladenov, V., Jayne, C., Iliadis, L. (eds.) EANN 2014. CCIS, vol. 459, pp. 113–122. Springer, Cham (2014). https://doi.org/10.1007/978-3-319-11071-4_11
27. Rahim, S.S., Palade, V., Jayne, C., Holzinger, A., Shuttleworth, J.: Detection of diabetic retinopathy and maculopathy in eye fundus images using fuzzy image processing. In: Guo, Y., Friston, K., Aldo, F., Hill, S., Peng, H. (eds.) BIH 2015. LNCS (LNAI), vol. 9250, pp. 379–388. Springer, Cham (2015). https://doi.org/10.1007/978-3-319-23344-4_37
28. Rahim, S.S., Palade, V., Shuttleworth, J., Jayne, C.: Automatic screening and classification of diabetic retinopathy and maculopathy using fuzzy image processing. Brain Inf. **3**, 249–267 (2016)
29. Rahim, S.S., Palade, V., Shuttleworth, J., Jayne, C., Omar, R.N.R.: Automatic detection of microaneurysms for diabetic retinopathy screening using fuzzy image processing. In: Iliadis, L., Jayne, C. (eds.) EANN 2015. CCIS, vol. 517, pp. 69–79. Springer, Cham (2015). https://doi.org/10.1007/978-3-319-23983-5_7
30. Kumar, T.A., Priya, S., Paul, V.: A novel approach to the detection of macula in human retinal imagery. Int. J. Signal Process. Syst. **1**(1), 23–28 (2013)
31. Mubbashar, M., Usman, A., Akram, M.U.: Automated system for macula detection in digital retinal images. In: Proceedings of the 2011 International Conference on Information and Communication Technologies, ICICT, pp. 1–5. IEEE, USA (2011)
32. Akram, M.U., Tariq, A., Khan, S.A., Javed, M.Y.: Automated detection of exudates and macula for grading of diabetic macular edema. Comput. Methods Programs Biomed. **114**, 141–152 (2014)
33. Sekhar, S., Al-Nuaimy, W., Nandi, A.K.: Automated localisation of optic disk and fovea in retinal fundus images. In: Proceedings of the 16th European Signal Processing Conference, pp. 1–5. IEEE, USA (2008)
34. Vimala, A.G.S.G., Kajamohideen, S.: Detection of diabetic maculopathy in human retinal images using morphological operations. Online J. Biol. Sci. **14**, 175–180 (2014)
35. Punnolil, A.: A novel approach for diagnosis and severity grading of diabetic maculopathy. In: Proceedings of the 2013 International Conference on Advances in Computing, Communications and Informatics, pp. 1230–1235. IEEE, New York (2013)

36. Tariq, A., Akram, M.U., Shaukat, A., Khan, S.A.: Automated detection and grading of diabetic maculopathy in digital retinal images. J. Digit. Imaging **26**(4), 803–812 (2013)
37. Siddalingaswamy, P.C., Prabhu, K.G.: Automatic grading of diabetic maculopathy severity levels. In: Mahadevappa, M., et al. (eds.) Proceedings of the 2010 International Conference on Systems in Medicine and Biology, pp. 331–334. Excel India Publishers, New Delhi (2010)
38. Hunter, A., Lowell, J.A., Steel, D., Ryder, B., Basu, A.: Automated diagnosis of referable maculopathy in diabetic retinopathy screening. In: Proceedings of 2011 Annual International Conference of the IEEE Engineering in Medicine and Biology Society, EMBC 2011, pp. 3375–3378. IEEE, USA (2011)
39. Chowriappa, P., Dua, S., Rajendra, A.U., Muthu, R.K.M.: Ensemble selection for feature-based classification of diabetic maculopathy images. Comput. Biol. Med. **43**(12), 2156–2162 (2013)
40. Goebel, R., et al.: Explainable AI: the new 42? In: Holzinger, A., Kieseberg, P., Tjoa, A.M., Weippl, E. (eds.) CD-MAKE 2018. LNCS, vol. 11015, pp. 295–303. Springer, Cham (2018). https://doi.org/10.1007/978-3-319-99740-7_21

Semi-automated Quality Assurance for Domain-Expert-Driven Data Exploration – An Application to Principal Component Analysis

Sandra Wartner[1(✉)], Manuela Wiesinger-Widi[1], Dominic Girardi[1],
Dieter Furthner[2], and Klaus Schmitt[3]

[1] RISC Software GmbH, Research Unit Medical Informatics, Hagenberg, Austria
{sandra.wartner,manuela.wiesinger-widi,dominic.girardi}@risc-software.at
[2] Department of Pediatrics and Adolescent Medicine, Salzkammergut-Klinikum,
Vöcklabruck, Austria
[3] Department of Pediatrics and Adolescent Medicine, Kepler University Hospital,
Linz, Austria

Abstract. Processing and exploring large quantities of electronic data
is often a particularly interesting but yet challenging task. Both the lack
of statistical and mathematical skills and the missing know-how of handling masses of (health) data constitute high barriers for profound data
exploration – especially when performed by domain experts. This paper
presents guided visual pattern discovery, by taking the well-established
data mining method Principal Component Analysis as an example. Without guidance, the user has to be conscious about the reliability of computed results at any point during the analysis (GIGO-principle). In
the course of the integration of principal component analysis into an
ontology-guided research infrastructure, we include a guidance system
supporting the user through the separate analysis steps and we introduce a quality measure, which is essential for profound research results.

Keywords: Principal Component Analysis · Data Quality ·
Guidance · Visual Analytics · Data mining · Doctor-in-the-Loop

1 Introduction

Due to the steadily rising amount of data in varying research domains, visual
data analytics is becoming increasingly important. In complex research domains
(such as biomedical research), deep integration of the domain expert into the data
analysis process is required [1]. A major technical obstacle for these researchers lies

This work was supported in part by the Austrian Research Promotion Agency (FFG
project number 851460) and by funds from the Strategic Economic and Research Program "Innovatives OÖ 2020" of the State of Upper Austria.

© IFIP International Federation for Information Processing 2019
Published by Springer Nature Switzerland AG 2019
A. Holzinger et al. (Eds.): CD-MAKE 2019, LNCS 11713, pp. 128–146, 2019.
https://doi.org/10.1007/978-3-030-29726-8_9

not only in handling, processing and analyzing complex research data [2], but also in giving chapter and verse for exploiting results. Since conclusions can be no better than the received input, users also have to be aware of this GIGO ("Garbage-In-Garbage-Out") principle when applying analytics methods.

With our work we aim to assist the domain expert in the whole process of data modeling, processing, analysis, and interpretation. In particular we are aspiring to increase interpretability and understanding of advanced analysis techniques, such as the Principal Component Analysis (PCA), which serves as an example for evaluating our approach. Preliminary work on this topic by Wartner et al. [3] has focused on the integration of basic PCA functionality, enabling its use by domain experts without assistance of a data scientist. Moreover, it has introduced the concept of quality of the result, by a preliminary selection of certain quality criteria, such as the sample size, the ratio between the number of observations and variables, and the properties of the correlation matrix. In this paper, an ascertainment of the quality criteria summarized and combined in an assessment scheme is presented. Finally, the guided PCA is applied on data from the MICA (Measurements for Infants, Children, and Adolescents) project which was performed in cooperation with the *Kepler University Hospital Linz*.

1.1   An Ontology-Based Research Platform

In order to address the issue of handling, processing, and analyzing complex research data, we have been working on an ontology-based research platform for domain-expert-driven data exploration and research. The key idea behind the platform is that, while being a completely generic system, it can be adapted to any specific research domain by modeling its relevant aspects (classes, attributes, relations, semantic rules, constraints, etc.) in the form of a domain ontology. The whole system adapts itself to this domain ontology at run-time and appears to the user like an individually developed system. Moreover, the elaborate structural meta-information about the research data is used to actively support the domain-expert in challenging tasks such as data integration, data processing, and finally data exploration.

For a more detailed description on the platform itself and the usage of the domain-ontology for data exploration the reader is kindly referred to [3–5].

1.2   The Nuts and Bolts of Principal Component Analysis

Multidimensional data can be hard to explore and visualize. Methods such as the Principal Component Analysis (PCA) are used for simplifying this challenging task by decreasing the dimensions of the data set. Dimensionality reduction aims to reduce the number of variables in the data set without significant loss of information. The new (fewer) variables – in the case of PCA called principal components – are linear combinations of the original variables, capturing most of the variation of the original data set. For an introduction to PCA see for example [6–8].

Based on the centered covariance or correlation matrix of numeric variables, PCA works as a solution to the eigenvalue problem. The newly obtained orthogonal (i.e., uncorrelated) axes constitute linear combinations of the original variables and are referred to as principal components. The corresponding eigenvalue is a measure of importance of the principal component, describing the amount of variation in the original data explained by the principal component. When projecting data into a lower dimensional space, the primary motivation is to preserve most of the variation – the direction with maximum variation is found in the first principal component, while successive principal components account for less variance than the previous ones. As a result the least important principal components are discarded. What we receive is a data set with a reduced number of variables. The proportion of a component's variation, respectively, sums up to the explained variance of the new system. Eventually, the procedure is completed as soon as a predefined percentage of the original system's variance or number of components has been reached.

There are different ways of visualizing PCA results. Score plots depict the transformed data points, and thus are used to find patterns and clusters in the data and to detect outliers within the model, i.e., observations which scatter far from the data center (see also Fig. 10). Ideally, the majority of data lie around the origin of the new coordinate system and spread just slightly. Further advantages arise when chronological sequences are to be analyzed. The analyst is able to determine when a process is getting out of control by tracking the time-related course of the transformed observations. In the event that the position of the observation is increasingly migrating from the origin and out of the control ellipse (see also Sect. 2.2), there is a high chance that one or more variables are taking on unfavorable values. Likewise, in non-sequential data (as in our experimental data), points of large deviation to the center of the hyperplane of the PCA model (and/or outside the control ellipse) differ more strongly from the rest of the observations. While score plots are used for examining the similar behavior of observations, loadings plots are used to investigate the influence of variables on a certain principal component (see also Fig. 9). A variable's position (weight) close to zero indicates little importance, whereas high weights emphasize the contribution to the component. For pattern recognition, the relative positions of the variables to each other also play a pivotal role (close locations imply high correlation and vice versa). Merging both, the sample scores and variable loadings, in one visualization, the resulting visualization is called a biplot. In order to find observations which are not explained well by the model, samples can be colored by the squared prediction error (SPE) to graphically represent the size of the residuals.

2   Quality Measure

To the author's knowledge, no previous work has proposed the assessment of PCA result quality by a single metric. As a result, we develop an expansive evaluation scheme, modifiable for versatile types of analysis. In fact, we differentiate between the quality of a criterion and that of the entire data set, where the latter comprises the assessment of all quality criteria. Furthermore, the procedure of quality assessment also comprises quality criteria not assigned a grade, such as linearity or normality. However, the system makes domain experts aware of checking these requirements.

2.1 A Polymorphic Evaluation Scheme

Firstly, each quality criterion is graded as *good*, *ok*, and *bad*. Secondly, the overall quality is determined according to the subsequent rules:

- If more than 50% of the quality criteria are graded as *good*, and no *bad* has been given, the result quality is graded as *good*.
- If at least 50% of the quality criteria are graded as *ok* and no *bad* has been given, *ok* is awarded.
- If at least one criterion is graded as *bad*, the quality of the result is marked as inadequate (*bad* quality). This is because when at least one quality test fails, severe side effects might arise in interpreting the results.

It is worth noting that the choice of 50% is not restrictive, but merely reflects the experience of the authors in certain studies. One major benefit of this evaluation scheme is its polymorphism, as each data mining method and statistical analysis has most diverse requirements to the underlying data structure and quality. A tailor-made solution can be achieved by cobbling together various key figures in order to meet method dependent requirements. In the following section, the quality criteria specifically used for determining the PCA result quality are described.

2.2 Quality Criteria and User-Interpretable Assessment Methods for Model Quality

This section provides an overview of the quality criteria specifically when applying PCA.

Sample Size and Ratio Between Number of Samples and Variables. A study by Osborne et al. [9] has shown that PCA results of data with both large sample sizes and high ratios had superior quality to those where the data had large sample sizes only. In the following table, as well as in the upper part of Fig. 7, we adapt their proposed sample guidelines to our assessment scheme, where n denotes the number of observations and r the ratio between the number of observations and variables (Table 1).

Table 1. Assessment scheme of the *sample size* quality criterion, where n denotes the number of observations. This classification is partially summarized from recommendations by [9].

Assessment	Sample size	Ratio
Good	$n \geq 500$	$r \geq 10$
Ok	$200 \leq n < 500$	$5 \leq r < 10$
Bad	$n < 200$	$r < 5$

Communalities. The communality r_x^2 of a variable x is computed as the sum of the squared correlations between the extracted principal components and the variable x. When using standardized data, the communality of a variable is computed as the sum of the squared *loadings* for this variable. It indicates how good the variable is explained by the extracted components. The closer the value to 1 is, the better the observed data for the variable is reflected in the model [10]. To the user, the communality for each selected variable is shown within the numeric output section (see last column in the result view section of Fig. 8).

Correlation. In a study conducted by Dziuban et al. [11], the importance of prior inspection of the correlation matrix, as well as the major interpretation pitfalls on examples of random data have been shown. It has been suggested that enough entries beyond the diagonal have to be greater or equal to 0.3 – raising a difficult question here in defining what is meant by "enough" [12]. To that end, the Kaiser-Meyer-Olkin test (KMO) has been introduced [13] to measure the sampling adequacy which gives, prior to analysis, an indication to the meaningfulness of applying PCA to the data set. It was later modified in Kaiser et al. [14] to the forms in Eqs. (1) and (2) to improve stability. A major advantage over other methods (e.g., the Bartlett's test) is the possibility of simultaneously interpreting individual features and the overall quality of the correlation matrix [11].

Let (r_{ij}) denote the correlation matrix, (s_{ij}) the inverse of the correlation matrix and (q_{ij}) the *partial correlation* (or anti-image correlation) matrix of the input variables, where

$$q_{ij} := -\frac{s_{ij}}{\sqrt{s_{ii}s_{jj}}}.$$

Then the overall measure of sampling adequacy (KMO) is defined as:

$$\mathrm{KMO} := \frac{\sum_i \sum_{j \neq i} r_{ij}^2}{\sum_i \sum_{j \neq i} r_{ij}^2 + \sum_i \sum_{j \neq i} q_{ij}^2} \tag{1}$$

The partial correlation estimates the relationship between two statistical variables while controlling for the effect of one or more other variables. Equation (2) provides the computation of the KMO for each variable separately:

$$\mathrm{KMO}(i) := \frac{\sum_{j \neq i} r_{ij}^2}{\sum_{j \neq i} r_{ij}^2 + \sum_{j \neq i} q_{ij}^2} \tag{2}$$

In contrast to traditional measures of sampling adequacy, which have been proven to be unstable for poor data, this instability has been corrected through normalization by Olkin, hence any result value must lie between zero and one. Following Kaiser [14], the assessment scheme is shown in the correlation part in Fig. 7. The closer the value is to 1, the more suitable the variables are (Table 2).

Table 2. Rule for the interpretation of the overall KMO measure after Kaiser [14] and the related assessments integrated in the quality measure.

Assessment	Evaluation	KMO
Good	Marvelous	≥ 0.9
	Meritorious	≥ 0.8
Ok	Middling	≥ 0.7
	Mediocre	≥ 0.6
Bad	Miserable	≥ 0.5
	Unacceptable	< 0.5

Linearity. Another procedure of human supported quality assessment involves the inspection of the structure the data follows [15]. Linearity is given in case there is constant spread in data, i.e., data is homoscedastic, and no (strong) outliers are detected in the data set. This quality criterion is not assigned any grade, but is rather to be examined by the user. A simple and effective way of investigation is offered by a scatter plot matrix, showing $n \cdot (n-1)/2$ plots, where n is the number of variables. The variables names are written in a diagonal line from top left to bottom right. Each bivariate plot delineates the association between two variables of the data set – desirably the vast majority should show linear relationship, i.e., the plot should look like a line. Non-linear patterns between variables can not be detected by PCA (see Fig. 1 for an example to interpret).

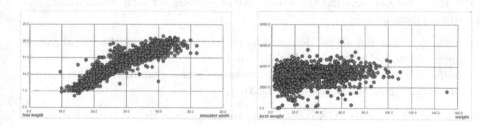

Fig. 1. Scatter plots showing the relationship between a pair of variables, each. In the first plot it is probably safe to say that there is a correlation between the variables, whereas the second plot does not show recognizable correlation.

Normally Distributed Data. Although PCA can be performed on data that is not normally distributed, it might overlook patterns since it only handles first and second order dependencies like mean and variance. If the data is not normally distributed, higher order dependencies might be present but are not detected by PCA. Furthermore, independence of the components is only guaranteed in case of normality. If the data is not normally distributed, other methods like independent component analysis (ICA) may give further insight [16]. Automatic grading of normality is not included in the combined quality metric, but it is recommended to the user to investigate normality plots or normality tests (see also Thode [17]) in the detailed quality description listing. To easily assess whether data fits a normal distribution or not (and to which extent), an additional scatter-chart related visualization is provided in the research infrastructure rather than normality tests. In this plot, the normal theoretical quantiles are plotted against each variable (see Fig. 2). According to the resulting curve characteristics, conclusions of the distribution can be drawn, including also supplementary information on kurtosis or skewness. If data follows a straight, linear pattern, it can be assumed that data is approximately normally distributed.

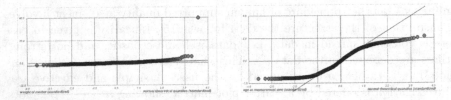

Fig. 2. The normality plots for the weight of the mother (on the left) and the age at measurement time (on the right). On the left-hand side graph, data follows approximately the normal distribution, except for the outlier on the top right corner. The right-hand side graph shows heavy tails, thus this variable doesn't follow the normal distribution.

Outliers. For determining the suitability of samples for inclusion, outliers can be distinguished as:

– *outliers within the model*, detected by the control ellipse by the visual assessment of unusual points (see Fig. 10), and,
– *outliers between the model and the measured data*, examined by the squared prediction error (SPE, see also Fig. 3).

Like all linear methods PCA is very sensitive to outliers. Hence, a lot of attention has to be given to proper outlier handling.

Control Ellipse. The control ellipse is an addition to the score plot. Its aim is to visually identify potential outliers within the model. The control ellipse is derived via Hotelling's T^2, which indicates if a certain observation conforms to the mean of observations. In other words, it measures if an observation is in control. According to [6],

$$T^2 := zL^{-1}z = \frac{k(n-1)}{n-k} F_{k,n-k,\alpha},$$

where z is the score vector of said observation, L is the diagonal matrix of eigenvalues, k the number of principal components, n the number of observations,

α the significance level and $F_{k,n-k,\alpha}$ the critical value of the F-distribution with respect to the parameters k and $n - k$. Since we only draw two-dimensional control ellipses for principal components PC_i and PC_j, this formula reduces to

$$T_{i,j}^2 = \frac{z_i^2}{\lambda_i} + \frac{z_j^2}{\lambda_j} = \frac{k(n-1)}{n-k} F_{k,n-k,\alpha},$$

with λ denoting the eigenvalues. Hence the control ellipse's center is 0 and its half-axes lengths HA(i) and HA(j) are

$$\mathrm{HA}(l) = \sqrt{\lambda_l F_{k,n-k,\alpha} \frac{k(n-1)}{n-k}}$$

for $l = i, j$. Most commonly used values for α are 0.05 or 0.01, depicting a control ellipse of 95% or 99%, respectively.

Squared Prediction Error. The squared prediction error (*SPE* or *Q-statistic*) for an observation is defined as the squared difference between its actual value and the value predicted by the model, i.e., the scalar product of the error vector of the predicted vs. the actual observation vector with itself. It gives an indication for the model fit (see Fig. 3).

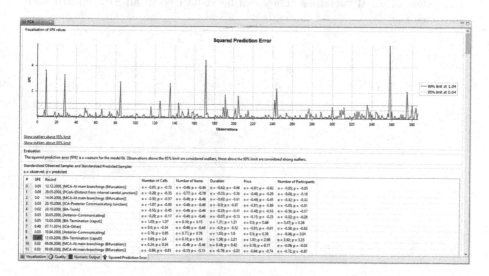

Fig. 3. The squared prediction error information board of the PCA result. The line chart shows that some SPE values are exceeding the 95% limit (orange horizontal line) or even the 99% limit (red horizontal line). The table below illustrates the numeric SPE values for each observation, numbered from 1 to n. According to the limit the record exceeds, the record is colored either orange or red. Further the observed and predicted values for each variable are listed. (Color figure online)

Contribution Plots. Contribution plots are a simple graphical way of investigating the contribution of a variable to the PCA model (see Miller et al. [18]). The aim here is to find the variable(s) that contribute most to unusual values detected in the model and to investigate how these values have been achieved. A variable's influence can either be positive or negative and vary in the strength of its contribution. Commonly, bar charts are used for illustration. In our system, those plots have been realized for score values and the squared prediction error of an observation.

Score Contribution Plot. The score of an observation is a vector whose dimension equals the number of principal components. The individual entries of a score vector are computed as a linear combination of the original variables, i.e., as a weighted sum of the original variables. The terms in this weighted sum are the contributions of the original variables to this observation with respect to the principal component and they can be visualized in a score contribution plot, which looks similar to the SPE contribution plot in Fig. 4.

SPE Contribution Plot. Every observation has an SPE value. This SPE value is the sum of squares of the entries in the error vector of the predicted vs. the actual observation vector. The individual summands of the SPE value each depend on an original variable and are called the SPE contributions of the individual original variables. They can be visualized in an SPE contribution plot (see Fig. 4).

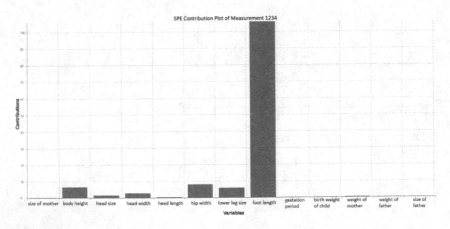

Fig. 4. A squared prediction error contribution plot example of the red marked measurement in Fig. 3 on the first principal component. As shown by the figure, the *foot length* contributes the most to this record. When going back to the raw data, this record shows a rather high value in the variable *foot length* compared to the other observations. (Color figure online)

It should be noted that the quality criteria listed above include a selection of the most important prerequisites for PCA quality assessment, but may not comprise all relevant prerequisites. If necessary, additional quality criteria may be embedded steadily into the proposed quality measure.

3 Results and Discussion

3.1 Data Set

The MICA (Measurements for Infants, Children, and Adolescents) project started in 2010 with the aim of acquiring detailed demographic and biometric data of children and thus to determine the body surface of children more accurately [19]. In cooperation with the *Kepler University Hospital Linz* about 3200 children aged 0 to 18 were measured by nurses in the outpatient department of the hospital. Those measurements contain more than 30 variables describing biometric variables, such as weight, length and circumference of the child's head, lower legs, hands and feet. Most of the variables are numeric, which have been included in the principal component analysis.

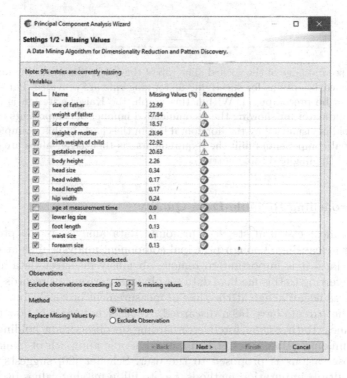

Fig. 5. On the first page of the wizard available numeric variables and the corresponding ratio of null values are shown. Though some of the variables are rated with a warning sign, we consider them to be sufficient for analysis. In this PCA run, missing values will be replaced by the variable mean.

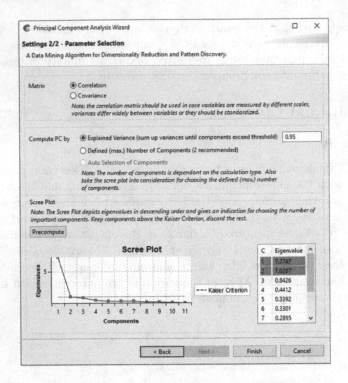

Fig. 6. The second page of the wizard. The top of the graph illustrates the matrix type selection. Beneath, options for specifying a stopping rule are provided. For support, a scree plot can be precomputed. Within this graph, the Kaiser criterion is illustrated as a second indicator for showing the recommended number of components to be kept. This threshold is visualized by the horizontal line in the chart. In this example, two out of 15 principal components fulfill the requirements, as their eigenvalues are exceeding the threshold (highlighted table entries).

3.2   Approaching Reliable Data Quality

One of the most essential steps in profound data analysis is the preliminary work of data cleansing. Though principal component analysis aims at detecting outliers in data, it is important to remove obviously incorrect data already in advance. Referring to this medical data set obviously erroneous records including values such as negative age at the time of measurement or zero-values in size or weight of the parents have been discarded. Another hot topic is how to handle missing values. Rather than forcible discarding an observation holding at least one single missing value, a softer and more dynamic approach of managing the affected observations is proposed to the user. Scheffer [20] suggests applying single or multiple imputation methods, i.e., to fill in missing values by including means, medians or modes, computed by all known values of the corresponding variable. The default setting in our implementation is the replacement of missing values by the variable mean. If missing values are not to be imputed, the system proposes an alternative configuration – variable and observation exclusion. Both exclusion thresholds can be adapted dynamically by the researcher. However, according to Nelson et al. [21], a higher threshold than 20% missing values leads to losses in performance when applying missing data algorithms (see also Figs. 5 and 6).

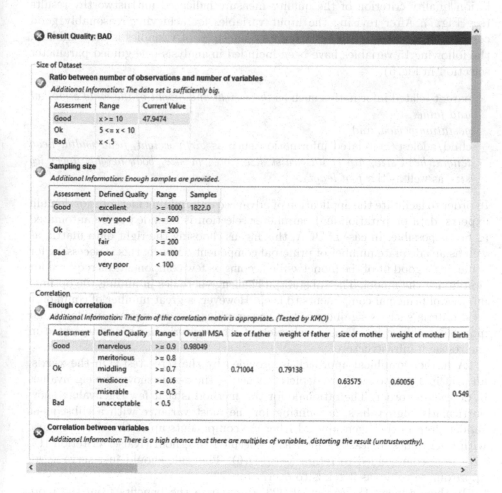

Fig. 7. Excerpt of the detailed quality description listing shown to the user. The overall quality is computed as *bad* (red highlighting on the top of the figure). Assessment scheme classification (*good, ok, bad*) of the actual computed key numbers are highlighted in light gray. As an additional information a short and comprehensible description is provided to the user. (Color figure online)

Regarding the medical data set, a vast majority of the variables describe measurements of the children. As a matter of fact, some parts of the body grow more or less at the same rate – resulting in similar proportions and therefore multiples in their correlation.

The aim of this evaluation primarily is to showcase the practical use of the quality measure, to illustrate how to derive new hypotheses from the data, and to compare and validate them with existing knowledge, rather than acquiring new medical knowledge or testing hypothesis on a confidence level.

In a first PCA run, including all of the measurement variables, the correlation quality criterion of the quality measure indicated untrustworthy results (see Fig. 7). After revising the input variables for achieving reasonably good correlation values, i.e., the occurrence of multiples of variables is very unlikely, the following 15 variables have been included in analysis (see guided parameter selection in Fig. 5):

- parent related information such as *the weight and height of the child's mother and father,*
- *gestation period,* and
- child/adolescent related information such as *birth weight, head width, head length, head size, hip width, waist size, forearm size, body height, lower leg size* as well as the *foot length.*

In order to facilitate the application of advanced data mining methods for domain experts, data preparation and parameter selection is intended to be automated as far as possible. In case of PCA, this means choosing the right data matrix as well as an adequate number of principal components. Hence, this is necessary for achieving a good fit of the model while keeping as few components as required for simplifying the model. The substantial challenge here lies in finding the optimal number of principal components to keep. However, a great number of commonly used criteria such as significance tests or graphical procedures are available and may be combined [6]. Thus, a selection of those criteria has been integrated in the research infrastructure.

A further graphical approach is provided by the scree test. On the x-axis, all principal components are depicted, whereas the y-axis shows the eigenvalues in descending order. The rationale for this method is that few eigenvalues show particularly high values, accounting for the most variance with a subsequent sudden drop in the eigenvalues. Either the components up to this drop or those with eigenvalues at least as high as a specified threshold (Kaiser criterion [13, 22]) are recommended to retain (see Fig. 6). This rule should just serve as an approximate value, as it tends to overfactor [23].

In their review, Hayton et al. [23] summarize the benefits from the more accurate factor retention method *parallel analysis.* The basic idea is to compare the eigenvalues of the sample data to those of a number of random generated data (exhibiting the same size and number of variables). What is expected here is, that observed eigenvalues of valid principal components are larger in comparison to the average of the eigenvalues of the parallel components. This method will be provided to users as a supplementary factor recommendation method.

Finally, all necessary input parameters are set to start PCA.

3.3   The Infinite Thirst for Knowledge

Interpreting a PCA output on a numeric base has proven to be complex and difficult for domain experts who are not trained in the fields of mathematics or statistics. When the result is computed, a result package including all relevant information and visualizations of the aforesaid sections is shown to the user.

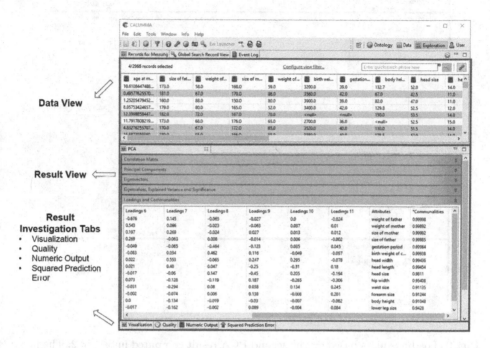

Fig. 8. Holistic view of the entire system and the integrated result of a PCA run

The holistic view is shown in Fig. 8. On top of the graph the data is shown. The bottom section (see Result View) opens subsequently to PCA computation, containing four tabs for result examination. On the present image, the numeric output tab is opened. In the visualization section, scores, loadings, and biplot are shown. The quality tab lists automatically computed quality criteria (see also Fig. 7). The last tab shows a view for the squared prediction error (see Fig. 3). In order to ensure correct results, all outputs of the implementation have been validated using the R language [24].

In order to facilitate access to interpretations, three types of plots described in Sect. 1.2 had already been integrated previously in the research infrastructure (see [3]).

In Fig. 9, loadings plots of the current PCA result are depicted. The first graph shows the transformed variables of the data set with the first principal component as the x-axis and the second principal component as the y-axis. Since rather all child related variables are located far from the origin, a strong

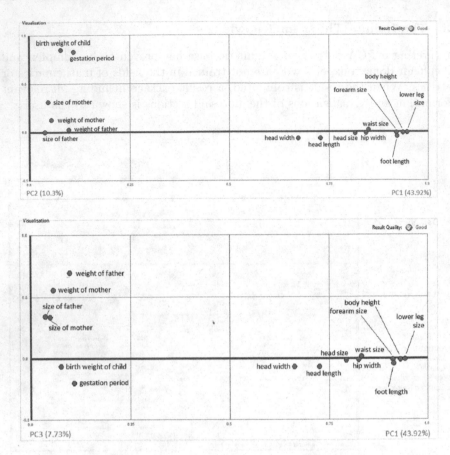

Fig. 9. Loadings plots based on the second PCA result computed in Sect. 3.2. The first graph illustrates the loadings plot of the first and the second principal component, characterizing the x-axis and y-axis. On the second graph, the x-axis and the y-axis depict the first and the third principal component. The location of the PCA-transformed variables plotted on the graph provides insight into the strength of correlation between the original variables and the influence of a variable on the specific component.

influence exists by those measurement variables on the first principal component. According to the graph, the child's proportions are correlating strongly with each other, as anticipated. When regarding the second principal component (y-axis), variables contributing to pregnancy and childbirth are described, as they delineate high weights according to this component. Based on our experimental data, due to their proximity, the *birth weight of the child* and the *gestation period* show high positive correlation, i.e., the longer a woman is pregnant, the heavier the child is at birth. It is also apparent that *maternal height* seems to have positive association with the *gestation period*, i.e., that the gestation length is increased for taller women. Literature research actually has shown that this slight influence has already been found in other experiments [25]. Additionally,

the third principal component has been evaluated as it gives an indication of
the parental influence on the child related data (see Fig. 9). According to the
positions of *parental body height and weight* divergent to *duration of pregnancy*,
influences of maternal and paternal properties are emerging (see also Morrison
et al. [26]).

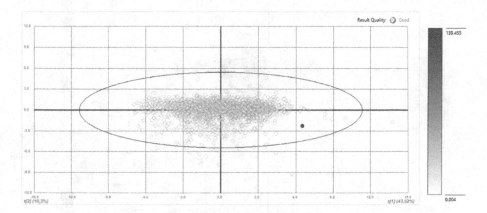

Fig. 10. The score plot (based on the second PCA result computed in Sect. 3.2) of
the first and the second principal component, characterizing the x-axis and y-axis.
The color scheme represents the SPE-values of each record. Small deviations (between
the actual and the predicted model) are colored white, whereas high differences are
highlighted in darker color. (Color figure online)

In Fig. 10 a score plot of the experimental data is shown. Each point is
colored after the squared prediction error. The color represents the distance of
the observation to the hyperplane – the darker a point, the higher the distance.
In this case, the specified observation is badly explained by the model and is
an indication for a potentially corrupted record. It might also be worth to take
a closer look at this specific record and to find out why it differs so greatly
from the remaining records. Generally, the interactive implementation of the
charts provides the connection from drawn points selected in the visualization
to the raw data. The corresponding records are then highlighted in the data
view (for an example of the data view see Fig. 8). Further investigations of the
experimental data showed the expected outcome and how to interpret PCA
results (see Fig. 11). Positive correlations between the growth of included body
regions and the stages of age are apparent in each of the plots.

Recall that projections by PCA are only applicable to numeric variables. In
case other data types such as categorical variables should be incorporated into
analysis, scores can be colored in the corresponding colors and therefore used,
e.g., for cluster analysis. Further analysis on comparing the behavior of various
subgroups follows the same procedure as introduced, by using subsets of interest
based e.g. on gender, ethnicity, or other characteristics. However, it is important

to bear in mind the possible bias in these responses as well as the fact that PCA merely can find linear patterns, it is not applicable for detecting non-linear relationships.

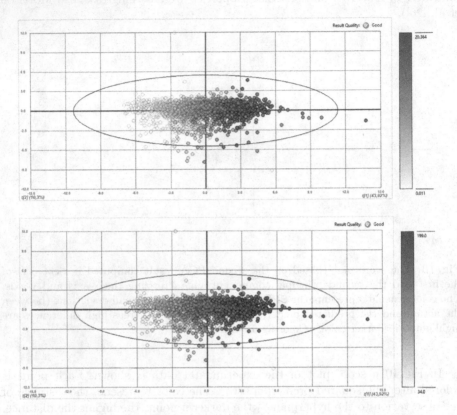

Fig. 11. Age graded score plot (first graph) and measurement graded score plot (second graph). The points are colored from white to blue, whereby the white points on the first graph represent children measured in early years, bluish points represent older test persons. On the second graph white points depict smaller children, the bluish points depict taller children. (Color figure online)

4   Concluding Remarks and Future Work

We integrated PCA with an augmented guidance system into an ontology-guided research platform – starting with the control for appropriate input variables and reasonable hyperparameters, going ahead with informing the user about the reliability of the received output and recommendations on how to approach a more trustworthy result, and providing interactive visualizations and result-related information for investigating data patterns and creating new hypotheses. Technically, we introduced a quality measure for PCA which supports the domain expert in checking if the data have the necessary quality and structure for a meaningful application of PCA, and in improving the data in order to increase its quality, e.g., gather more data, remove essentially duplicate variables, etc. We also showed how this guidance works by applying it to the MICA data set. Overall, it is crucial to reach higher interpretability from data mining techniques, so that humans can understand the path from data to results and – by far more important – the meaning and reliability of the result. Further developments of the linkage between ontologies and probabilistic machine learning is a hot future topic and may lead to profound contributions in terms of explainable AI [27]. Future work includes an usability study with medical doctors to evaluate how good the guidance system works in practice. We also plan to integrate other methods augmented with a guidance system like mixed factor analysis, regression and different clustering methods. Another component to include will be support for the user in the choice of the correct statistical test for a problem, depending on (amongst other) what the user wants to examine, the number of variables or the data type of the chosen variables. Therefore the user will be asked the necessary questions in a wizard in advance and a selection of possible tests will be given as a result.

References

1. Roddick, J.F., Fule, P., Graco, W.J.: Exploratory medical knowledge discovery: experiences and issues. SIGKDD Explor. Newslett. **5**, 94–99 (2003)
2. Anderson, N.R., et al.: Issues in biomedical research data management and analysis: needs and barriers. J. Am. Med. Inform. Assoc. **14**(4), 478–488 (2007)
3. Wartner, S., Girardi, D., Wiesinger-Widi, M., Trenkler, J., Kleiser, R., Holzinger, A.: Ontology-guided principal component analysis: reaching the limits of the doctor-in-the-loop. In: Renda, M.E., Bursa, M., Holzinger, A., Khuri, S. (eds.) ITBAM 2016. LNCS, vol. 9832, pp. 22–33. Springer, Cham (2016). https://doi.org/10.1007/978-3-319-43949-5_2
4. Girardi, D., Dirnberger, J., Giretzlehner, M.: An ontology-based clinical data warehouse for scientific research. Saf. Health **1**(1), 1–9 (2015)
5. Girardi, D., et al.: Interactive knowledge discovery with the doctor-in-the-loop: a practical example of cerebral aneurysms research. Brain Inform. **3**(3), 133–143 (2016)
6. Jackson, J.: A User's Guide to Principal Components. Wiley, New York (1991)
7. Rencher, A.: Methods of Multivariate Analysis. Wiley Series in Probability and Statistics. Wiley, Hoboken (2002)
8. Kessler, W.: Multivariate Data Analysis for Pharma-, Bio- and Process Analytics. WILEY-VCH Verlag GmbH & Co. KGaA, Weinheim (2007)
9. Osborne, J.W., Costello, A.B.: Sample size and subject to item ratio in principal components analysis. Pract. Assess. Res. Eval. **9**(11), 8 (2004)
10. Beaumont, R.: An Introduction to Principal Component Analysis & Factor Analysis Using SPSS 19 and R (psych Package), April 2012
11. Dziuban, C.D., Shirkey, E.C.: When is a correlation matrix appropriate for factor analysis? some decision rules. Psychol. Bull. **81**(6), 358 (1974)
12. Tabachnick, B.G., Fidell, L.S., Osterlind, S.J.: Using Multivariate Statistics. Allyn and Bacon, Boston (2001)
13. Kaiser, H.F.: A second generation little jiffy. Psychometrika **35**(4), 401–415 (1970)

14. Kaiser, H.F., Rice, J.: Little Jiffy Mark IV. Educ. Psychol. Measur. **34**, 111–117 (1974)
15. Jackson, D.A., Chen, Y.: Robust principal component analysis and outlier detection with ecological data. Environmetrics **15**(2), 129–139 (2004)
16. Kim, D., Kim, S.-K.: Comparing patterns of component loadings: principal component analysis (PCA) versus independent component analysis (ICA) in analyzing multivariate non-normal data. Behav. Res. Meth. **44**, 1239–1243 (2012)
17. Thode, H.C.: Testing for Normality. CRC Press, Boca Raton (2002)
18. Miller, P., Swanson, R.E., Heckler, C.E.: Contribution plots: a missing link in multivariate quality control. Appl. Math. Comput. Sci. **8**(4), 775–792 (1998)
19. Thumfart, S., et al.: Proportionally correct 3D models of infants, children and adolescents for precise burn size measurement (187). Ann Burns Fire Disasters **28**, 5–6 (2015)
20. Scheffer, J.: Dealing with missing data. Res. Lett. Inf. Math. Sci. **3**, 153–160 (2002)
21. Nelson, P.R., Taylor, P.A., MacGregor, J.F.: Missing data methods in PCA and PLS: score calculations with incomplete observations. Chemometr. Intell. Lab. Syst. **35**(1), 45–65 (1996)
22. Kaiser, H.F.: The Varimax Method of Factor Analysis. Unpublished Doctoral Dissertation, University of California, Berkeley (1956)
23. Hayton, J.C., Allen, D.G., Scarpello, V.: Factor retention decisions in exploratory factor analysis: a tutorial on parallel analysis. Organ. Res. Methods **7**(2), 191–205 (2004)
24. R Core Team, R: A Language and Environment for Statistical Computing. R Foundation for Statistical Computing, Vienna, Austria (2014)
25. Myklestad, K., Vatten, L.J., Magnussen, E.B., Salvesen, K.Å., Romundstad, P.R.: Do parental heights influence pregnancy length?: a population-based prospective study, HUNT 2: BMC Pregnancy Childbirth **13**(1), 33 (2013)
26. Morrison, J., Williams, G., Najman, J., Andersen, M.: The influence of paternal height and weight on birth-weight. Aust. N. Z. J. Obstet. Gynaecol. **31**(2), 114–116 (1991)
27. Holzinger, A., Kieseberg, P., Weippl, E., Tjoa, A.M.: Current advances, trends and challenges of machine learning and knowledge extraction: from machine learning to explainable AI. In: Holzinger, A., Kieseberg, P., Tjoa, A.M., Weippl, E. (eds.) CD-MAKE 2018. LNCS, vol. 11015, pp. 1–8. Springer, Cham (2018). https://doi.org/10.1007/978-3-319-99740-7_1

Ranked MSD: A New Feature Ranking and Feature Selection Approach for Biomarker Identification

Ghanshyam Verma[1,2]([✉]) [iD], Alokkumar Jha[1,2], Dietrich Rebholz-Schuhmann[3], and Michael G. Madden[1,2] [iD]

[1] Insight Centre for Data Analytics, National University of Ireland Galway,
Galway, Ireland
{ghanshyam.verma,alokkumar.jha}@insight-centre.org,
michael.madden@nuigalway.ie
[2] School of Computer Science, National University of Ireland Galway,
Galway, Ireland
[3] ZB MED - Information Center for Life Sciences, University of Cologne,
Cologne, Germany
rebholz@zbmed.de

Abstract. In the era of big data when a huge amount of data is continuously being generated, it is common for situations to arise where the number of samples is much smaller than the number of features (variables) per sample. This phenomenon is often found in biomedical domains, where we may have relatively few patients, compared to the amount of data per patient. For example, gene expression data typically has between 10,000 and 60,000 features per sample. A separate issue arises from the "right to explanation" found in the European General Data Protection Regulation (GDPR), which may prevent the use of black-box models in applications where explainability is required. In such situations, there is a need for robust algorithms which can identify the relevant features from experimental data by discarding irrelevant ones, yielding a simpler subset that facilitates explanation. To address these needs, we have developed a new algorithm for feature ranking and feature selection, named *Ranked MSD*. We have tested our proposed approach on two real-world gene expression data sets, both of which relate to respiratory viral infections. This Ranked MSD feature selection algorithm is able to reduce the feature set size from 12,023 genes (features) to 65 genes on the first data set and from 20,737 genes to 31 genes on the second data set, in both cases without any significant loss in disease prediction accuracy. In an alternative configuration, our proposed algorithm is able to identify a small subset of features that gives better accuracy than that of the full feature set. Our proposed algorithm can also identify important biomarkers (genes) with their importance score for a particular disease and the identified top-ranked biomarkers can play a vital role in drug discovery and precision medicine.

Keywords: Machine learning · Respiratory viral infection ·
Feature ranking · Feature selection · Classification · Explainable AI

© IFIP International Federation for Information Processing 2019
Published by Springer Nature Switzerland AG 2019
A. Holzinger et al. (Eds.): CD-MAKE 2019, LNCS 11713, pp. 147–167, 2019.
https://doi.org/10.1007/978-3-030-29726-8_10

1   Introduction

It has been observed that the use of ML (machine learning) algorithms has been increased in healthcare applications that deeply impact the life of patients [24]. The term "black-box model" is used for those ML models that fail to explain their predictions in a way that humans can understand and make some meaningful conclusions for decision making. According to Rudin [17] due to the lack of proper explanation and transparency of black box models, there might be severe consequences of using them for decision making specifically in health, finance and in the domains where people are directly involved. Therefore, rather than using black-box models without a proper explanation if we use models that are explainable or use secondary analyses to generate explanations from black-box models, then it can surely help in better decision making, particularly in the medical domain where medical professionals want to understand how and why a machine decision has been made. Moreover, algorithms that can facilitate meaningful explanations could enhance the trust of medical professionals in future AI or ML based systems [11].

In supervised machine learning, a classification algorithm is learned by applying it to a set of training samples or instances, where each instance contains a vector of attribute values (also called features) and a class [16]. For example, in genomics, the features are generally genes (of which there are typically thousands) and the class label might denote whether or not a patient is infected. Here the problem is that we have thousands of genes and all the genes are not relevant for a particular problem or disease. We are only interested in very few most important genes which can be targeted for further study or drug discovery. Therefore, it is very important to identify and select the very few most important genes. One way to solve this problem is to use an appropriate feature selection technique. Most of the machine learning algorithms are designed in such a way that they learn which are the most important features to use for decision making. In theory, they should never select irrelevant and unhelpful features. But there is a difference between theory and practice. In practice, irrelevant or distracting features often confuse machine learning systems and lead to deterioration of classification accuracy [31]. Having a large set of irrelevant features require excessive computational time and memory space. Moreover, a large number of irrelevant features make it very hard to interpret the representation of the target concept. Because of these bad effects of irrelevant features, it is common to perform feature selection before applying any learning algorithm.

When we want to perform feature selection, there are two different broad types of approaches. The first type are *filter* methods, which make an assessment of feature importance based on general characteristics of the data, using criteria that are independent of the subsequent machine learning algorithm. The second type are *wrapper* methods. Wrapper methods start with an empty set of features, and iteratively add/remove features until an optimal feature set is found [31]. An important property of filter methods is that they can assign a score to all the features, based on which features can be ranked in the desired order. This is useful when we need to select the top few features for further analysis. On the other hand, wrapper methods do not, in general, have any mechanism to rank features.

In this work, our overall goal is to identify the important biomarkers or genes for a particular disease, and assign an importance score to each biomarker. Therefore, the focus of this paper is on filter methods, and wrapper methods are not applicable. In this paper, we propose a feature ranking algorithm named *Ranked MSD* which gives a ranked list of all the features with their importance score. We also propose two more feature selection algorithms $\mathcal{F}_{equal}$ and $\mathcal{F}_{best}$. $\mathcal{F}_{equal}$ gives most strongly relevant features and $\mathcal{F}_{best}$ gives all the relevant features by discarding irrelevant features. The overall approach can identify the most important biomarkers and can help in explaining the predictions.

The rest of the paper is structured as follows. In Sect. 2, we describe related work. Section 3 describes the two real-world data sets used to perform experiments. In Sect. 4, we explain the proposed algorithms. In Sects. 5 and 6, we present the overall experimental design and methodology used. In Sect. 7, we discuss how we can explain predictions using our approach. In Sect. 8, we discuss results in detail with comparative analysis. In Sect. 9, we present the identified biomarkers using proposed algorithms. In Sect. 10, we discuss the significance of identified biomarkers, and finally we conclude in Sect. 11.

2   Related Work

A basic and natural way to interpretability is to provide explanations of an ML model's predictions in the form of input features [14]. This is the reason most of the work that tried to explain the predictions of black-box models used in some sense the features that have some influence on the class of interest [14,23,26]. Riberio et al. proposed an approach called LIME [23] which can explain the prediction of a classifier by providing a small list of features that either contribute to the prediction or are evidence against the prediction. In our work, using Ranked MSD approach we are also suggesting a small list of features with their importance score that can explain the predictions of a classifier. This feature importance score also denotes class discriminative power of that feature. For computing contribution of features, LIME uses K-Lasso an approach based on Lasso [8] and we are using our proposed approach which is explained in Algorithm 1. Filter methods can be used to compute contribution or importance of features. Most of the filter methods use feature ranking as a principle mechanism for feature selection [9]. Feature ranking is a type of preprocessing that ranks features in ascending or descending order of their relevance to the class label based on a computed score for each variable or feature. A suitable threshold is then used to select the top ranked features [4]. Filter methods are not dependent on the choice of the classifier or predictor. However, under certain assumptions, it may produce optimal solution for a given predictor [28]. One of the most important properties of a good feature is that it contains useful information of the different possible classes in the data. This property is known as feature relevance [16], and relates to the usefulness of a feature in discrimination of classes.

In the following sub-sections, we will discuss two state-of-the-art filter methods, against which we will compare our proposed feature ranking algorithm.

2.1   Correlation Criteria

As described by Guyon et al. [9], the Pearson correlation coefficient can be defined as:

$$R(i) = \frac{Cov\,(X_i, Y)}{\sqrt{Var\,(X_i)\,Var\,(Y)}} \tag{1}$$

Here Cov denotes covariance and Var denotes the variance. X_i denotes the ith feature vector and Y denotes the outcome. $R(i)$ represents the fraction of the total variance around the mean value $\bar{y}$, therefore, the $R(i)^2$ can be used as a variable ranking criterion, with which we can rank features in ascending or descending order. $R(i)^2$ can be used for two-class classification, for which each class label is mapped to a given value of Y, e.g., 0 & 1.

2.2   Information Gain

Another well-known feature selection approach is Information Gain, which can be classed as an information theoretic ranking criterion [4,9,16]. It is based on Shannon's definition of entropy which can be represented as:

$$H\,(Y) = -\sum_y p\,(y)\,log\,(p\,(y)) \tag{2}$$

This formula represents the information content or uncertainty in any variable Y. Now if we observe a new variable X, then the conditional entropy can be represented by the following formula:

$$H\,(Y|X) = -\sum_x \sum_y p\,(x, y)\,log\,(p\,(y|x)) \tag{3}$$

This formula says that by observing a variable X, the uncertainty of the output or variable Y is reduced. Now the formula for Information Gain IG can be represented as:

$$IG(Y, X) = H(Y) + H(X) - H(Y|X) \tag{4}$$

Here $H(Y)$ is the information content or uncertainty of class variable Y, the second term $H(X)$ is the information content of observed variable X and $H(Y|X)$ is the conditional entropy.

3   The Respiratory Viral Data Sets

We have conducted experiments on two real-world data sets, both of which are related to respiratory viral infections. The first data set is collected from 7 Respiratory Viral Challenge studies which is available for open access on Gene

Expression Omnibus (GEO) using accession number GSE73072[1]. This first data set consists of a total of 151 human volunteers. All the volunteers were healthy when they enrolled for the study. After enrolment in the study, all subjects were inoculated with one of the 4 viruses (H1N1, H3N2, HRV, RSV). Their blood samples were taken at different pre-defined time-points, thus delivering gene expression profiles from non-infected individuals as well as from infected ones [19]. The details related to labels and other additional details of this data set can be found on GEO (accession number GSE73072). From the start, total 151 subjects were enrolled in these 7 challenge studies, however, we have to exclude 47 subjects from the study because those subjects' gene expression data are either inconsistent (faulty) or missing, and faulty data can be misleading and harmful while model-building. Detailed information about the excluded subjects can be found in a paper that used this dataset before [30].

The second dataset contains gene expression profiles of 133 adults whose samples are taken in three different seasons - Fall, Winter and Spring. Baseline samples are taken at the time of enrolment of volunteers (Fall season). Day 0, Day 2, Day 4, Day 6 and Day 21 samples are taken during the winter season (Influenza season). Samples of all the volunteers are taken again in Spring season. For each volunteer, samples are taken at up to seven time points before, during, and after the occurrence of illness (Influenza and other acute respiratory viral infections). Among those seven time points, the samples taken before illness (baseline), at day 21 and during Spring season are healthy samples and rest of the samples are infected samples as they are taken during the illness (day 0, day 2, day 4 and day 6). A total of 890 microarray samples were collected. Any samples that failed Quality Control were excluded from the study ($N = 10$), leaving 880 high-quality arrays from which the subsequent analysis was conducted. Out of these 880 samples, 373 samples are healthy samples and rest of 507 samples are infected samples [33]. There were in total 47,254 probe IDs in each microarray sample from which 20,737 probe IDs have unique gene mapping; therefore, we left with a total of 20,737 genes for further analysis. This data set is also openly accessible on GEO via accession number GSE68310[2].

The first dataset contains in total 12,023 genes and the second data contains in total 20,737 genes; however, a large number of genes have little or no contribution in finding the progression of a particular disease, so it is crucial to find that small number of genes which actually provide diagnostic signals and contribute the most at the time of a particular disease progression. It is also important to understand their importance for that disease prediction and for finding treatment targeting those genes. In this work, using the proposed *Ranked MSD* feature selection algorithm, we are interested in finding the strongly relevant features (genes) which are potential biomarkers and contributing the most in respiratory viral disease prediction. Our software implementing our algorithm is open-source and freely available; R code for it can be accessed here: https://github.com/researher/Ranked_MSD.

---

[1] https://www.ncbi.nlm.nih.gov/geo/query/acc.cgi?acc=GSE73072.
[2] https://www.ncbi.nlm.nih.gov/geo/query/acc.cgi?acc=GSE68310.

4   Proposed Algorithms

We have developed a feature ranking algorithm and two additional schemes (Algorithms 2 and 3) for feature subset selection. One can observe these three algorithms as one but here we have designated them as 3 individual algorithms for simplicity of explanation; their purpose is to rank all the features, identify *strongly relevant* and *relevant* features respectively for a particular problem.

Algorithm 1. Ranked MSD Feature Ranking Algorithm

Input: The training data $Y_{tr} = [G_1, G_2, ..., G_m]_{n \times m}$, where n = number of samples, m = total number of features; a vector of class labels $C = [C_1, C_2, ..., C_p]$, where p = total number of classes and $p \geq 2$

Output: $\mathcal{RF}_{All}$ = Ranked list of all the m features

1: Create *Reference* vector $R_{1 \times m}$, where R is the feature wise mean of all the samples belong to class C_1 of Y_{tr}
2: Create *Target* matrix $T_{j \times m}$, where j = number of samples in T and T contains all the samples from Y_{tr} except class C_1
3: Compute Mean Squared Difference (MSD) for each feature
$$MSD_{1 \times m} = \frac{\sum_1^j \left(T_{j \times m} - R_{1 \times m}\right)^2}{j} \qquad \triangleright \text{ } MSD \text{ is a named vector that contains}$$
feature names and MSD values.
4: Compute $\mathcal{RF}_{All}$ by arranging the MSD in the descending order of their values.
5: Return $\mathcal{RF}_{All}$

The name of the proposed feature ranking algorithm is *Ranked MSD* which needs only training data in input and returns a ranked list of all the features ($\mathcal{RF}_{All}$) with their importance score. In the first step of Algorithm 1, we create a *Reference* vector $R_{1 \times m}$ which contains the feature wise mean of all the samples belong to a particular class say class one or base class. In our application, all these samples of class one belong to the negative class (all healthy samples) thus the *Reference* vector R represents the gene expression values of healthy samples on an average. In other words, the *Reference* vector R can be used as a representative of the base class. In the second step we create a *Target* matrix $T_{j \times m}$ which contains all the samples from training data except class one i.e. healthy samples. In the third step, we compute the Mean Squared Difference (MSD) for every feature (gene) from the formula shown in Algorithm 1. This MSD serves as a scoring function for the proposed algorithm. This scoring function preserves the difference between healthy gene expression values and infected gene expression values and gives the highest score to the highly differentially expressed gene and so on. Then in the next step, we compute the ranked list of all the features ($\mathcal{RF}_{All}$) by arranging the computed MSD score in descending order of their values. Here the MSD vector is a named vector which has feature names as heading and computed scores as values. Therefore when we arrange them in descending order of their values the most important genes are the top ones.

Once all the features are ranked, we may wish to select a subset of relevant features from ($\mathcal{RF}_{All}$). A simple solution is to take ranked features one by one sequentially and evaluate them by training a classifier until we get the best accuracy. In the worst case, this method has high time complexity of order $O(m * T)$ where m is the total number of features and T is time complexity to train a classifier which depends on the classification algorithm used (and may have higher-order complexity). The proposed Algorithms 2 and 3 provide a less time-consuming solution for this problem in comparison to the sequential search. To understand the Algorithms 2 and 3, first, we need to define the following terms.

Definition 1: (Feature Set with First Statistically Equal Accuracy) - $\mathcal{F}_{equal}$ is the feature set that has minimum number of top variables or genes (subset of full feature set) and gives accuracy which is statistically (according to a t-test) equivalent to the accuracy of using the full feature set.

Definition 2: (Feature Set with Best Accuracy) - $\mathcal{F}_{best}$ is a subset of the full feature set which gives the best possible accuracy.

Algorithms 2 and 3 are designed in such a way that they can give us $\mathcal{F}_{equal}$ and $\mathcal{F}_{best}$ respectively, since depending on the problem that we wish to solve, we may need to identify strongly relevant features or all relevant features: $\mathcal{F}_{equal}$ contains *strongly relevant* features, such as the most important biomarkers if applied to the biomedical domain. $\mathcal{F}_{best}$ gives us the feature set with best accuracy, which therefore contains all relevant features (including the strongly relevant features as a subset). To find irrelevant features we just need to remove all relevant features produced by $\mathcal{F}_{best}$ from the list of all the features $\mathcal{RF}_{All}$ ($\mathcal{RF}_{All}$ - $\mathcal{F}_{best}$).

In our application, we are interested in identifying potential biomarkers. Therefore, we use $\mathcal{F}_{equal}$ because it gives us the smallest optimal feature set without any significant loss in disease prediction accuracy. This observation is backed up by the results obtained (See Sect. 8).

Algorithm 2 identifies and returns $\mathcal{F}_{equal}$. Finding $\mathcal{F}_{equal}$ using Algorithm 2 is much less time consuming than a full linear search through the list of features, as it recursively applies binary search to find candidate entries for $\mathcal{F}_{equal}$. Every time the *SearchFEqual* function is called, it calculates the accuracy using top $M - 1, M$ and $M + 1$ features and performs t-tests to find whether or not they are statistically equal to the accuracy of full feature set. A t-test might show statistical equivalence with multiple feature sets in a range, but we want the feature set that has the minimum number of features. There are 8 possibilities based on 3 feature subsets $M - 1, M$ and $M + 1$ and 2 options which are statistically equal or not. These 8 possibilities are $000, 001, 010, 011, 100, 101, 110, 111$ where 1 denotes that the statistically equal accuracy found and 0 denotes not found using a particular feature subset. For example, if $M - 1$ is 0, M is 0 and $M + 1$ is 1 means we found the $\mathcal{F}_{equal}$. So there are 2 possibilities for $\mathcal{F}_{equal}$ that is 001 and 011 and if these are true it returns the $\mathcal{F}_{equal}$, otherwise, it checks for other possibilities. In the remaining 6 possibilities, if 000 is true then we move to the right side else in rest of the other cases we move to the left side as $\mathcal{F}_{equal}$ would

Algorithm 2. Feature set with $\mathcal{F}_{equal}$ accuracy

Input: The training data $Y_{tr} = [G_1, G_2, ..., G_m]_{n \times m}$ and Ranked list of all the m features ($\mathcal{RF}_{All}$), Significance level (α).

Output: $\mathcal{F}_{equal}$ ▷ $\mathcal{F}_{equal}$ = Feature Size with First Statistically Equal Accuracy

1: **function** SEARCHFEQUAL(Y_{tr}, $\mathcal{RF}_{All}$, $L = 1$, $R = m$, α)
2:     $M = ceiling(\frac{L+R}{2})$
3:     Train the desired classifier using top $M - 1$, M and $M + 1$ features from $\mathcal{RF}_{All}$
4:     Perform ***t-test*** between $Acc(M - 1, M, M + 1)$ and $Acc(m)$ individually
5:     **if** ((p-value(M-1) $< \alpha$) & (p-value(M) $< \alpha$) & (p-value(M+1) $> \alpha$)) **then**
6:         $\mathcal{F}_{equal} = \mathcal{RF}_{All}[M + 1]$
7:         Return $\mathcal{F}_{equal}$
8:     **else**
9:         **if** ((p-value(M-1) $< \alpha$) & (p-value(M) $> \alpha$) & (p-value(M+1) $> \alpha$)) **then**
10:             $\mathcal{F}_{equal} = \mathcal{RF}_{All}[M]$
11:             Return $\mathcal{F}_{equal}$
12:         **else**
13:             **if** ((p-value(M-1) $< \alpha$) & (p-value(M) $< \alpha$) & (p-value(M+1) $< \alpha$)) **then**
14:                 SEARCHFEQUAL(Y_{tr}, $\mathcal{RF}_{All}$, $L = M$, $R = R$, α)
15:             **else**
16:                 SEARCHFEQUAL(Y_{tr}, $\mathcal{RF}_{All}$, $L = L$, $R = M$, α)
17:             **end if**
18:         **end if**
19:     **end if**
20: **end function**

be in the left side. The time complexity of finding $\mathcal{F}_{equal}$ using Algorithm 2 is $O(log_2 m * 3T)$ where m is the total number of features, T is the time complexity to train a classifier, and the constant 3 can be neglected.

While calling the *SearchFEqual* function, we have to pass value of α as one of the parameters. Here α, the decision-making significance threshold, denotes the probability of type-I error that we are willing to accept in a particular experiment during Null Hypothesis Significance Test (NHST) and it determines the probability of a type-II error (β) for a study [21]. A type-I error occurs when we reject the null hypothesis incorrectly and a type-II error occurs when we fail to reject the null hypothesis when the alternative hypothesis is true. It is not possible to remove both the errors at a given time because if we decrease the probability of type-I error, it increases the probability of type-II error due to the nonlinear but negative and monotonic nature of the relationship between α and β. In general, a low value of α should be chosen if it is important to avoid a type-I error and a low value of β if the research question makes it particularly important to avoid a type-II error [1]. In our case while finding $\mathcal{F}_{equal}$ our objective makes it particularly important to avoid a type-II error, therefore, we have to choose low value of β and as we know both α and β are connected: we can't lower one without raising the level of other. To find $\mathcal{F}_{equal}$, we are performing

---

Algorithm 3. Feature set with $\mathcal{F}_{best}$ accuracy

Input: The training data $Y_{tr} = [G_1, G_2, ..., G_m]_{n \times m}$ and Ranked list of all the m features ($\mathcal{RF}_{All}$)

Output: $\mathcal{F}_{best}$ ▷ $\mathcal{F}_{best}$ = Feature set that leads to best accuracy

1: $i = size[\mathcal{F}_{equal}]$ ▷ Start from $i = 2$ if not computing $\mathcal{F}_{equal}$
2: $Overall\_Best\_Acc = 0, Best\_Acc = Acc(m)$
3: **while** $i \leq m$ **do**   ▷ m = total number of features
4:     $L = i, R = i * 2$
5:     **while** $(L + 1) < R$ **do**
6:         $M = ceiling(\frac{L+R}{2})$
7:         Train the desired classifier using top M features from $\mathcal{RF}_{All}$
8:         **if** $(Acc(M) > Best\_Acc)$ **then**
9:             $Best\_Acc = Acc(M)$
10:            $Best\_Feature\_Size = M$
11:            $R = M$
12:        **else**
13:            $L = M$
14:        **end if**
15:    **end while**
16:    **if** $(Best\_Acc > Overall\_Best\_Acc)$ **then**
17:        $Overall\_Best\_Acc = Best\_Acc$
18:        $\mathcal{F}_{best} = \mathcal{RF}_{All}[Best\_Feature\_Size]$
19:    **end if**
20:    $i = i * 2$
21: **end while**
22: Return $\mathcal{F}_{best}$

---

repeated t-tests and to avoid the chances of type-II error we have used high value of alpha ($\alpha = 0.05$). A value of alpha is considered low if it is around 0.01.

Algorithm 3 takes training data (Y_{tr}) and ranked list of all the features ($\mathcal{RF}_{All}$) in input and returns $\mathcal{F}_{best}$. This algorithm makes the assumption that the feature-ranking algorithm is able to rank the features successfully to give best solutions, otherwise, it may give a sub-optimal solution. If we are interested in $\mathcal{F}_{equal}$ and have already calculated $\mathcal{F}_{equal}$ then it starts from $i = size[\mathcal{F}_{equal}]$, otherwise it starts from $i = 2$, increment by $i = i * 2$ and searches the full feature space. To find $\mathcal{F}_{best}$, it searches for the feature subset that gives the best accuracy by applying binary search within each i and $i * 2$ number of features and the feature size which gives overall best accuracy will be stored into $\mathcal{F}_{best}$. The $\mathcal{F}_{best}$ gives all the features which are relevant for a particular problem.

5   Experimental Design

The overall experimental design is illustrated in Fig. 1. We explain it by taking the example of Dataset 1. In Dataset 1, we have a total of 12023 features and 2042 samples. The data is divided into separate training and test sets. In all experiments, 80% of the data is used to train the classifiers and the remaining 20% is kept as a hold-out test set, using stratified sampling. To build the ML model for each algorithm, we estimate model parameters over the training data using 10-fold cross-validation, repeated 3 times. Only the training data is used for feature selection. The proposed *Ranked MSD* feature selection algorithm is applied to rank the features and two feature subsets $\mathcal{F}_{equal}$ and $\mathcal{F}_{best}$ are obtained. Two existing feature selection methods, correlation criteria and information gain, were also applied to compare with our proposed method. Four well-known ML classifiers are trained using selected features after applying feature selection techniques and without applying any feature selection techniques, and performance evaluation is carried out as shown in Fig. 1.

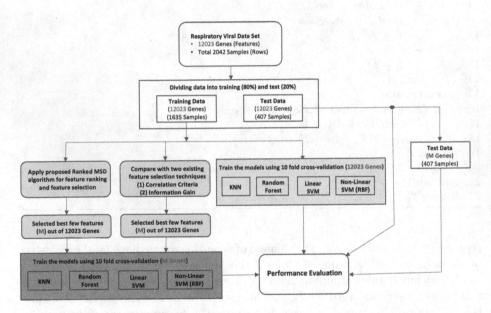

Fig. 1. Overall experimental design for evaluation of proposed Ranked MSD Algorithm.

6 Methodology

In this section, we briefly explain the methodology used for the evaluation of proposed Ranked MSD feature selection algorithm and potential biomarker identification. It is well known that no single ML algorithm is best for all kind of datasets, so we tested a selection of different ML approaches. The best performing classifier is then used for biomarker identification.

First, we used a very simple algorithm, k-NN, which is an instance-based learning algorithm; see for example [6]. k-NN is an important algorithm in the sense that it can give us good explanations if we have few features or a way to reduce our feature set to the most important features [20]. Moreover, it can be used to set a base to compare the results and to see the improvements yielded by more complex algorithms. We also used the Random Forest algorithm which is an ensemble technique [7]. We then employed both linear SVM [3] and SVM with RBF kernel which has inbuilt capability to learn pattern from high dimensional data [25]. We have used R programming language version 3.4.1 for coding [22].

6.1 k-Nearest Neighbour (k-NN)

The k-NN utilizes the nearest neighbours of a data sample for prediction. The k-NN has two stages, the first stage is the determination of the nearest neighbours i.e. the value of k and the second is the prediction of the class label using those neighbours. The "k" nearest neighbours are selected using a distance metric. We have used Euclidean distance for our experiments. There are various ways to use this distance metric to determine the class of the test sample. The most straightforward way is to assign the class that the majority of k-nearest neighbours have. In the present work, the optimum value of k is searched over the range of $k = 1$ to 30.

6.2 Random Forest

The Random Forest algorithm constructs an ensemble of many classification trees [18,27]. Each classification tree is created by selecting a bootstrap sample from the whole training data and a random subset of variables with size denoted as $mtry$ is selected at each split. We have used the recommended value of $mtry$: $(mtry = \sqrt{(number\ of\ genes)})$ [7]. The number of trees in the ensemble is denoted as $ntree$. We have used $(ntree) = 10,001$ so that each variable can reach a sufficiently large likelihood to participate in forest building.

6.3 Support Vector Machine (SVM)

Assume that we have given a training set of instance-label pairs $(x_i, y_i); \forall i \in \{1, 2, ..., l\}$ where $x_i \in \mathbb{R}^n$ and $y \in \{1, -1\}^l$, then the SVM [3,10,12] can be formulated and solved by the following optimization problem:

$$
\begin{aligned}
\min_{w,b,\xi_i} \quad & \tfrac{1}{2}w^T w + C \sum_{i=1}^{l} \xi_i, \\
\text{subject to} \quad & y_i \left(w^T \phi(x_i) + b\right) \geq 1 - \xi_i, \\
& \xi_i \geq 0.
\end{aligned}
\tag{5}
$$

Here the parameter $C > 0$ is the penalty parameter of the error term [12] and $\xi_i \forall i \in \{1, 2, ..., l\}$ are positive slack variables [3]. For linear SVM, we did a search for best value of parameter C for a range of values $(C = 2^{-7}, 2^{-3}, ..., 2^{15})$ and the one with the best 10-fold cross validation accuracy has finally been chosen.

We also used SVM with RBF kernel which is a non-linear kernel. There are four basic kernels that are frequently used: linear, polynomial, sigmoid, and

RBF. We picked the RBF kernel, as recommended by Hsu et al. [12]. It has the following form:

$$K\left(\boldsymbol{x}_i, \boldsymbol{x}_j\right) = \exp\left(\frac{-\|\boldsymbol{x}_i - \boldsymbol{x}_j\|^2}{2\sigma^2}\right); \frac{1}{2\sigma^2} > 0.$$

We performed a grid-search over the values of C and σ using 10-fold cross validation. The different pairs of (C, σ) values are tried in the range of $(C = 2^{-7}, 2^{-3}, ..., 2^{15}; \sigma = 2^{-25}, 2^{-13}, ..., 2^3)$ and the values with the best 10-fold cross validation accuracy are picked for the final model building.

7   Explaining Predictions

Similar to the LIME [23], we also believe that it is possible to explain predictions of any classifier by explaining the contribution of important features that led to those predictions. Providing the explanation for an individual prediction is relatively easy and can be achieved by explaining the contribution of important features for that particular prediction. It is relatively hard to provide global interpretation, however, it can be achieved by either explaining a set of representative predictions of each class or explaining all the predictions as a whole [23]. Here, we are providing global interpretation by explaining all the predictions using a 3D-plot. Figure 2(B) is showing the contribution of the 3 most important features (genes) suggested by the proposed Ranked MSD approach. Figure 2 contains the held-out test set predictions of GSE68310 gene expression data set using least important 3 genes (left) and using most important 3 genes (right).

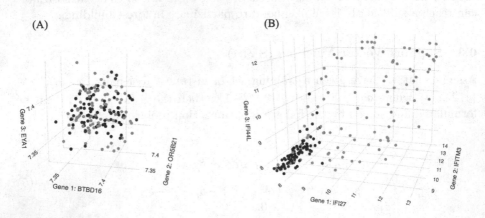

Fig. 2. Providing global interpretation of SVM with RBF Kernel model by plotting test data predictions. (A) 3D-plot of least important 3 genes which fail to achieve class separability and do not give any explanation (B) 3D-plot of most important 3 genes ranked by proposed algorithm. Using these 3 most important genes it is possible to achieve a greater class separability thus helping in explaining predictions. Green dots denote the healthy test samples and Red dots denote infected test samples. The axis denotes the gene expression values of the corresponding gene. (Color figure online)

More clear and persuasive visual explanations can be provided if contributions of all the important feature can be plotted all together. We leave this exploration for future work.

8 Results

We experimentally obtained the 10-fold cross-validation accuracy at full feature set, $\mathcal{F}_{equal}$ and $\mathcal{F}_{best}$ by applying proposed Ranked MSD algorithm on two datasets. We also compared the performance of proposed algorithm with two existing algorithms using four classifiers: k-NN, Random Forest, linear SVM, and SVM with RBF Kernel. The results from both datasets can be seen in Tables 1 and 2. To show the performance of the proposed algorithm in comparison to existing algorithms, we have plotted graphs of feature size versus 10-fold cross-validation accuracy using the four classifiers for both the data sets (see Figs. 3 and 4). The shaded region in the figures showing the standard deviation calculated over 10 fold-cross validation accuracies (repeated 3 times so 30 accuracies in total).

Table 1. Comparison between the performance of proposed Ranked MSD and other feature selection algorithms using four well-known ML algorithms trained on full feature set, $\mathcal{F}_{equal}$ and $\mathcal{F}_{best}$ feature set of the first dataset. Here $\mathcal{F}_{equal}$ is feature size which gives statistically equal accuracy to that of full feature set and $\mathcal{F}_{best}$ is feature size which gives best accuracy.

Feature ranking algorithm	ML model	Total features	Accuracy (all features)	$\mathcal{F}_{equal}$	Accuracy $\mathcal{F}_{equal}$	$\mathcal{F}_{best}$	Accuracy $\mathcal{F}_{best}$
Ranked MSD	KNN	12023	89.17%	22	88.60%	110	91.68%
Correlation Criteria	KNN	12023	89.17%	70	88.73%	1472	90.78%
Information Gain	KNN	12023	89.17%	8950	87.95%	11976	89.23%
Ranked MSD	Linear SVM	12023	91.52%	367	91.05%	2560	93.17%
Correlation Criteria	Linear SVM	12023	91.52%	561	90.68%	3040	92.62%
Information Gain	Linear SVM	12023	91.52%	11961	91.39%	12013	91.52%
Ranked MSD	Random Forest	12023	88.97%	45	88.32%	544	91.17%
Correlation Criteria	Random Forest	12023	88.97%	352	88.44%	3040	89.42%
Information Gain	Random Forest	12023	88.97%	11271	88.85%	11930	88.99%
Ranked MSD	SVM with RBF Kernel	12023	93.3%	65	92.25%	128	93.3%
Correlation Criteria	SVM with RBF Kernel	12023	93.3%	327	92.46%	1916	93.39%
Information Gain	SVM with RBF Kernel	12023	93.3%	6387	92.54%	11895	93.32%

Based on the results obtained, it can be concluded that the scoring function used in the Ranked MSD algorithm is successfully able to rank the features in descending order of their importance, because we are able to see the increase in accuracy when we are adding the top-ranked features. For example, in case of SVM with RBF kernel (see Table 1), the $\mathcal{F}_{equal}$ using proposed Ranked MSD algorithm gives 92.25% accuracy using top 65 genes (strongly relevant) whereas total 12023 genes give 93.3% accuracy. Here the benefit of $\mathcal{F}_{equal}$ is that it can hold the potential biomarkers for the respiratory viral infection because it finds a small number of strongly relevant features which contribute to reaching accuracy of 92.25%. $\mathcal{F}_{best}$ yields 93.3% accuracy using the top 128 genes from $\mathcal{RF}_{All}$ and there is no improvement in accuracy if we add more genes, which shows that the rest of the genes after top 128 genes are irrelevant for this classifier. A similar behaviour can be observed for the second data set (see Table 2 and Fig. 4).

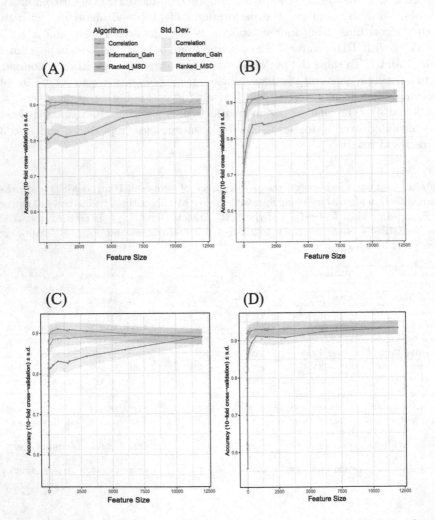

Fig. 3. Comparing performance of Ranked MSD algorithm with other existing feature selection techniques on first data set using (A) KNN (B) Linear SVM (C) Random Forest and (D) SVM with RBF Kernel.

In the area of drug discovery, we wish to target the smallest number of important genes. In such cases, the use of F_{equal} is valuable because we don't want to include all 12,023 or 20,737 genes as potential targets for drug discovery but those 65 genes (Dataset 1) or 31 genes (Dataset 2) which contribute to reaching the F_{equal} accuracy (See Tables 1 and 2).

The other optimal feature subset, F_{best} can be used according when needed (See Tables 1 and 2), in cases where one requires the best possible accuracy, while allowing a larger number of features to be selected; in such cases F_{best} provides all the features that are relevant. The standard deviation of repeated 10 fold cross-validation accuracies is not significantly high which suggests that the algorithm is able to produce stable results.

Table 2. Comparison between the performance of proposed Ranked MSD and other feature selection algorithms using four well-known ML algorithms trained on full features set, F_{equal} and F_{best} feature set of the second dataset. Here F_{equal} is feature size which gives statistically equal accuracy to that of full feature set and F_{best} is feature size which gives best accuracy.

Feature ranking algorithm	ML-model	Total features	Accuracy (all features)	F_{equal}	Accuracy F_{equal}	F_{best}	Accuracy F_{best}
Ranked MSD	KNN	20737	80.7%	2	79.82%	48	82.55%
Correlation Criteria	KNN	20737	80.7%	2	80.64%	224	83.05%
Information Gain	KNN	20737	80.7%	13	78.39%	632	83.53%
Ranked MSD	Linear SVM	20737	86.44%	45	85.82%	1725	90.16%
Correlation Criteria	Linear SVM	20737	86.44%	8	84.87%	6108	88.74%
Information Gain	Linear SVM	20737	86.44%	90	84.91%	1920	88.73%
Ranked MSD	Random Forest	20737	83.03%	10	81.42%	1664	85.44%
Correlation Criteria	Random Forest	20737	83.03%	7	81.83%	83	85.48%
Information Gain	Random Forest	20737	83.03%	93	81.75%	332	85.09%
Ranked MSD	SVM with RBF Kernel	20737	86.44%	31	84.35%	1239	90.07%
Correlation Criteria	SVM with RBF Kernel	20737	86.44%	8	84.83%	6136	88.97%
Information Gain	SVM with RBF Kernel	20737	86.44%	88	85.15%	1920	88.83%

As the results show, our *Ranked MSD* algorithm is able to achieve significantly higher accuracy using very few genes compared to two well-known feature selection approaches, which indicates that our algorithm is selecting more highly informative genes than the other approaches. Based on the results of these experiments, we can conclude that the proposed Ranked MSD algorithm outperforms the existing correlation-based and entropy-based feature selection methods on investigated datasets. For further details, additional figures can be found in a supplementary file at this link: https://figshare.com/articles/Ranked_MSD/8312402.

9   Biomarker Identification

In this section, we show the top 18 important biomarkers (see Table 3) which are obtained from taking the intersection of $\mathcal{F}_{equal}$ genes suggested from the best-performing classifier. We have selected SVM with RBF kernel as the best performing classifier, because it gives the consistently best accuracy with smallest optimal feature set of the classification algorithms evaluated. For Dataset 1, the $\mathcal{F}_{equal}$ size is 65 and for Dataset 2, the $\mathcal{F}_{equal}$ size is 31. The intersection of both optimal feature sets is 18, as illustrated in Fig. 5(b). Table 3 lists the 18

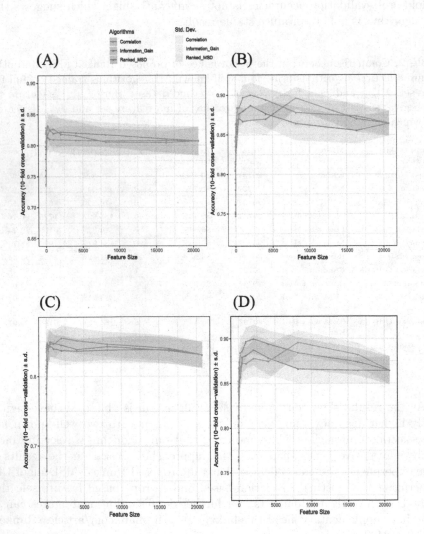

Fig. 4. Comparing performance of Ranked MSD algorithm with other existing feature selection techniques on second data set using (A) KNN (B) Linear SVM (C) Random Forest and (D) SVM with RBF Kernel.

biomarkers that are common to both datasets, with their importance scores as given by our *Ranked MSD* algorithm. The combined importance score of a biomarker is the average of its importance score from Dataset 1 and Dataset 2. These 18 biomarkers are found to be the most important ones for the progression of respiratory viral infection as they are common best biomarkers for both the respiratory viral data sets and play an important role in the discrimination of infected samples from non-infected ones.

10   Biological Significance of Biomarkers

To determine biological significance of identified 18 common biomarkers (see Table 3), we performed Molecular Enrichment Analysis (MEA) developed as an extension of Gene Set Enrichment Analysis (GESA) [29]. The biomarkers shown in Table 3 work as seed genes for the MEA analysis and the results of the enrichment can be seen in Fig. 5(a). We used KEGG pathway database [13] to perform the MEA analysis. Biomarkers retrieved only through gene expression can lead to non-relevant signatures and irrelevant phenotypes [5], therefore, a

Table 3. Top 18 biomarkers obtained from taking intersection of $\mathcal{F}_{equal}$ genes suggested by best performing classifier from both the data sets with their importance score given by the proposed Ranked MSD algorithm.

Sr. no.	Gene symbol	Importance score data set 1	Importance score data set 2	Combined importance score
1	IFI27	7.440807172	8.744873626	8.092840399
2	RSAD2	7.781213672	3.788627429	5.78492055
3	IFI44L	6.574818892	4.978166941	5.776492916
4	RPS4Y1	9.091357246	2.008783926	5.550070586
5	ISG15	4.063142667	4.578217378	4.320680022
6	IFI44	5.014484011	2.860260613	3.937372312
7	IFITM3	2.495110022	5.359004043	3.927057032
8	HERC5	4.04471719	3.211724492	3.628220841
9	MX1	2.98390575	3.508595477	3.246250613
10	LY6E	2.249102841	3.834706217	3.041904529
11	IFIT3	3.441885872	2.50066336	2.971274616
12	OAS3	3.705077795	2.176990862	2.941034328
13	IFIT2	3.015140745	2.840130002	2.927635374
14	IFI6	3.192691447	2.236622559	2.714657003
15	OASL	3.303346964	1.954874216	2.62911059
16	HBG2	2.275345902	2.923786916	2.599566409
17	OAS2	2.345796307	2.446741233	2.39626877
18	XAF1	2.675522172	1.975049326	2.325285749

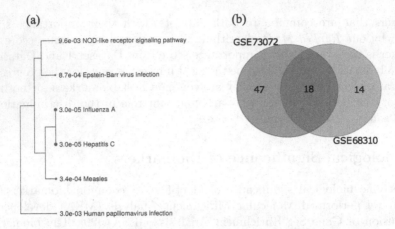

Fig. 5. (a) MEA analysis of seed genes and associated KEGG terms. The larger blue dots show higher enrichment with seed genes. (b) Venn-diagram of important genes obtained from data set 1 (GSE73072) and data set 2 (GSE68310). Total 18 genes are overlapping genes which are final biomarkers. (Color figure online)

biological significance analysis is essential. The reproducible and clinically attainable results provided by us, proves the authenticity of biomarkers and can be a great fit in precision medicine era. The 18 gene retrieved through our analysis yields Influenza A and Hepatitis C with enrichment score of 3.0e-05. It has been reported earlier that any gene signature profound in Influenza A also elevates the expression profile of Hepatitis C with similar intensity [2]. Further, Influenza A and measles are viruses that both cause respiratory symptoms thus enrichment of measles provide appropriate phenotype for our 18 gene signature [32]. Now, looking into the pathway from Fig. 5 (a) the enriched pathways NOD-like receptor signalling pathway is associated with higher immunity. Thus any targeted study through these 18 biomarkers will provide a clinical acceptable therapy for viral diseases specially in the case of Influenza. Thus it can be inferred that our biomarker panel covers immune response [15] with disease progression and provides a cohesive platform for precision medicine.

11   Conclusions and Future Work

In this work, we have aimed to tackle the issues that arise when the number of samples is much smaller than the number of features (commonly referred to as $n << p$). It becomes very hard to interpret the target concept in these situations. In addition, irrelevant features often confuse machine leaning systems and lead to deterioration of classification accuracy. Also, recent GDPR issues may make it difficult to use black-box models particularly in business and medicine. To address these problems, we have proposed a feature ranking algorithm named *Ranked MSD* with two additional algorithms to identify *strongly relevant* features and *relevant* features by discarding irrelevant features. Our experimental results show that the proposed *Ranked MSD* algorithm outperforms two well-known feature ranking methods, Correlation Criteria and Information Gain, and thereby can help with better disease prediction. Moreover, we have identified 18 biomarkers which are common biomarker across the two datasets that we have analysed and which have been identified as strongly relevant features by our approach. The importance of these 18 genes are confirmed with the four classifiers as they all yield improvements in accuracy using these top genes. To determine the biological significance of these 18 genes, we performed Molecular Enrichment Analysis (MEA); the results show that these biomarkers are strongly related to the target disease, and can therefore be considered as potential targets for drug discovery, and could play an important role in precision medicine.

In this work, we have demonstrated our proposed approach by applying it on datasets related to respiratory viral infections only. In future work, we aim to perform analyses on more datasets of different diseases and domains. We also aim to incorporate useful information that is openly available in the form of biomedical knowledge graphs.

Acknowledgements. This publication has emanated from research conducted with the financial support of Science Foundation Ireland (SFI) under Grant Number SFI/12/RC/2289, co-funded by the European Regional Development Fund.

References

1. Banerjee, A., Chitnis, U., Jadhav, S., Bhawalkar, J., Chaudhury, S.: Hypothesis testing, type i and type ii errors. Ind. Psychiatry J. **18**(2), 127 (2009)
2. Bolen, C.R.: The blood transcriptional signature of chronic hepatitis c virus is consistent with an ongoing interferon-mediated antiviral response. J. Interferon Cytokine Res. **33**(1), 15–23 (2013)
3. Burges, C.J.: A tutorial on support vector machines for pattern recognition. Data Min. Knowl. Disc. **2**(2), 121–167 (1998)
4. Chandrashekar, G., Sahin, F.: A survey on feature selection methods. Comput. Electr. Eng. **40**(1), 16–28 (2014)
5. Cun, Y., Fröhlich, H.: Biomarker gene signature discovery integrating network knowledge. Biology **1**(1), 5–17 (2012)
6. Cunningham, P., Delany, S.J.: k-nearest neighbour classifiers. Multiple Classifier Syst. **34**, 1–17 (2007)
7. Díaz-Uriarte, R., De Andres, S.A.: Gene selection and classification of microarray data using random forest. BMC Bioinform. **7**(1), 3 (2006)
8. Efron, B., Hastie, T., Johnstone, I., Tibshirani, R., et al.: Least angle regression. Ann. Stat. **32**(2), 407–499 (2004)
9. Guyon, I., Elisseeff, A.: An introduction to variable and feature selection. J. Mach. Learn. Res. **3**, 1157–1182 (2003)
10. Guyon, I., Weston, J., Barnhill, S., Vapnik, V.: Gene selection for cancer classification using support vector machines. Mach. Learn. **46**(1), 389–422 (2002)
11. Holzinger, A., Biemann, C., Pattichis, C.S., Kell, D.B.: What do we need to build explainable AI systems for the medical domain? arXiv preprint arXiv:1712.09923 (2017)

12. Hsu, C.W., Chang, C.C., Lin, C.J.: A practical guide to support vector classification (2010)
13. Kanehisa, M., Goto, S.: Kegg: kyoto encyclopedia of genes and genomes. Nucleic Acids Res. **28**(1), 27–30 (2000)
14. Kim, B., et al.: Interpretability beyond feature attribution: Quantitative testing with concept activation vectors (TCAV). In: Dy, J., Krause, A. (eds.) Proceedings of the 35th International Conference on Machine Learning. Proceedings of Machine Learning Research, vol. 80, pp. 2668–2677. PMLR, StockholmsmÄssan, Stockholm Sweden, 10–15 July 2018. http://proceedings.mlr.press/v80/kim18d.html
15. Kim, Y.K., Shin, J.S., Nahm, M.H.: Nod-like receptors in infection, immunity, and diseases. Yonsei Med. J. **57**(1), 5–14 (2016)
16. Kohavi, R., John, G.H.: Wrappers for feature subset selection. Artif. Intell. **97**(1–2), 273–324 (1997)
17. Li, O., Liu, H., Chen, C., Rudin, C.: Deep learning for case-based reasoning through prototypes: a neural network that explains its predictions. In: Thirty-Second AAAI Conference on Artificial Intelligence (2018)
18. Liaw, A., Wiener, M.: Classification and regression by randomForest. R News **2**(3), 18–22 (2002). http://CRAN.R-project.org/doc/Rnews/
19. Liu, T.Y., et al.: An individualized predictor of health and disease using paired reference and target samples. BMC Bioinform. **17**(1), 47 (2016)
20. Molnar, C., et al.: Interpretable machine learning: A guide for making black box models explainable. Christoph Molnar, Leanpub (2018)
21. Mudge, J.F., Baker, L.F., Edge, C.B., Houlahan, J.E.: Setting an optimal α that minimizes errors in null hypothesis significance tests. PLoS ONE **7**(2), e32734 (2012)
22. R Core Team: R: A Language and Environment for Statistical Computing. R Foundation for Statistical Computing, Vienna, Austria (2013)
23. Ribeiro, M.T., Singh, S., Guestrin, C.: Why should I trust you?: explaining the predictions of any classifier. In: Proceedings of the 22nd ACM SIGKDD International Conference on Knowledge Discovery and Data Mining, pp. 1135–1144. ACM (2016)
24. Rudin, C.: Stop explaining black box machine learning models for high stakes decisions and use interpretable models instead. Nat. Mach. Intell. **1**, 206–215 (2019)
25. Scholkopf, B., et al.: Comparing support vector machines with Gaussian kernels to radial basis function classifiers. IEEE Trans. Sig. Process. **45**(11), 2758–2765 (1997)
26. Selvaraju, R.R., Cogswell, M., Das, A., Vedantam, R., Parikh, D., Batra, D.: Grad-cam: visual explanations from deep networks via gradient-based localization. In: Proceedings of the IEEE International Conference on Computer Vision, pp. 618–626 (2017)
27. Statistics, L.B., Breiman, L.: Random forests. In: Machine Learning, pp. 5–32 (2001)
28. Stork, E., Duda, R., Hart, P., Stork, D.: Pattern Classification. Academic Internet Publishers, New York (2006)
29. Subramanian, A., et al.: Gene set enrichment analysis: a knowledge-based approach for interpreting genome-wide expression profiles. Proc. Nat. Acad. Sci. **102**(43), 15545–15550 (2005)
30. Verma, G., Jha, A., Rebholz-Schuhmann, D., Madden, M.G.: Using machine learning to distinguish infected from non-infected subjects at an early stage based on viral inoculation. In: Auer, S., Vidal, M.-E. (eds.) DILS 2018. LNCS, vol. 11371, pp. 105–121. Springer, Cham (2019). https://doi.org/10.1007/978-3-030-06016-9_11

31. Witten, I.H., Frank, E., Hall, M.A., Pal, C.J.: Data Mining: Practical Machine Learning Tools and Techniques. Morgan Kaufmann, Burlington (2016)
32. Zachariah, P., et al.: Vaccination rates for measles, mumps, rubella, and influenza among children presenting to a pediatric emergency department in New York city. J. Pediatr. Infect. Dis. Soc. **3**(4), 350–353 (2014)
33. Zhai, Y., et al.: Host transcriptional response to influenza and other acute respiratory viral infections-a prospective cohort study. PLoS Pathog. **11**(6), e1004869 (2015)

How to Improve the Adaptation Phase of the CBR in the Medical Domain

Ľudmila Pusztová[1]([✉]), František Babič[1], Ján Paralič[1],
and Zuzana Paraličová[2]

[1] Department of Cybernetic and Artificial Intelligence,
Faculty of Electrical Engineering and Informatics,
Technical University of Košice, Letná 9, 042 00 Košice, Slovak Republic
{ludmila.pusztova.2, frantisek.babic,
jan.paralic}@tuke.sk
[2] Faculty of Medicine, University of Pavol Jozef Šafárik,
Trieda SNP 1, 040 11 Košice, Slovak Republic
zuzana.paralicova@unlp.sk

Abstract. This paper reviews the Case-based reasoning (CBR) approach and its usability in the medicine and presents a new concept on how to improve its adaptation phase. We use the CBR as a supporting method for decision support like diseases diagnostics or therapy identification. We investigated existing approaches, studies, and research works to solve one of the most critical problems in the CBR cycle - adaptation, which is often done manually by the experts in the relevant field. Based on the findings and our experiences with medical diagnostics through suitable data analytical methods, we proposed a new solution to solve this challenge. This approach is based on a comparison of the stored decision rules with the new one related to the current case. This comparison can result in three alternative states: (1) case base contains a similar case, and relevant rule can be applied. (2) The new case is very different from the stored ones, so the input from participated experts is needed, and a new rule will be stored. (3) The new case is partially similar satisfying adaptability conditions, in such a situation we adopt related decision rule to the new conditions under the supervision of the expert. We plan to experimentally test and verify this concept within available medical samples from our previous experiments.

Keywords: Case-based reasoning · Medical diagnostics · Data analysis

1 Introduction

Information and communication technologies are increasingly being deployed in the medical domain to support various activities on all sides of the relevant processes. This situation is also related to the ever-increasing amount of data that needs to be processed and analyzed. For doctors, it´s hard to consider a higher volume of data in the diagnosis procedure, or in determining the right treatment.

The evidence of the World Health Organization (WHO) from 2015 indicates that up to 5% of patients had an incorrect diagnosis. The diagnostic procedure represents a complicated process, in which it is essential to have the right information available for

© IFIP International Federation for Information Processing 2019
Published by Springer Nature Switzerland AG 2019
A. Holzinger et al. (Eds.): CD-MAKE 2019, LNCS 11713, pp. 168–177, 2019.
https://doi.org/10.1007/978-3-030-29726-8_11

the right people at the right time. If analytical support models are available for doctors, they will help them consider all the contexts and important hidden patterns in the data. They will also reduce the time needed to make a decision. The input and continuous participation of the experts is an essential part of the analytical projects. In some cases, we can capture and store the expert's knowledge in a suitable formal way.

Many researchers apply various intelligent techniques to create decision support systems or models to help the doctor determine the correct patient diagnosis and enable them to design the best treatment for their current health condition.

Case-based reasoning (CBR) approach was proposed by Schank in 1982 to solve identified problems related to the decision support systems like knowledge elicitation, adaptation, or maintenance [1]. Since medicine requires experts with a mixture of knowledge and experience, case-oriented methods should be very efficient, mainly because reasoning with cases corresponds with the typical decision-making process of physicians. Also, incorporating new cases means automatically updating parts of the changeable knowledge [2].

In our previous research works focusing on the diagnostics of the various diseases like metabolic syndrome, mild cognitive impairment, heart or brain attacks, we typically extracted different models and hidden knowledge through suitable analytical methods. In most cases, these were relatively small samples with records up to 500. It means that the adaptation of the generated decision rules to new examples was simple, e.g., we spent some time re-generate the new ones. However, in this paper, we want to focus on a different situation, when the knowledge base contains a large number of rules, and it is important to decide about possible updates effectively.

The paper is organized as follows: the first section introduces our motivation and the topic, the second one presents the Case-based reasoning in the medical domain, identified challenges and new proposed approach for adaptation phase. The conclusion summarizes the main points and outlines future work.

1.1   Case-Based Reasoning

The CBR methodology has attracted significant attention because the basic idea of reusing experience to solve previous problems is a powerful and often used way of addressing people's issues. In CBR terminology, a case usually means a problem situation that one needs to resolve.

Doyle et al. in 1998 [3] described the CBR as a problem-solving paradigm in many ways significantly different from other major artificial intelligence approaches. But the situation has changed during the last years. Computational analogy-making and CBR are closely related areas. Analogy-making involves at least several subprocesses like building representation, retrieval from a base for the analogy, mapping onto the target, validation, and learning from the experience [4]. Other approaches rely only on general knowledge of the problem area, but CBR can use specific knowledge about previous problematic situations [5]. CBR reasoning has an incremental character [6]: it means that whenever the problem is resolved, the new experience will be retained and immediately available for future use of problem-solving.

Therefore, solving each problem in the CBR cycle consists of 4 phases such as retrieve, reuse, revise, and retain. We identified several models in existing literature like the Hunt model, the Allen model, the model by Kolodner and Leake [7], and the R4 model proposed by Aamodt and Plaza [8]. The R4 model is one of the most used and defines the CBR cycle with the following four primary steps.

Finding and retrieving the most similar cases is done at the RETRIEVE step. According to several authors [9, 10], this phase is one of the most important. It includes a case-finding process based on their similarity. For this purpose, we can typically use the nearest neighbor search, inductive approaches, knowledge approaches, Bayesian network, clustering Euclidian distance, or other similarity measures. In many practical applications, it is often difficult to distinguish the REUSE and REVISE steps because many researchers associate them into one phase called ADAPT (adaptation) [11]. In this step is the case used again, and the proposed solution is checked. In the last step - RETAIN - is performed preservation (storage) of the learned case for future use. There are several approaches to achieving this goal like retaining only the solution of the previous problem or the new one. In many cases, this retention process leads to uncontrolled growth of the case's base, which consequently causes the system performance to deteriorate in terms of speed [12].

2   Case-Based Reasoning in the Medical Domain

CBR reasoning process is medically accepted and getting increasing attention from the medical domain. In 1988, 1989 and 1991 were organized three CBR seminars by the American Defense Research Agency (DARPA), which officially marked the beginning of CBR discipline.

The authors in [13] present a summary of 21 studies that dealt with medical CBR systems. They described a list of methods used in each CBR step and the success rate in system verification. In the RETRIEVE phase, the most used methods are Euclidian distance, nearest neighborhood, similarity function, and weight set ranked by a decision tree. In the REUSE phase, authors used neural nets, fuzzy rules, stepwise regression, manual reuse, but most systems do not use any technique. The REVISE phase performed either manually, or do not use any specific method. In RETAIN phase are cases stored manually or not at all. The most used evaluation methods are k-fold cross-validation, leave-one-out strategy, conditional probability, AUC curve, statistical frequency, and correlation.

Many studies [11, 14] have attempted to investigate existing medical CBR systems since 1987. The most systems were developed to solve a specific disease; most systems perform as prototypes and not as the final product. Another visible trend was the successful hybridization of CBR with various computational methods. According to [11], in 32 systems out of 76, CBR was used in combination with other techniques. Also, out of 76 systems, in 51 systems, automatic adaptation is completely avoided, so they only work as retrieval systems.

The use of CBR in the medical field is currently reviving. The knowledge base of medical knowledge is continually changing; sometimes, there is more than one solution; doctors have different approaches and medicines. The fact that the CBR system methodology very much resembles the doctor's thinking process suggests the successful use of CBR in medicine [15]. The main advantage of CBR in this field is the possibility of adapting the knowledge base [16], which is a significant aspect of decision making in the medical field.

2.1 Existing Limitations

Although the use of the CBR method appears to be successful, there are some limitations. In the medical domain, the number of similar cases is often extremely high, and this fact causes a complex generalization [14]. High memory/storage requirements and time-consuming retrieval accompany CBR systems utilizing large case bases and can take significant processing time to find similar cases in case-base. CBR systems have problems with handling noisy data. Unsuccessful assessment of such noise may result in the same problem being unnecessarily stored numerous times. In turn, this implies inefficient storage and retrieval of cases. The number of systems using the full CBR cycle (retrieve, adaptation, retain) is still very low. However, the most critical problem in the successful implementation of CBR techniques in medical systems is the problem of the adaptation step. D'Aquin et al. [16] note that this step is a relatively complex process because it has to address the lack of relevant patient information, the usability, consequences of the decision, the proximity of decision thresholds and the need to consider patients in different ways. Schmidt et al. indicate that introducing the adaptation step in the CBR system was a challenging step in medicine [17]. Most CBR systems which don't apply the adaptation step, can't solve some new problems, and thus their accuracy is unconvincing in critical areas [5]. The adaptation phase is, therefore limited to planning tasks [18].

2.2 Adaptation Step Problem

The study [13] mentioned that medical CBR systems solve the problem of adaptation in two ways. Most systems avoid an adaptation problem by applying only the RETRIEVE step in the CBR cycle, while others are trying to resolve it. One of the first medical expert systems CASEY [19] attempted to solve the adapting problem through rule-based domain theory. Knowledge acquisition is a barrier to the development of rule-based systems; therefore, the development of adaptation rules has never become a successful technique in medical CBR systems [17]. Some of the newer systems successfully used adaptation using computational techniques, e.g., eXiT * CBR.v2 [20] revises and reuses cases using genetic algorithms; EquiVox developed by Henriet et al. [21] performs adaptation using artificial neural networks.

The studies [5, 6] solved the adaptation problem by the creation of a hybrid CBR system integrating CBR (case) and RBR (rule) reasoning. This system automatically applied the adaptation process using adaptive rules.

In the study [5], after the resolution of the new case, the knowledge base was expanded, and the adaptation and reasoning rules were updated. To achieve integration into REUSE step was added a new process called REASON, which applied the reasoning rules to get a solution if the REUSE and ADAPT process failed to find a solution. They first applied the CBR and after that, RBR to the available data. The authors used multiple cross-validations to evaluate accuracy. The developed prototype achieved an average accuracy of 99.53% on the diagnosis of thyroid disease and 99.33% on breast cancer diagnosis (accuracy by other systems ranged from 80% to 97%).

The study [6] provided a hybrid system to help healthcare professionals in early diagnose on cancer patients. In the proposed approach, CBR was used as the primary reasoning process, and RBR was used to improve part of the process. For this research, they gathered real data about patients with gastrointestinal cancer. To evaluate accuracy, they also used multiple cross-validations. The results showed increased diagnosis accuracy by 22.92% compared to the use of a single CBR method.

Salem and El Bagouras [22] have proposed a hybrid adaptation model that combines transformational and hierarchical adaptation techniques with artificial neural networks and factors for the diagnosis of thyroid cancer. Zubi and Saad [23] used combined data mining techniques with neural networks for early diagnosis of lung cancer. For the diagnosis of breast cancer, Keles, Keles, and Yavuz [24] used neuro-fuzzy rules, while Sharaf-el-Deen et al. [25] introduced a hybrid approach that also combined CBR and RBR reasoning.

We can see that authors tried to solve the adaption problem in three ways: avoiding the adaptation problem by using CBR systems only for RETRIEVE step; the use of computational techniques such as genetic algorithms and artificial neural networks; creating a hybrid CBR system that integrates CBR and RBR reasoning.

2.3   New Proposal How to Support the Adaptation

Figure 1 presents graphically our approach on how to support the adaptation phase in the CBR cycle. The assumption is a list of decision rules generated by suitable machine learning algorithms stored in case base: IF conditions THEN consequences (target value, expected diagnosis).

The CBR cycle starts with the RETRIEVE step as a response to a new example without target diagnosis. The new case is compared with existing ones from the case base by an inference mechanism. We will calculate the distances between cases with similarity metrics like Euclidean, Manhattan, or Hamming distance. The result of the comparison can be one of the three alternatives:

1. The mechanism will find an identical case to the new one. The target diagnosis will be the same as for the existing one.
2. The mechanism will not find a match; all stored cases are significantly different. This situation requires re-generate the current rules based on the original set of records extended by the latest case classified by the expert.
3. The mechanism will find partly similar cases with different target values. Therefore, the CBR cycle will continue with other steps like REUSE, REVISE, and RETAIN.

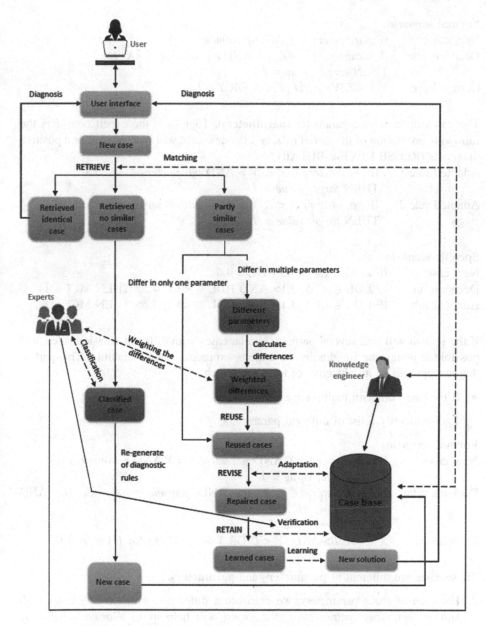

Fig. 1. The new approach for adaptation

Before the cycle continues, we will investigate the differences:

- If the cases differ only in one condition (parameter) on the left side of the rules, the expert will consider possible adjusting of it. After several iterations, we will be able to create a separate knowledge base with the knowledge from the experts and will be able to do this step in a semi-automatic way. An example:

Formal scenario:

New case: IF $parameter_1 = X$ AND $parameter_2 = Y$

Decision rule: IF $parameter_1 \in$ <Z, V> AND $parameter_2 \in$ <K, L>
                   THEN *target value = 1*

Comparison: $X \in$ <Z, V> AND (Y < K OR Y > L)

The cases differ in one parameter (**parameter$_2$**). Therefore, the expert considers the following adaptation of the stored rule, and the new case will be classified as a positive diagnosis (REUSE-REVISE-RETAIN).

Adapted rule 1:   IF $parameter_1 \in$ < Z, V > AND $parameter_2 \in$ <K, Y>
                   THEN *target value = 1*

Adapted rule 2:   IF $parameter_1 \in$ < Z, V > AND $parameter_2 \in$ <Y, L>
                   THEN *target value = 1*

Specific scenario:

New case: IF LDL = 1.8 AND HDL = 4.6

Decision rule: IF LDL $\in$ <1.5, 3.1> AND HDL $\in$ <3.1, 4.5> THEN MCI = 1

Adapted rule: IF LDL $\in$ <1.5, 3.1> AND HDL $\in$ <3.1, 4.6> THEN MCI = 1

If the system will find several partially similar cases with different decision rules, it is possible to assign the weights by the experts expressing their suitability. This part of the concept will be an objective of further research.

- If the cases differ in multiple parameters:

  1. We identify a list of different parameters.

Formal scenario:

New case: IF $parameter_1 = X$ AND $parameter_2 = Y$ AND $parameter_3 = Z$
                   THEN *target value = 1*

Decision rule: IF $parameter_1 \in$ <A, B> AND $parameter_2 \in$ <C, D> AND
                   $parameter_3 \in$ <E, F>
                   THEN *target value=1*

Comparison: $X \in$ <A, B> AND (Y < C OR Y > D) AND (Z < E OR F > D)

These cases are different in **parameter$_2$** and **parameter$_3$**.

2. For each of these parameters, we calculate a difference with existing cases with suitable similarity metric. Next, the expert will help us to allocate weights by importance for particular differences.
3. The parameters with high weights will be adapted to the most similar case, and the target class will be determined (REUSE-REVISE-RETAIN).

Specific scenario:

New case: IF LDL = 1.8 AND HDL = 4.6 AND BMI = 34 THEN MCI = 1

Decision rule: IF LDL $\in$ <1.5, 3.1> AND HDL $\in$ <3.1, 4.5> AND BMI $\in$ <25.1, 29.3> THEN MCI = 1

Adapted rule 1:   IF LDL $\in$ <1.5, 3.1> AND HDL $\in$ <3.1, 4.6> AND BMI $\in$ <25.1, 34> THEN MCI = 1

                   etc.

3   Conclusion

The CBR methodology has attracted significant attention because the basic idea of reusing experience to solve previous problems looks very attractive. It can use specific knowledge about past problematic situations solving. The number of medical systems using the full CBR cycle (retrieve, adaptation, retain) is still very low. The most critical issue is the successful adaptation step. We propose a new concept to solve this issue. We found the inspiration in the research of professor Holzinger research group called interactive machine learning (iML) with a human-in-the-loop. This approach leads to algorithms that can interact with both computational agents and human agents and can optimize their learning behavior through these interactions [26, 27].

For this purpose, we use a combination of data analysis methods and CBR extending by communication with an expert, which helps us determine the importance of the parameters, their settings and the determination of the suitable adaptation.

In future work, we will focus on experimentally testing and verification of the proposed approach on the available medical data samples.

Acknowledgment. The work was partially supported by the Slovak Grant Agency of the Ministry of Education and Academy of Science of the Slovak Republic under grant no. 1/0493/16 and The Slovak Research and Development Agency under grants no. APVV-16-0213 and APVV-17-0550.

References

1. López-Fernández, H., Fdez-Riverola, F., Reboiro-Jato, M., et al.: Using CBR as design methodology for developing adaptable decision support systems. In: Jao, C.S. (ed.) Efficient Decision Support Systems - Practice and Challenges From Current to Future. IntechOpen, Croatia (2011)
2. Schmidt, R., Montani, S., Bellazzi, R., Portinale, L., Gierl, L.: Cased based reasoning for medical knowledge-based systems. Int. J. Med. Informatics **64**(2), 355–367 (2001)
3. Doyle, M., Hayes, C., Cunningham, P., Smith, B.: CBR Net: smart technology over a network. Department of Computer Science, Trinity College Dublin (1998)
4. Kokinov, B., French, R.M.: Computational models of analogy-making. In: Nadel, L. (ed.) Encyclopedia of Cognitive Science, vol. 1, pp. 113–118. Nature Publishing Group, London (2003)

5. Sharaf-El-Deen, D.A., Moawad, A.F., Khalifa, M.E.: A new hybrid case-based reasoning approach for medical diagnosis systems. J. Med. Syst. **38**, 9 (2014). https://doi.org/10.1007/s10916-014-0009-1

6. Saraiva, R.M., Bezerra, J., Perkusich, M., Almeida, H., Siebra, C.: A Hybrid Approach Using Case-Based Reasoning and Rule-Based Reasoning to Support Cancer Diagnosis: A Pilot Study. MedInfo (2015). https://doi.org/10.3233/978-1-61499-564-7-862

7. Leake, D.B.: Case-Based Reasoning: Experiences, Lessons, and Future Directions. MIT Press, Cambridge (1996)

8. Aamodt, A., Plaza, E.: Case-based reasoning: Foundational Issues, Methodological Variations, and System Approaches. In: AI Communications, vol. 7, pp. 39–59. OSI Press (1994)

9. De Mantaras, R.L., et al.: Retrieval, reuse, revision, and retention in case-based reasoning. Knowl. Eng. Rev. **20**(03), 215–240 (2005)

10. Pal, S.K., Shiu, S.C.: Foundations of Soft Case-Based Reasoning. Wiley, Chichester (2004)

11. Holt, A., Bichindaritz, I., Schmidt, R., Perner, P.: Medical applications in case-based reasoning, September 2005. https://doi.org/10.1017/s0269888906000622

12. Lawanna, A., Daengdej, J.: Methods for case maintenance in case-based reasoning. Int. J. Comput. Inform. Eng **4**, 10–18 (2010)

13. Blanco, X., Rodríguez, S., Corchado, J.M., Zato, C.: Case-based reasoning applied to medical diagnosis and treatment. In: Omatu, S., Neves, J., Rodriguez, J.M.C., Paz Santana, J. F., Gonzalez, S.R. (eds.) Distributed Computing and Artificial Intelligence. AISC, vol. 217, pp. 137–146. Springer, Cham (2013). https://doi.org/10.1007/978-3-319-00551-5_17

14. Choudhury, N.: A survey on case-based reasoning in medicine. October 2018. https://doi.org/10.14569/ijacsa.2016.070820

15. Macura, R.T., Macura, K.: Case-based reasoning: opportunities and applications in health care. Artif. Intell. Med. **9**(1), 1–4 (1997)

16. D'Aquin, M., Lieber, J., Napoli, A.: Adaptation knowledge acquisition: a case study for case-based decision support in oncology. Comput. Intell. **22**(3–4), 161–176 (2006)

17. Schmidt, R., Montani, S., Bellazzi, R., Portinale, L., Gierl, L.: Cased-based reasoning for medical knowledge-based systems. Int. J. Med. Informatics **64**(2), 355–367 (2001)

18. Pous, C., et al.: Modeling reuse on case-based reasoning with application to breast cancer diagnosis. In: Dochev, D., Pistore, M., Traverso, P. (eds.) AIMSA 2008. LNCS (LNAI), vol. 5253, pp. 322–332. Springer, Heidelberg (2008). https://doi.org/10.1007/978-3-540-85776-1_27

19. Koton, P.: A medical reasoning program that improves with experience. Comput. Methods Programs Biomed. **30**(2), 177–184 (1989)

20. Pla, A., LóPez, B., Gay, P., Pous, C.: Distributed case-based reasoning tool for medical prognosis. Decis. Support Syst. **54**(3), 1499–1510 (2013)

21. Henriet, J., Leni, P.E., Laurent, R., Salomon, M.: Case-based reasoning adaptation of numerical representations of human organs by interpolation. Expert Syst. Appl. **41**(2), 260–266 (2014)

22. Salem, A.B.M. Bagoury, B.M.E.L.: A case-based adaptation model for thyroid cancer diagnosis using neural networks. In: Society, Florida Artificial Intelligence Research, Cairo, Egypt, pp. 155–159 (2003)

23. Zubi, Z.S., Saad, R.A.: Using some data mining techniques for early diagnosis of lung cancer. In: Recent Researches in Artificial Intelligence, Knowledge Engineering, and Data Base, Tripoli, Libya, pp. 32–37 (2011)

24. Keles, A., Keles, A., Yavuz, U.: Expert system based on neuro-fuzzy rules for diagnosis breast cancer. Expert Syst. Appl. **38**(5), 5719–5726 (2011)

25. Sharaf-elDeen, D.A., Moawad, I.F.: A breast cancer diagnosis system using hybrid case-based approach. Int. J. Comput. Appl. **72**, 9–14 (2013)
26. Holzinger, A.: Interactive machine learning (iML). Informatik Spektrum **39**(1), 64–68 (2016)
27. Holzinger, A.: Interactive machine learning for health informatics when do we need the human in the loop. Brain Inf. **3**(2), 119–131

Machine Learning for Family Doctors: A Case of Cluster Analysis for Studying Aging Associated Comorbidities and Frailty

František Babič[1(✉)], Ljiljana Trtica Majnarić[2,3], Sanja Bekić[2], and Andreas Holzinger[4]

[1] Department of Cybernetics and Artificial Intelligence,
Faculty of Electrical Engineering and Informatics,
Technical University of Košice, Letná 9,
042 00 Košice, Slovak Republic
`frantisek.babic@tuke.sk`
[2] Department of Internal Medicine, Family Medicine and the History
of Medicine, Faculty of Medicine, Josip Juraj Strossmayer University of Osijek,
Josipa Huttlera 4, 31000 Osijek, Croatia
`ljiljana.majnaric@mefos.hr, sanja.bekicl@gmail.com`
[3] Department of Public Health, Faculty of Dental Medicine and Health,
Josip Juraj Strossmayer University of Osijek, Crkvena 21, 31000 Osijek, Croatia
`ljiljana.majnaric@fdmz.hr`
[4] Institute for Medical Informatics/Statistic, Medical University Graz,
Auenbruggerplatz 2/V, 8036 Graz, Austria
`andreas.holzinger@medunigraz.at`

Abstract. Many problems in clinical medicine are characterized by high complexity and non-linearity. Particularly, this is the case with aging diseases, chronic medical conditions that are known to tend to accumulate in the same person. This phenomenon is known as multimorbidity. In addition to the number of chronic diseases, the presence of integrated geriatric conditions and functional deficits, such as walking difficulties, of frailty (a general weakness associated with weight and muscle loss and low functioning) are important for the prediction of negative health outcomes of older people, such as hospitalization, dependency on others or pre-term mortality. In this work, we identified *how* frailty is associated with clinical phenotypes, which most reliably characterize the group of older patients from our local environment: the general practice attenders. We have performed cluster analysis, based on using a set of anthropometric and laboratory health indicators, routinely collected in electronic health records. Differences found among clusters in proportions of prefrail and frail versus non-frail patients have been explained with differences in the central values of the parameters used for clustering. Distribution patterns of chronic diseases and other geriatric conditions, found by the assessment of differences, were very useful in determining the clinical phenotypes derived by the clusters. Once more, this study demonstrates the most important aspect of any machine learning task: the quality of the data!

Keywords: Aging comorbidities · Complexity · Data-driven clustering

© IFIP International Federation for Information Processing 2019
Published by Springer Nature Switzerland AG 2019
A. Holzinger et al. (Eds.): CD-MAKE 2019, LNCS 11713, pp. 178–194, 2019.
https://doi.org/10.1007/978-3-030-29726-8_12

1 Introduction

The general trend towards an ageing population in our western society is still unbroken, and life expectancy is even rising [1]. It poses enormous future challenges for the medical profession in general and particularly for the daily practice of family doctors (general practitioners, GP's).

Aging is often associated with the burden of chronic diseases [2]. However, some older patients with chronic diseases are more vulnerable than others in the development of negative health outcomes, including hospitalization, falls, disability, low physical and mental functioning, dependency on others and most of all to pre-term mortality [3]. The clinical signs of this vulnerable state have been identified by the gerontologists and include the following features: decreased muscle mass and strength, general weakness, slow gait speed, impaired balance and generally low activity [4]. It is termed frailty and explained by the reduced reserve capacity of various physiological systems, due to the aging process and accumulation of chronic medical conditions [5]. There are numerous diagnostic criteria for assessing frailty divided into the two principal groups: the tests focusing on physical functions of frailt and a more comprehensive approach, based on the Cumulative Deficit Model of Frailty (the Frailty Index), which accounts for a broad range of medical, cognitive, psychological, and functional deficits [6].

Frailty is a new concept, and precise theory is still missing. There are several assessment tools (or predictive models), but it may not be the best solution. One of the most widely used tests for assessing physical frailty is the Fried Phenotypic Model of Frailty [4]. It is based on the assessment of the small number of measurable components, including slow walking speed, low grip strength (measured by the hand grip dynamometer), self-reported exhaustion, unintended weight loss, and low physical activity. The credibility of this instrument relies on the fact that it has been derived from data of the large epidemiologic study. In line with the proposed evolutive and dynamic nature of frailty, this instrument also includes a prefrailty state [4, 5]. Gradation is based on disorder counting, so that 1–2 disorders indicate prefrailty and 3–5 indicate frailty [4].

In general, determinants of frailty may be divided into three groups: (1) chronic diseases and conditions associated with disability, such as falls and walking difficulties; (2) markers of physiological systems disturbation; and (3) behavioural and societal factors, such as nutrition, low socioeconomic status and low education [7]. Some chronic diseases were found in epidemiological studies as being more often associated with frailty, than some others, including diabetes, cardiovascular disease (CVD), malignant disease, chronic obstructive pulmonary disease (COPD), and chronic renal failure [2, 8, 9].

In the fundamental study on frailty, Fried et al. have shown that the risks for the development of prefrailty and frailty increase in parallel with the number of chronic diseases and the number of pathophysiology disorders, described with measures of anaemia, inflammation, micronutrient deficits, metabolic regulatory factors, body composition and neuromuscular function [5]. This study has provided a rationale for non-linearity in the development of frailty and heterogeneity of older people concerning prefrailty and frailty.

In this study, we have attempted to go a step forward and to show how prefrail and frail older people are distributed within the "naturally formed groups". These clusters are defined by the set of parameters which indicate significant pathophysiology disorders associated with frailty. We used only easily available data from General Practice (GP) electronic health records (eHRs) and patient self-reports. To identify the most important chronic diseases and geriatric disabling conditions that are associated with particular clusters, we have assessed their differences among the clusters.

Clustering methods are still rarely used in medical research. However, they have proved to be appropriate for grouping patients with chronic medical conditions and comorbidities in situations with overlapping between patients usually. The solving of such problems is like copying with the system's complexity. These methods allow insights into the "natural" grouping of patients, that is, in a case when theoretical assumptions of the way of their grouping (classification) are low. The main point which we have taken into account when deciding to use the clustering or the classification methods, was "the quantity" of theory, or how much the postulated hypothesis is convincing. The clustering methods allow a higher degree of uncertainty than the classification methods or take a larger, still unknown context. Also, they may be used when we wish to reconcile the old hypothesis to provide a new paradigm. In contrast to this, the classification methods need a strongly grounded theory and are, consequently, more predictive, than the clustering methods.

We assume that by using this methodology approach, it would be possible to identify clinical patterns that in some local population are mostly associated with frailty. The final aim is to implement the efficacious strategy for recognizing older people, GP patients, who are at increased risk for negative health outcomes. We believe that this study will add value to the requirements that screening of older people on frailty become the standard procedure in PC [10]. In general, using predictive machine learning enables the discovery of potential risk factors which provides the family doctor with information on a probable patient outcome and to react promptly and to avert likely adverse events in advance [11].

1.1   Related Work

To our knowledge, studies with similar approaches to our have not been published to date. The only study that we could find was based on using single measures of physical performance and cognitive function impairment tests, to identify clusters [12]. A similar paper, by ourselves, has been prepared on the smaller sample and with a larger set of parameters, is now under review in the journal devoted to analytical, clinical research, but is not in conflict with this conference paper.

Bertini et al. proposed two predictive frailty models for subjects older than 65 years old by exploiting information from 12 socio-clinical databases available in the Municipality of Bologna [13]. The authors take into account many diagnoses and functional conditions that may be impaired (decreased). They also noted that the frailty has not yet emerged as a well-defined clinical or social concept. Clegg et al. tried to develop and validate an electronic frailty index (eFI) using routinely available primary care electronic health record data [14]. For this purpose, they used anonymized data from a total of 931 541 patients aged 65–95. The eFI enables identification of older people who are fit, and those with mild, moderate, and severe frailty.

Both presented works represent a relatively tricky and lengthy method for daily usage. By clustering methods, we can narrow this high heterogeneity of patients (based on different combinations of many diseases and disorders) and limit it to only several subgroups. These subgroups are clusters defined according to a less number of the main features. Also, the results of some large-scale studies, including older people with multiple chronic medical conditions, may not be applied directly to the local situation, because of the large variability between populations. Thus, small datasets are appropriate for use, when there is a need that the results of research inform local healthcare providers.

2 Data Analysis

A retrospective analytical study was conducted during 2018, in the General Practice (GP) setting, in the town of Osijek, eastern Croatia, during a six-month follow-up. Only GP attendees, old 60 years and more, and not those on home care, were included in the study. Patients were assessed at their regular encounters and were recruited if they gave their written informed consent. Exclusion criteria included: acute medical conditions, worsening of chronic conditions, dementia and active chemo- or biological therapies.

Our dataset contained 261 records characterized by 10 numerical variables (89 males and 172 females). We present these variables as average and standard deviation (SD) and as the median and interquartile range (Table 1).

Table 1. Descriptive statistics of the numerical variables.

Variable	Min	Max	Median (interquartile range)	Mean (standard deviation)
Age (years)	60	90	71.00 (10.00)|	71.16 (6.43)
BMI (kg/m^2)	14.33	47.05	29.71 (5.78)	30.21 (4.71)
Waist circumference (cm)	50.00	148.00	99.00 (16.00)	99.88 (12.58)
Mid arm circumference (cm)	18.00	45.00	31.00 (5.00)	31.38 (3.66)
Fasting glucose (mmol/L)	3.60	16.20	5.60 (1.70)	6.24 (1.91)
Total cholesterol (mmol/L)	2.90	9.70	53.70 (1.70)	5.73 (1.33)
LDL cholesterol (mmol/L)	1.20	8.90	3.50 (1.50)	3.53 (1.18)
Glomerular filtration rate (mL/min/1.73 m^2)	24.00	191.00	86.00 (37.00)	87.25 (26.42)
Haemoglobin (g/L)	54.00	177.00	137.00 (15.00)	137.00 (13.49)
Haematocrit (g/L)	0.22	1.00	0.42 (0.04)	0.42 (0.05)

The quality of the data matters more than the selected model or the chosen algorithm; therefore, data-pre-processing is the most important step for all machine learning tasks.

It is rare in the practice that the number of clusters is known at the beginning of the experiments. One possibility of how to identify the most suitable number of clusters (k value in the case of the K-Means algorithm) is Elbow method [15, 16]. This method provides a graphical visualization and uses the percentage of variance explained as a function of the number of clusters.

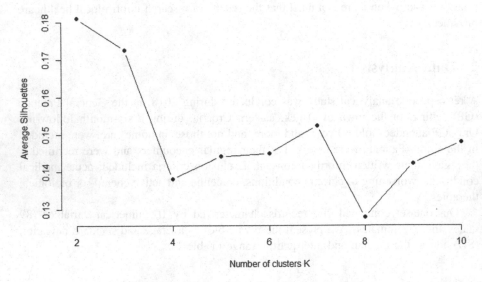

Fig. 1. The results of Average silhouette method.

The idea is to run this method on the dataset for a range of values of k (for example k from 1 to 10), and for each value of k calculate the total within-clusters sum of squared errors. On the graph, the first clusters will add much information (explain a lot of variances), but at some point, the marginal gain will drop, giving an angle in the graph [15]. This point represents the expected number of clusters. The Elbow method also has some limitation, such as the elbow is no always unambiguously identified. In this case, we can use the Average silhouette method calculating how well each object lies within its cluster. The optimal number of clusters k is the one that maximizes the average silhouette over a range of possible values for k [17]. We calculated the maximal values for the 2 or 3 possible clusters (Fig. 1). Based on the data characteristics, we finally choose the 3 as k-value.

For clustering, we selected the K-Means algorithm as a very popular technique to partitioning data sets with numerical attributes [18]. It is an unsupervised learning algorithm constructing a partition of a data of n objects into a set of k clusters. The k-value has to be specified at the beginning. Each cluster is represented by its centre. Next, we were looking for differences among them in other features (expressed by the categorical variables – gender, diagnoses, etc.)

We applied the K-Means algorithm to pre-processed data:

- The dataset contained only numerical variables like age, bmi, waist Circumference (wc), mid-arm circumference (mac), fasting glucose (glu_f), cholesterol (chol), low-density lipoprotein (ldl), glomerular filtration rate (gfr), hemoglobin (hb) and hematocrit (htc).
- All variables were normalized based on the z-score standardization (transforms all the variables to a mean value of 0 and a standard deviation of 1).

3   Results

We constructed 3 clusters visualized in Fig. 2, distinguished by the signs. The Principal Component Analysis generates the plot, and an eclipse is drawn around each cluster (but not represented a boundary). The data points are plotting according to the first two principal components coordinate.

The centres of the clusters are characterized by the relevant value of each input variable, see Table 2.

Table 2. The clusters centres.

Number of cases	Clusters	age	bmi	wc	mac	glu_f	chol	ldl	gfr	hh	htc
66	1	67.560	30.950	100.154	32.467	5.619	6.798	4.469	96.275	139.835	0.437
92	2	71.113	34.641	112.968	34.145	7.900	4.650	2.556	105.758	143.161	0.435
103	3	74.213	27.031	92.134	28.875	5.806	5.455	3.306	169.028	131.129	0.402

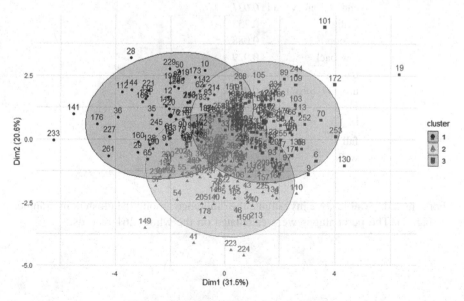

Fig. 2. The plot of 3 constructed clusters.

We evaluated the differences between clusters. We started with the proportion of prefrail and frail vs. non-frail patients in our dataset. Cluster 2 contains the highest number of non-frail patients (51). On the other hand, the numbers of prefrail or frail patients in the first and second clusters are relatively similar. The highest number of prefrail patients characterizes the third cluster. The gender radio of cluster1 is relatively balanced (1:1); the next two have a different one 2.5:1.

The Kruskal-Wallis test by rank is a non-parametric alternative to the one-way ANOVA test when we have more than two groups [19, 20]. As the p-value is less than the significance level 0.05, we can conclude that there are significant differences between the clusters (Table 3). For this purpose, we used the following list of variables:

- The number of diagnoses of chronic diseases (som com).
- Selected diagnoses of chronic diseases: diabetes mellitus (dm), chronic obstructive pulmonary disease (copd) or asthma, cardio-vascular disease (cvd), including coronary heart disease (chd) or cardio-vascular disease or peripheral arterial disease (pad), malignant disease (malig), osteoporosis (osteop), low back pain (low back), osteoarthritis (oa).
- Geriatric syndromes other than frailty: urogenital disease or urinary incontinence (urogenit incont), visus impairment (visus), hearing loss (hear), falls (with or without bone fracture) (fall nf, f), walk difficulties (walk).

Table 3. Differences between clusters - Kruskal-Wallis test by rank.

Variable (clusters)	Kruskal-Wallis rank sum test
som com	*0.003*
dm	*0.00000001*
copd_asthma	0.690
chd_coron_cv_pad	*0.021*
malig	0.753
osteop	0.088
low back	0.929
oa	0.254
urogent_incont	0.449
visus	0.609
hear	*0.045*
fall	0.083
walk	*0.025*

For selected variables, we investigate the difference among the clusters in graphical form (Fig. 3). The percentages were calculated on the whole 261 records.

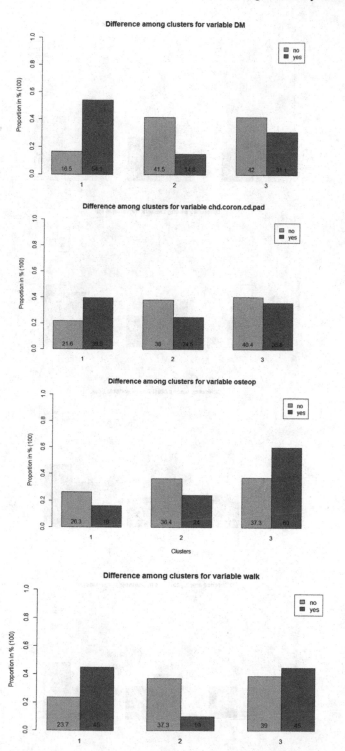

Fig. 3. Particular chronic medical conditions among clusters.

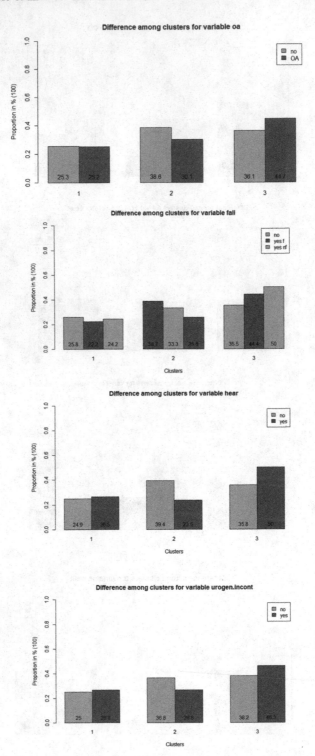

Fig. 3. (*continued*)

The proportion of chronic diseases and various impairments also differs among the clusters (Figs. 4 and 5).

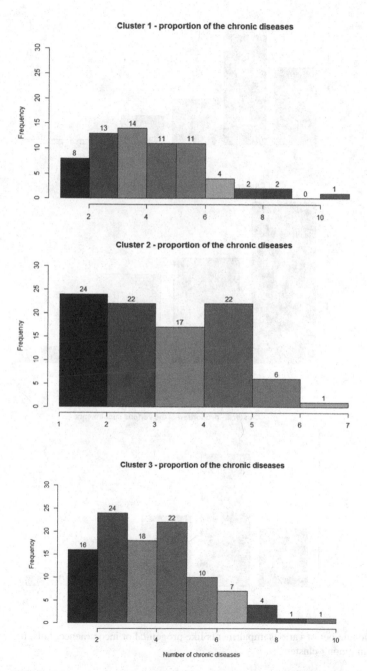

Fig. 4. The number of diagnoses of chronic diseases among clusters.

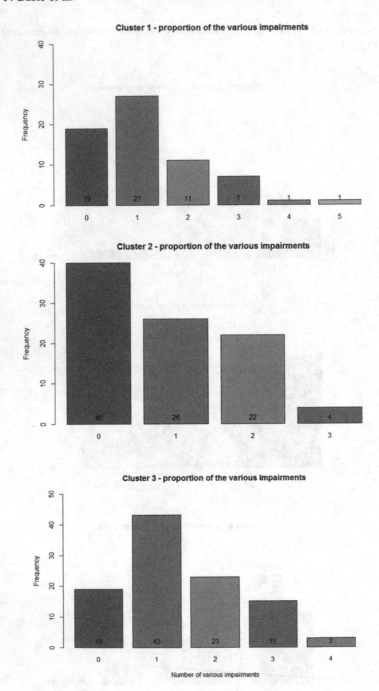

Fig. 5. The number of various impairments like urogenital or incontinence, falls, hear, walk and chronic pain among clusters.

We used the Kruskal-Wallis test also for an evaluation of the difference between the three target categories of the patients: prefrail, frail, and non-frail (Table 4).

Table 4. The difference between prefrail, frail, and non-frail patients.

Variable (prefrail/frail vs non-frail)	Kruskal-Wallis rank sum test
som com	0.0003
dm	0.954
copd_asthma	0.315
chd_coron_cv_pad	0.065
malig	0.527
osteop	0.005
lumb	0.127
oa	0.006
urogent_incont	0.202
visus	0.891
hear	0.012
fall	0.002
walk	0.00000000006

4 Discussion

We used a small set of laboratory and anthropometric measures, to describe the participants' health status, and the clustering procedure, to identify major clinical phenotypes within the group of older PC patients from the local community (Tables 1 and 2). Surprisingly, the values of the collected parameters have shown a large degree of diversity (Table 1). On the other hand, this characteristic is in line with non-linear age-related dysregulation of multiple body systems, which stays is in the background of the clinical expression of multimorbidity [3, 5].

The three clusters have been identified, each containing 66, 92, and 103 patients (Table 2). They overlap to some degree with each other, which is usual in people with multiple comorbidities and accounts for their exceptional heterogeneity (Fig. 2) [2, 3].

These characteristics, in combination with heterogeneity and overlapping, are emphasized by the distribution patterns of prefrail and frail patients in the clusters. Namely, in all three clusters, there are also non-frail, prefrail, and frail patients, but presented with different proportions, their patterns depending on characteristics of patients in the clusters.

Analysis of the clusters has shown that patients in the clusters No. 1 and No. 2 are of a lower age than those in the cluster No. 3 (67.5, 71.1 and 74.2 years, respectively), and characterized with high values of bmi and wc measures (30.95 vs 34.64 and 100.15 vs 112.96), which according to the knowledge indicate obesity, in the case of the cluster No. 2, even extreme obesity (Table 2) [21].

These two clusters have similar proportions of prefrail and frail vs. non-frail patients. This ratio is somewhat highe in cluster No. 1. Despite these similarities, patients in these two clusters also show some important differences, such as significantly higher proportions of patients diagnosed with diabetes and CVD, in the cluster No. 1, which according to the evidence are two conditions strongly associated with both, obesity and frailty (Table 3) (Fig. 3) [22, 23]. This fact can explain a higher proportion of patients in the cluster No. 1 who have reported subjective difficulties in a walk, indicating a higher level of physical disability of patients in this cluster, compared to the cluster No. 2 (Table 3) (Fig. 3). Some recently published papers highlight the role of measuring gait speed in the screening of patients with CVD who are in particular vulnerability for negative outcomes of hospitalization and surgery [24].

The differences between the clusters No. 1 and 2 may also be associated with differences in participation of males vs. females. The participation of the males in the cluster No. 1 is relatively higher, than in cluster No. 2, and compared to the whole sample. Regarding these results, it is known that women, in general, gain obesity, and subsequently diabetes and CVD, later than men in the life course, yet in the age after menopause [25]. This fact may explain our results that women, who make a prevalent part of patients in cluster No. 2, are characterized by extreme obesity (Table 2). But they are still not dominated with diabetes, as it is the case with obese patients in cluster No. 1 (Fig. 3). Extreme obesity has been recognized as a possible cause of prefrailty and frailty, although this fact is in contradiction to what is, in general, considered under the term of frailty, including weakness, muscle loss and unintended weight loss [4, 26].

The cluster No. 3 contains the oldest patients (the prevalent age 74.2 years), who, distinctly from patients in the other two clusters, have reduced renal function (as indicated with the parameter gfr = 69.02 vs. 96.27 and 105.75) (Table 2). According to the evidence, frailty is strongly associated with progressive renal impairment, which in this study is highlighted with the highest proportion of prefrail and frail vs non-frail patients, found in cluster No. 3 (a ratio of about 2.6 vs 1.4 and 1.3), compared to the other two clusters [27].

Other characteristics of patients in this cluster, which are markedly different from those in other two clusters, include the lack of obesity (indicated with values of the parameters bmi = 27.03 and wc = 92.13) and reduced muscle mass (indicated with the parameter mac = 28.87 vs. 32.46 and 34.14) [28].

A clinical phenotype that relies on the description with these parameters can be characterized as muscle wasting and frailty, reflecting the both, highly developed levels of frailty and significantly reduced renal function [4, 29]. Other characteristics of patients in this cluster, including lower (than in other two clusters) values of the parameter hb, indicating anemia, and the parameter htc, indicating decreased blood viscosity, as well as moderate (not elevated) values of the parameters fglu and chol, indicating the metabolic status, fasting glucose and total serum cholesterol, are in line with this proposed clinical phenotype [29, 30].

When comorbidities associated with particular clusters were analyzed, a tendency for disorders accumulation has been recognized in cluster No. 3, compared to the other two clusters. It is indicated with the highest proportions, in this cluster, of patients having 3 or more diagnoses of chronic diseases (frequencies 45:46:63) (Fig. 4). This tendency is even more emphasized when integrated geriatric conditions are considered, including walking difficulties, falls, hearing loss, urinary incontinence and chronic pain (frequencies 47:52:84) (Fig. 5). Of particular disorders, the most prevalent disorders in cluster No. 3 were osteoporosis, falls, and hearing loss with significant differences and osteoarthritis and urinary incontinence as non-significant (Table 3) (Fig. 3). The findings indicating functional disorders accumulation in people of older age, with high levels of frailty expression, can find their support in the growing body of evidence [31]. What is emphasized in this study is the key role of renal function impairment in the development of higher levels of frailty expression. This message is in line with the concept of unsuccessful aging, the course of aging that has been turned towards the development of renal function decline, multimorbidity, functional deficits, and frailty [29].

The non-linearity in frailty development, on distribution patterns of chronic medical conditions in the clusters, is visible in Tables 3 and 4. Significant differences among clusters were found in diagnoses of diabetes and CVD (Table 3). When prefrail and frail vs non-frail patients were considered, significant differences were found in diagnoses of osteoporosis and osteo-arthritis. Both conditions are known to be directly associated with frailty and difficulties in a walk (Table 4). The patient group consisted of only prefrail and frail patients, in a great part resembles characteristics of the cluster No. 3, where osteoporosis, osteo-arthritis and other physical dysfunctions have shown a tendency to accumulate (Figs. 3 and 5).

5 Conclusions and Future Outlook

For medical research, we usually select a standard, already proved a machine learning method, to ensure the reproducibility of the results. What is innovative in this paper is a research approach, that is, a use of a combination of ML and statistical and graphical methods, for solving a complex medical problem. This research approach requires a significant input of a medical researcher in designing research. The clusters can be considered as a first level analytical method, the results of which can inform more consistent future models.

There is a problem in medical research of small dataset conditions, such as the need for research within one health institution or when the size of the dataset is constrained by the complexity and a high cost of large-scale experiments. To meet these constraints, many theoreticians in ML methods, try to adapt these methods to be accurate and appropriate for use in small datasets [32].

By using a set of simple parameters from the general practitioners both electronic health records and patient self-reports, together clustering method, they could discover new insights into the main clinical phenotypes of the group of older patients from the local community and their associated rates of prefrail and frail patients. It is an important problem-solving task, characterized by non-linearity and high complexity, which requires a data-driven analytical approach. The setting of the general practitioner creates an ideal place for conducting such research. It provides access to a huge amount and variety of medical data in combination with his broad implicit knowledge of the GP. Moreover, this study showed again that *data-quality* matters most.

There is an urgent need for closer collaboration between medical experts, particularly general practitioners, and machine learning experts. If the GP's want to use the full capability of machine learning, there is a need soon to include it to the daily routine workflows of the GP's. These calls for simple to use Human-AI Interaction, fostering to understand the data within the context of a medical problem and to support decision making under the constraint of increasing workload and time pressure. Consequently, such methods must be trustworthy, and this requires explainability on demand. To reach such a level of explainable medicine it needs much future research in explainability [33] and causability [34].

Acknowledgements. This work was partially supported by the Slovak Grant Agency of the Ministry of Education and Academy of Science of the Slovak Republic under grant no. 1/0493/16 and The Slovak Research and Development Agency under grants no. APVV-16-0213 and APVV-17-0550.

Ethics Statement. The study was conducted in accordance with the Declaration of Helsinki and was approved by the Ethics Committee of the Faculty of Medicine, JJ Strossmayer University, Osijek (Ethics approval number 641-01/18-01/01).

References

1. Kleinberger, T., Becker, M., Ras, E., Holzinger, A., Müller, P.: Ambient intelligence in assisted living: enable elderly people to handle future interfaces. In: Stephanidis, C. (ed.) UAHCI 2007. LNCS, vol. 4555, pp. 103–112. Springer, Heidelberg (2007). https://doi.org/10.1007/978-3-540-73281-5_11
2. Barnett, K., Mercer, S.W., Norbury, M., et al.: Epidemiology of multimorbidity and implications for health care, research and medical education: a cross-sectional study. Lancet **38**, 37–43 (2012)
3. Onder, G., Palmer, K., Navickas, R., et al.: Time to face the challenge of multimorbidity. A European perspective from the joint action on chronic diseases and promoting healthy ageing across the life cycle (JA-CHRODIS). Eur. J. Intern. Med. **26**, 157–159 (2015)
4. Fried, L.P., Ferrucci, L., Darer, J., et al.: Untagling the concepts of disability, frailty and comorbidity: implications for improved targeting and care. J. Gerontol. **59**, 255–263 (2004)
5. Fried, L.P., Qian-Li, X., Cappola, A.R., et al.: Nonlinear multisystem physiological dysregulation associated with frailty in older women: implications for etiology and treatment. J. Gerontol. Ser. A Biol. Sci. Med. Sci. **64**(10), 1049–1057 (2009)
6. Rockwood, K., Andrew, M., Mitniski, A.: A comparison of two approaches to measuring frailty in elderly people. J. Gerontol. Ser. A Biol. Sci. Med. Sci. **62**, 738–743 (2007)
7. Lang, P.O., Michel, J.P., Zekry, D.: Frailty syndrome: a transitional state in a dynamic process. Gerontology **55**, 539–549 (2009)
8. Hanlon, P., Nicholl, B.I., Dinesh, J.B., et al.: Frailty and pre-frailty in middle-aged and older adults and its association with multimorbidity and mortality: a prospective analyses of 493 737 UK biobank participants. Lancet Public Health **3**, e323–e332 (2018)
9. Nixon, A.C., Bampouras, T.M., Pendleton, N., et al.: Frailty and chronic kidney disease: current evidence and continuing uncertainties. Clin. Kidney J. **11**(2), 236–245 (2018)
10. Wallace, E., Salisbury, C., Guthrie, B., Lewis, C., Fahey, T., Smith, S.M.: Managing patients with multimorbidity in primary care. BMJ **350**, h176 (2015)

11. Hassler, A.P., Menasalvas, E., García-García, F.J., Rodríguez-Manas, L., Holzinger, A.: Importance of medical data preprocessing in predictive modeling and risk factor discovery for the frailty syndrome. BMC Med. Inform. Decis. Mak. **19**(1), 33 (2019)
12. Bandelow, S., Xu, X., Xiao, S., Hogervorst, E.: Cluster analysis of physical and cognitive ageing patterns in older people from Shanghai. Diagnostics **6**(11), 2–13 (2016)
13. Bertini, F., Bergami, G., Montesi, D., et al.: Predicting frailty condition in elderly using multidimensional socioclinical databases. Proc. IEEE **106**(4), 723–737 (2018)
14. Clegg, A., Bates, C., Young, J., et al.: Development and validation of an electronic frailty index using routine primary care electronic health record data. Age Ageing **45**(3), 353–360 (2016)
15. Ketchen, D.J., Shook, C.L.: The application of cluster analysis in strategic management research: an analysis and critique. Strateg. Manag. J. **17**(6), 441–458 (1996)
16. Kodinariya, T.M., Makwana, P.R.: Review on determining number of cluster in K-means clustering. Int. J. Adv. Res. Comput. Sci. Manag. Stud. **1**(6), 90–95 (2013)
17. Kaufman, L., Rousseeuw, P.J.: Finding Groups in Data: An Introduction to Cluster Analysis. Wiley, Hoboken (1990)
18. MacQueen, J.B.: Some methods for classification and analysis of multivariate observations. In: Proceedings 5th Berkeley Symposium on Mathematical Statistics and Probability, pp. 281–297 (1967)
19. Kruskal, W.: Use of ranks in one-criterion variance analysis. J. Am. Stat. Assoc. **47**(260), 583–621 (1952)
20. Corder, G.W., Foreman, D.I.: Nonparametric Statistics for Non-statisticians, pp. 99–105. Wiley, Hoboken (2009)
21. Rydén, L., Grant, P.J., Anker, S.D., et al.: ESC guidelines on diabetes, pre-diabetes and cardiovascular diseases developed in collaboration with the EASD: the Task Force on diabetes, pre-diabetes and cardiovascular diseases of the European Society of Cardiology (ESC) and developed in collaboration with the European Association for the Study of Diabetes (EASD). Eur. Heart J. **34**, 3035–3087 (2013)
22. Sinclair, A.J., Rodriguez-Mañas, L.: Diabetes and frailty: two converging conditions? Can. J. Diab. **40**(1), 77–83 (2016)
23. Afilalo, J., Karunananthan, S., Eisenberg, M.J., Alexander, K.P., Bergman, H.: Role of frailty in patients with cardiovascular disease. Am. J. Cardiol. **103**(11), 1616–1621 (2009)
24. Chen, M.A.: Frailty and cardiovascular disease: potential role of gait speed in surgical risk stratification in older adults. J. Geriatr. Cardiol. **12**(1), 44–56 (2015)
25. Chae, C.U., Derby, C.A.: The menopausal transition and cardiovascular risk. Obstet. Gynecol. Clin. N Am. **38**, 477–488 (2011)
26. Porter Starr, K.N., McDonald, S.R., Bales, C.W.: Obesity and physical frailty in older adults: a scoping review of intervention trial. J. Am. Med. Dir. Assoc. **15**, 240–250 (2014)
27. Cook, W.L.: The intersection of geriatrics and chronic kidney disease: frailty and disability among older adults with kidney disease. Adv. Chronic Kidney Dis. **16**(6), 420–429 (2009)
28. Karniya, K., Masuda, T., Matsue, Y., et al.: Complementary role of arm circumference to body mass index in risk stratification in heart failure. JACC Heart Fail. **4**, 265–273 (2016)
29. Walker, S.R., Wagner, M., Tangri, N.: Chronic kidney disease, frailty and successful aging: a review. J. Renal Nutr. **24**, 364–370 (2014)
30. Fried, L.P., Tangen, C.M., Walston, J., et al.: Frailty in older adults: evidence for a phenotype. J. Gerontol. A Boil. Sci. Med. Sci. **56**, 146–156 (2001)
31. Davidson, J.G.S., Guthrie, D.M.: Older adults with a combination of vision and hearing impairment experience higher rates of cognitive impairment, functional dependence and worse outcomes, across a set of quality indicators. J. Aging Health **31**, 1–24 (2017)

32. Shaikhina, T., Khovanova, N.: Handling limited datasets with neural networks in medical applications: a small-data approach. Artef Intell. Med. **75**, 51–63 (2017)
33. Holzinger, A.: From machine learning to explainable AI. In: 2018 World Symposium on Digital Intelligence for Systems and Machines (DISA), pp. 55–66 (2018)
34. Holzinger, A., Langs, G., Denk, H., Zatloukal, K., Mueller, H.: Causability and explainability of AI in medicine. Wiley Interdisc. Rev. Data Min. Knowl. Discovery (2019)

Knowledge Extraction for Cryptographic Algorithm Validation Test Vectors by Means of Combinatorial Coverage Measurement

Dimitris E. Simos[1]([✉]), Bernhard Garn[1], Ludwig Kampel[1], D. Richard Kuhn[2], and Raghu N. Kacker[2]

[1] SBA Research, 1040 Vienna, Austria
{dsimos,bgarn,lkampel}@sba-research.org
[2] National Institute of Standards and Technology, Gaithersburg, MD, USA
{kuhn,raghu.kacker}@nist.gov

Abstract. We present a combinatorial coverage measurement analysis for test vectors provided by the NIST Cryptographic Algorithm Validation Program (CAVP), and in particular for test vectors targeting the AES block ciphers for different key sizes and cryptographic modes of operation. These test vectors are measured and analyzed using a combinatorial approach, which was made feasible via developing the necessary input models. The extracted model from the test data in combination with combinatorial coverage measurements allows to extract information about the structure of the test vectors. Our analysis shows that some test sets do not achieve full combinatorial coverage. It is further discussed, how this retrieved knowledge could be used as a means of test quality analysis, by incorporating residual risk estimation techniques based on combinatorial methods, in order to assist the overall validation testing procedure.

Keywords: Combinatorial measurement ·
Cryptographic applications · Data analysis · Knowledge extraction

1 Introduction

The implementation of cryptographic algorithms is a demanding task, involving various fields of computer science and software engineering. Accordingly, the testing of cryptographic applications is a complex task, at the same time being of utmost importance, as the relevance of requirements, user expectations and standards for security and privacy grow in modern information society. Thorough testing and measurement of mission critical systems – such as medical, transportation or cryptographic systems – is of vital and crucial importance as recent studies have shown [15, 23–25].

© IFIP International Federation for Information Processing 2019
Published by Springer Nature Switzerland AG 2019
A. Holzinger et al. (Eds.): CD-MAKE 2019, LNCS 11713, pp. 195–208, 2019.
https://doi.org/10.1007/978-3-030-29726-8_13

The *Cryptographic Algorithm Validation Program* (CAVP) [19] by the *National Institute of Standards and Technology* (NIST) provides validation testing of FIPS-approved and NIST-recommended cryptographic algorithms and their individual components. Cryptographic algorithm validation is a prerequisite of *cryptographic module validation*, which is the subject of the *Cryptographic Module Validation Program* (CMVP) [20] established at NIST in 1995. The CMVP is a joint effort between NIST and the Canadian Centre for Cyber Security, a branch of the Communications Security Establishment. FIPS 140-2 [18] precludes the use of unvalidated cryptography for the cryptographic protection of sensitive or valuable data within Federal systems in USA.

As of this writing, the CAVP tests block ciphers including the *Advanced Encryption Standard* (AES) [16], among others. In *The Advanced Encryption Standard Algorithm Validation Suite* (AESAVS) [1] the testing requirements for different modes of implementations of the AES algorithm are specified.

Recently, the Secretary of Commerce approved Federal Information Processing Standards Publication (FIPS) 140-3, *Security Requirements for Cryptographic Modules* [22], which supersedes FIPS 140-2 and will come effective on September 22, 2019. FIPS 140-3 aligns with ISO/IEC 19790:2012(E) [6] and includes modifications of the Annexes that are allowed to the CMVP, as a validation authority. As of this writing, the corresponding documents have not been released yet.[1] In [1] it is noted that the testing performed within the AESAVS uses statistical sampling meaning that only a small number of the possible cases are tested. Nevertheless, AESAVS states to provide testing of an *implementation under test* (IUT) to determine the correctness of the algorithm implementation.

In recent years, big data analysis has become a focus of research in information technologies and information processing, reinforced by and also advancing the current interest in machine learning and artificial intelligence, see [5]. In a branch of software testing called *combinatorial testing* (CT) [12], combinatorial methods have been used to analyze *test sets* in term of *combinatorial coverage*, which can be interpreted as a means to extract knowledge. In this work, we analyze the test data used in the AESAVS in terms of combinatorial coverage. To this end, we transformed the data into an appropriate model which enabled the combinatorial analysis of the test sets. The combinatorial measurement quantifies the parameter-value interactions executed during testing and in doing so provides a structural analysis of the test data. Within a software testing context, extracting the knowledge about potentially left out combinations has been used to estimate the residual risk that remains after testing [11]. Moreover, a comparison of exhaustive testing with combinatorial testing for cryptographic software [15] showed that covering arrays were able to detect all errors found in exhaustive testing, using a test set 700 times smaller. Thus, it is useful to evaluate the level of combinatorial coverage, since the CAVP tests cannot be exhaustive. Full combinatorial coverage, for an appropriate level of t, suggests a strong capacity for error detection. Available tools not only can compute these

---

[1] According to [21], NIST plans to release drafts for public comment in mid-2019 and final publication of those documents will occur by September 22, 2019.

measurements, but also have the functionality to present the results in different ways which are easily intelligible for the human eye.

Contribution. In this paper, we perform a study of the combinatorial coverage of various test sets, originating from the AES algorithm validation suite. A featured model extraction is made feasible via a transformation of the test vectors into test sets. These are then used as a basis for the analysis provided by combinatorial coverage measurement tools. We use visualization techniques that arise from the combinatorial coverage measurement to display the results and interpret the extracted knowledge as first steps towards recommendations for future software validation endeavors.

The paper is structured as follows. In Sect. 2 we give some preliminaries. We present the derived combinatorial model for our analysis in Sect. 3 and present our results in Sect. 4. We discuss implications of these findings in Sect. 5 and conclude the paper in Sect. 6.

2   Background Information

In this section, we provide some necessary preliminaries that will be used throughout the paper. We summarize important properties about AES and how its testing is specified in AESAVS in Sect. 2.1 and introduce CT, including employed combinatorial concepts, in Sect. 2.2.

2.1   AES and AESAVS

AES. The AES algorithm [16] is a symmetric block cipher that can encrypt and decrypt data. The AES algorithm is capable of using cryptographic keys of 128, 192, and 256 bits to encrypt and decrypt data in blocks of 128 bits. NIST has approved several modes of the approved block ciphers in a series of special publications [17].

AESAVS. The Advanced Encryption Standard algorithm validation suite [1] is designed to test the following modes of operation [17]:

- ECB, which stands for electronic codebook mode
- CBC, which stands for cipher block chaining mode
- OFB, which stands for output feedback mode
- CFB, which stands for cipher feedback mode with the following variants:
  - CFB1 (CFB, where the length of the data segment is 1 bit, $s = 1$)
  - CFB8 (CFB, where the length of the data segment is 8 bits, $s = 8$)
  - CFB128 (CFB, where the length of the data segment is 128 bits, $s = 128$)
- Counter (Counter mode is tested by selecting the ECB mode)

Note that it is not necessary for validation for every mode implemented to support the same key sizes and ciphering directions [1]. To initiate a validation process of the AESAVS, a vendor submits an application to an accredited laboratory requesting the validation of their implementation. The AESAVS is designed for testing of an IUT at locations remote to the AESAVS using communications via REQUEST and RESPONSE files. The test data is provided to an IUT in REQUEST files. The IUT processes this data and creates a corresponding RESPONSE file, which in turn will be verified.

AESAVS specifies three categories of tests: the *Known Answer Test* (KAT), the *Multi-block Message Test* (MMT), and the *Monte Carlo Test* (MCT). The KAT category is further split into four types: GFSbox, KeySbox, Variable Key and Variable Text. The MMT is designed to test the ability of the implementation to process multi-block messages, which may require chaining of information from one block to the next. For each supported mode, ten messages are supplied with lengths of i times the blocklength, for $1 \leq i \leq 10$. Each MCT ciphers 100 pseudorandom texts, where these texts are generated using an algorithm depending on the mode of operation being tested.

2.2   CT and CCM

CT [12] is an efficient black-box software testing methodology for effective software testing at lower cost. It is based on an *input parameter model* (IPM) of the *system under test* (SUT[2]) that models its input or configuration space, by identifying finitely many parameters that can take finitely many values each [4]. In CT, the defining property of t-way test sets is the coverage of all *t-way interactions* of parameter-value assignments for any combination of t parameters, for a specific value of t. Informally, a t-way interaction can be described as a parameter value assignment for exactly t parameters. The key insight underlying the empirically observed effectiveness of CT results from a series of studies by NIST [2, 3, 7–10, 25]. NIST research showed that most software bugs and failures are caused by one or two parameter interactions, with progressively fewer by three or more. These findings have important implications for software testing, because it means that testing these few parameter-value combinations can provide strong assurances. Based upon that, a hypothesis has been formulated – which is referred to as the *interaction rule* – stating that most failures are induced by single factor faults or by the joint combinatorial effect (interaction) of two factors, with progressively fewer failures induced by interactions between three or more factors [12].

CT methods can also be applied to an existing *legacy test sets*, where an existing test set is used as a basis and analyzed in terms of combinatorial coverage. Subsequently, should higher or complete t-way coverage be desired than exhibited in the legacy test set, it is possible to create additional tests specifically covering those missing interactions. The union of all test cases coming from the legacy test set and the newly created ones then achieves the desired coverage properties. This approach is an alternative to creating combinatorial test sets newly from scratch.

---

[2] In this paper, we use the terms SUT and IUT interchangeably.

Measuring the achieved level of combinatorial coverage can help in estimating the degree of risk that remains after testing; meaning that if a high level of coverage has been achieved (e.g., more than 90%), then presumably the risk is small, but if the coverage is much lower, then the risk may be substantial [11].

To address the need for such measurements, NIST has developed suitable methods and tools to quantify the achieved combinatorial overage of test sets [11]. We briefly describe combinatorial coverage by means of an example and refer the reader to [12] for further information. Consider given an SUT that is modelled by five binary parameters A,B,C,D,E that can take the values 0 or 1. A 3-way interaction for this SUT is specified by a combination of three of the five parameters, together with a specification of a value for each parameter, e.g. $(A = 0, B = 1, E = 1)$ is one 3-way interaction. In total, for such an IPM, there are $2^3 \cdot \binom{5}{3} = 80$ different 3-way interactions. Consider now a test set comprised of the following four test vectors:

	A	B	C	D	E
test_1:	1	1	1	1	0
test_2:	1	0	1	0	1
test_3:	0	0	0	1	1
test_4:	0	1	1	0	1

We see that the 3-way interaction $(A = 0, B = 1, E = 1)$ is *covered* by test_4, i.e. the parameters A,B,E take the values 0,1,1 in this test vector, respectively. From the overall 80 3-way interactions for this IPM, the four test vectors cover 39 different 3 way interaction, in other words, the combinatorial coverage measurement of these vectors yields a 3-way coverage (also called *total 3-way coverage* in [11]) of 48.75%. To summarize, to perform combinatorial coverage measurement, one requires a test set together with an IPM against which we can measure the t-way coverage of the test set.

3   Modelling and Measuring Combinatorial Coverage of AESAVS Test Data

Our analysis concerns, for a given test set file of the AESAVS, the achieved combinatorial coverage of the binary-transformed extracted hex-values of the given keys in the individual test vectors. We start with an example for the data extraction, before we detail how complete files containing test data are transformed and analyzed.

The test data for a specific configuration (category of test, mode of operation and key size) are provided in REQUEST files. The RESPONSE files contain the same data as the REQUEST files with the addition of the ciphertext for encryption or plaintext for decryption. The generic structure of a single test vector in a RESPONSE file is as follows:

– an AES key of size 128, 192 or 256 bits, denoted by KEY, which is to be used for encryption or decryption. The mode of operation is further encoded into the filename of the test data;

- an initialization vector (if applicable to the mode of operation), denoted by IV;
- a sample plaintext, denoted by PLAINTEXT;
- the corresponding ciphertext, denoted by CIPHERTEXT;
- where the order of plaintext/ciphertext or ciphertext/plaintext indicates the ciphering direction.

To make our approach more tangible, consider the test data provided in the file CBCMCT192.rsp, which specifies test vectors for CBC mode of operation with a key size of 192 bits for the category of MCT:

```
COUNT = 51
KEY = 3461389779e6debf3e58d02175a33cd46663812b73b66082
IV = 88687bf1375300b8412cf10e35f6a0b1
PLAINTEXT = 03c1f719854c00e5a16c302e25621807
CIPHERTEXT = cf5d505c14e1e272634b4ad58b6ef3d9
```

The COUNT variable simply indicates the ordinal number of the test vector in this file.

For our analysis of the test data provided by the CAVP, we focused on the combinatorial measurement of the keys used for testing. Hence, we extract the hexadecimal value that instantiates the key used in the AES implementation. This value is translated to a binary vector of length 128, 192 or 256, depending on the chosen key size.

A test set consists of test vectors for both encryption and decryption. In our analysis, we *aggregated* the binary vectors in two different files depending on their origin, e.g. one for encryption and one for decryption. For each of these two resulting sets of test vectors, we carry out a combinatorial analysis in two steps:

1. Extraction of an IPM,
2. Combinatorial coverage measurement based on this IPM.

In the first step we determine for each parameter, that models the key, the set of values it takes over the course of the whole test set being executed. Thus, we extract an IPM for the AES key, from the test vectors. These extracted models contain either 128, 192, or 256 parameters. Depending on the considered test set, these parameters are unary or binary. In the second step, we measure the combinatorial 2-, 3- and 4-way coverage, of the test vectors against the IPM obtained in the first step. In our study we used the Combinatorial Coverage Measurement Tool (CCMtool) [14], developed by NIST and the Centro Nacional de Metrologia of Mexico, for both of the just described steps. Other combinatorial coverage measurement tools include the CAmetrics tool [13] which provides for additional visualization and combinatorial metrics.

We make this process more explicit by means of the following example where we consider again the CBCVarKey192 AES validation test set. This set contains 192 test vectors for testing encryption, from which we extract the values of the keys and transform them to binary vectors, which constitutes a test set of 192 binary vectors of length 192. From these vectors we extract an IPM consisting of 192 parameters. In this specific case, the first parameter is unary, only taking the value 1 and the remaining 191 parameters are binary, taking the values 0 or 1. Finally we measure the combinatorial coverage of the 192 test vectors with regard to this IPM. The test set covers 54817 out of 72962 2-way interactions and 4626975 out of 9217660 3-way interactions, i.e. it achieves 75.15% 2-way coverage and 50.2% 3-way coverage.

4 Measurement Results

For the AES KAT Vectors, AES MCT Sample Vectors and AES MMT Sample Vectors, we measured the total 2-way through 4-way coverage, separately considering the keys for encryption of plaintext and decryption of ciphertext.

4.1 AES KAT

The vectors extracted from the AES KAT test sets are the same for the different modes (ECB, CBC, OFB, CFB1, CFB8, CFB128) when considering AES versions of the same key size. Further, the keys for encryption and decryption are the same. Thus, we do not specify the mode when we refer to a set of test vectors, e.g. {Mode}GFSbox128 refers to CBCGFSbox128 as well as CFB1GFSbox128 and further do not distinguish between encryption and decryption. The size of each AES test set can be seen in Table 1, below.

Table 1. AES KAT test set sizes (for encryption or decryption).

	128	192	256
{Mode}GFSbox	7	6	5
{Mode}KeySbox	21	24	16
{Mode}VarKey	128	192	256
{Mode}VarTxt	128	128	128

Now, from the AES test sets we extracted the following IPMs:

- IPM({Mode}GFSbox): 1^{128} (all unary)
- IPM({Mode}KeySbox): $1^1, 2^{127}$ (first parameter unary, others binary)
- IPM({Mode}VarKey): 2^{128} (all binary)
- IPM({Mode}VarTxt): 1^{128} (all unary)

The results of our coverage measurement are depicted in Fig. 1 to give an comprehensive overview. Moreover, in Table 2 we detail the results of the 3-way coverage measurement for the different test vectors of length 128.

The results of the coverage measurement visualized in Fig. 1 need to be interpreted carefully. For the case of {Mode}VarTxt and {Mode}GFSbox, the coverage is 100%, simply because the IPM consists only of unary parameters. The vectors of the test sets where the extracted IPMs are not trivial, achieve lower t-way coverage.

Table 2. 3-way coverage of vectors for 128 length against extracted IPM.

	Extracted IPM	# tuples	# tuples covered	Coverage %
IPM({Mode}GFSbox128)	1^{128}	341376	341376	100%
IPM({Mode}KeySbox128)	2^{128}	2731008	2575694	94.3%
IPM({Mode}VarKey128)	$1^1, 2^{127}$	2699004	1357503	50.3%
IPM({Mode}VarTxt128)	1^{128}	341376	341376	100%

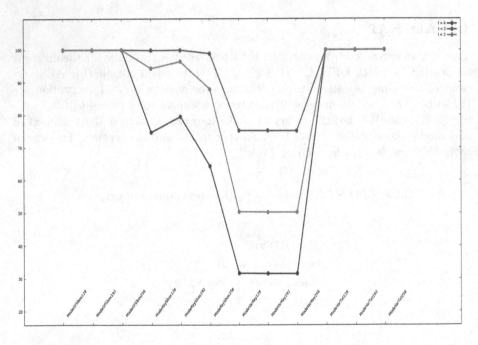

Fig. 1. 2-way, 3-way and 4-way coverage of AES KAT test sets.

4.2   AES MCT

The vectors extracted from the MCT test sets contained 200 vectors for each mode (ECB, CBC, OFB, CFB1, CFB8, CFB128), which are split into two test sets for encryption and decryption, as before. Again, we extract the values for the keys from these vectors. For different modes the keys are instantiated differently and also the keys in the test vectors for encryption differ from the keys in the test vectors for decryption. For all modes and both test sets - for encryption and decryption - the IPMs extracted from the sets of keys consist of only binary parameters. Figures 2 and 3 show the results of our coverage measurements for 2-way, 3-way, and 4-way coverage. On the x-axes we denote the various AES modes and the key sizes and on the y-axes the percentage of t-way coverage. The figures show that for both, encryption and decryption, the AES keys achieve full 2-way coverage and almost full 3-way coverage (the lowest percentage across all modes and sizes being 99.9994% for encryption, and 99.9996% for decryption). The keys also have good 4-way coverage, staying above 99.80%, except for the case of CFB128MCT128 achieving 99.77% 4-way coverage.

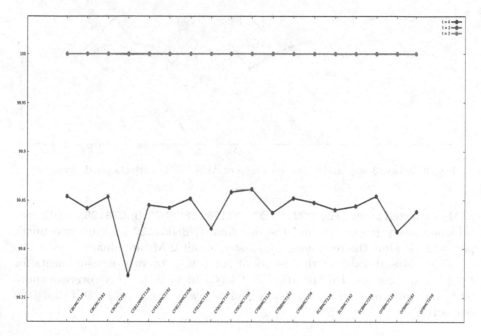

Fig. 2. 2-way, 3-way and 4-way coverage of AES MCT test sets for encryption.

4.3   AES MMT

The vectors extracted from the MMT test sets contain 20 vectors, where again for each mode (ECB, CBC, OFB, CFB1, CFB8, CFB128) the test vectors are split in two sets for encryption and description. As before, for different modes the keys are instantiated differently and the keys for encryption differ from the keys for decryption. When extracting the IPMs from these test sets, we retrieve IPMs containing mostly binary parameters, but some IPMs extracted from sets of test vectors also contain unary parameters. To be more specific, from the encryption test sets, the IPMs extracted from the vectors for the modes CFB1MMT192, ECBMMT256, CBCMMT128, CBCMMT256 and CFB128MMT192 contain one unary parameter, while the remaining ones are binary; and for decryption the

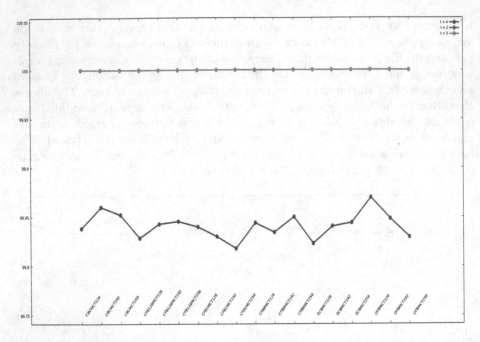

Fig. 3. 2-way, 3-way and 4-way coverage of AES MCT test sets for decryption.

IPMs extracted from CFB1MMT192, ECBMMT128, CBCMMT128, CFB128MMT192 contain one unary parameter and the one from CFB8MMT128 contains two unary parameters, while the remaining parameters in all IPMs are binary.

Figures 4 and 5 depict the results of our t-way coverage measurements for $t \in \{2, 3, 4\}$, showing that the MMT test sets achieve high 2-way coverage above 90%, but only medium to low 3-way and 4-way coverage below 80% and 50% respectively.

5   Discussion Related to Testing

The combinatorial coverage measurement analysis in the previous section shows that some of the extracted and transformed keys from the AESAVS test sets do not exhibit full t-way combinatorial coverage for some values of t. This finding has some implications for the currently specified testing requirements in AESAVS.

First, we already pointed out in the introduction that in the AESAVS document [1], it is noted that the testing performed within the AESAVS uses statistical sampling to generate the test data. With our measurement approach, we are now able to assess the result of the *statistical sampling* with respect to the key space used to generate the test sets in terms of the achieved combinatorial t-way coverage.

Second, some works in the software testing literature have linked achieved combinatorial t-way coverage to the residual risk that remains after testing [11],

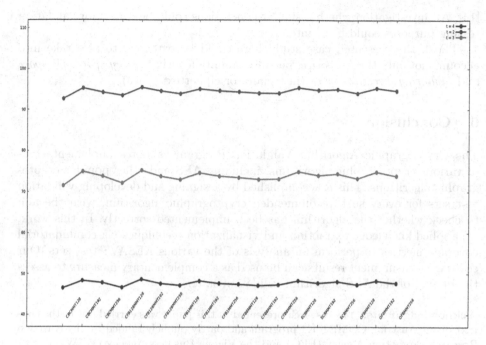

Fig. 4. 2-way, 3-way and 4-way coverage of AES MMT test sets for encryption.

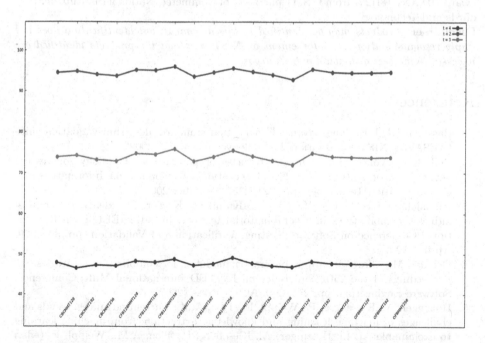

Fig. 5. 2-way, 3-way and 4-way coverage of AES MMT test sets for decryption.

[14]. An investigation whether similar conclusions could be drawn for validation testing purposes could be of interest.

Third, the presented case study here could be extended to also take into account not only the *key space*, but also simultaneously the *key space*, *IV space* and *ciphering direction space* (i.e., plain- or ciphertext space).

6   Conclusion

The Cryptographic Algorithm Validation Program validates implementations of various cryptographic algorithms, including AES and other popular cryptographic algorithms. This is accomplished by designing and developing validation test sets for every such recommended cryptographic algorithm, with the aim to check whether the algorithm has been implemented correctly. In this work, we applied knowledge extraction and visualization techniques via combinatorial coverage metrics to perform an analysis of the various AESAVS test sets. Our coverage measurement results can be used as a complementary measure to assess the quality of the AES algorithm validation suite.

Acknowledgments. The research presented in this paper was carried out in the context of the Austrian COMET K1 program and partly publicly funded by the Austrian Research Promotion Agency (FFG) and the Vienna Business Agency (WAW).

Moreover, this work was performed partly under the following financial assistance award 70NANB18H207 from U.S. Department of Commerce, National Institute of Standards and Technology.
Disclaimer: *Products may be identified in this document, but identification does not imply recommendation or endorsement by NIST, nor that the products identified are necessarily the best available for the purpose.*

References

1. Bassham III, L.E.: The advanced encryption standard algorithm validation suite (AESAVS). NIST Information Technology Laboratory (2002)
2. Bell, K.Z., Vouk, M.A.: On effectiveness of pairwise methodology for testing network-centric software. In: 2005 International Conference on Information and Communication Technology, pp. 221–235, December 2005
3. Ghandehari, L.S.G., Lei, Y., Xie, T., Kuhn, R., Kacker, R.: Identifying failure-inducing combinations in a combinatorial test set. In: 2012 IEEE Fifth International Conference on Software Testing, Verification and Validation, pp. 370–379, April 2012
4. Grindal, M., Offutt, J.: Input parameter modeling for combination strategies. In: Proceedings of the 25th conference on IASTED International Multi-Conference: Software Engineering, pp. 255–260. ACTA Press (2007)
5. Holzinger, A., Kieseberg, P., Weippl, E., Tjoa, A.M.: Current advances, trends and challenges of machine learning and knowledge extraction: from machine learning to explainable AI. In: Holzinger, A., Kieseberg, P., Tjoa, A.M., Weippl, E. (eds.) CD-MAKE 2018. LNCS, vol. 11015, pp. 1–8. Springer, Cham (2018). https://doi.org/10.1007/978-3-319-99740-7_1

6. ISO/IEC JTC 1/SC 27: Information technology - Security techniques - Security requirements for cryptographic modules (2012). https://www.iso.org/standard/52906.html. Accessed 01 Apr 2019
7. Kuhn, D.R., Kacker, R.N., Lei, Y.: Estimating t-way fault profile evolution during testing. In: 2016 IEEE 40th Annual Computer Software and Applications Conference (COMPSAC), vol. 2, pp. 596–597, June 2016
8. Kuhn, D.R., Okum, V.: Pseudo-exhaustive testing for software. In: 2006 30th Annual IEEE/NASA Software Engineering Workshop, pp. 153–158, April 2006
9. Kuhn, D.R., Reilly, M.J.: An investigation of the applicability of design of experiments to software testing. In: 27th Annual NASA Goddard/IEEE Software Engineering Workshop 2002 Proceedings, pp. 91–95, December 2002
10. Kuhn, D.R., Wallace, D.R., Gallo, A.M.: Software fault interactions and implications for software testing. IEEE Trans. Softw. Eng. **30**(6), 418–421 (2004)
11. Kuhn, D.R., Kacker, R.N., Lei, Y.: Combinatorial coverage as an aspect of test quality. CrossTalk **28**(2), 19–23 (2015)
12. Kuhn, D., Kacker, R., Lei, Y.: Introduction to Combinatorial Testing. Chapman & Hall/CRC Innovations in Software Engineering and Software Development Series. Taylor & Francis (2013)
13. Leithner, M., Kleine, K., Simos, D.E.: CAmetrics: a tool for advanced combinatorial analysis and measurement of test sets. In: 2018 IEEE International Conference on Software Testing, Verification and Validation Workshops (ICSTW), pp. 318–327 (2018)
14. Mendoza, I.D., Kuhn, D.R., Kacker, R.N., Lei, Y.: CCM: a tool for measuring combinatorial coverage of system state space. In: 2013 ACM/IEEE International Symposium on Empirical Software Engineering and Measurement, pp. 291–291. IEEE (2013)
15. Mouha, N., Raunak, M.S., Kuhn, D.R., Kacker, R.: Finding bugs in cryptographic hash function implementations. IEEE Trans. Reliab. **67**(3), 870–884 (2018)
16. National Institute of Standards and Technology: ADVANCED ENCRYPTION STANDARD (AES) (2001). https://nvlpubs.nist.gov/nistpubs/FIPS/NIST.FIPS.197.pdf. Accessed 01 Apr 2019
17. National Institute of Standards and Technology: Recommendation for Block Cipher Modes of Operation: Methods and Techniques (2001). https://nvlpubs.nist.gov/nistpubs/Legacy/SP/nistspecialpublication800-38a.pdf. Accessed 01 Apr 2019
18. National Institute of Standards and Technology: SECURITY REQUIREMENTS FOR CRYPTOGRAPHIC MODULES (2001). https://nvlpubs.nist.gov/nistpubs/FIPS/NIST.FIPS.140-2.pdf. Accessed 01 Apr 2019
19. National Institute of Standards and Technology: Cryptographic algorithm validation program (2019). https://csrc.nist.gov/projects/cryptographic-algorithm-validation-program. Accessed 01 Apr 2019
20. National Institute of Standards and Technology: Cryptographic module validation program (2019). https://csrc.nist.gov/projects/cryptographic-module-validation-program. Accessed 01 Apr 2019
21. National Institute of Standards and Technology: FIPS 140–3 Development (2019). https://csrc.nist.gov/projects/fips-140-3-development#schedule. Accessed 01 Apr 2019
22. National Institute of Standards and Technology: SECURITY REQUIREMENTS FOR CRYPTOGRAPHIC MODULES (2019). https://nvlpubs.nist.gov/nistpubs/FIPS/NIST.FIPS.140-3.pdf. Accessed 01 Apr 2019

23. Simos, D.E., Kleine, K., Voyiatzis, A.G., Kuhn, R., Kacker, R.: TLS cipher suites recommendations: a combinatorial coverage measurement approach. In: 2016 IEEE International Conference on Software Quality, Reliability and Security (QRS), pp. 69–73. IEEE (2016)
24. Simos, D.E., Kuhn, R., Voyiatzis, A.G., Kacker, R.: Combinatorial methods in security testing. Computer **49**(10), 80–83 (2016)
25. Wallace, D.R., Kuhn, D.R.: Failure modes in medical device software: an analysis of 15 years of recall data. Int. J. Reliab. Qual. Saf. Eng. **08**(04), 351–371 (2001)

An Evaluation on Robustness and Utility of Fingerprinting Schemes

Tanja Šarčević[ID] and Rudolf Mayer[(✉)][ID]

SBA Research, Vienna, Austria
{tsarcevic,rmayer}@sba-research.org

Abstract. Fingerprinting of data is a method to embed a traceable marker into the data to identify which specific recipient a certain copy of the data set has been released to. This is crucial for releasing data sets to third parties, especially if the release involves a fee, or if the data contains sensitive information due to which further sharing and potential subsequent leaks should be discouraged and deterred from. Fingerprints generally involve distorting the data set to a certain degree, in a trade off to preserve the utility of the data versus the robustness and traceability of the fingerprint. In this paper, we will thus compare several approaches for fingerprinting for their robustness against various types of attacks, such as subset or collusion attacks. We further evaluate the effects the fingerprinting has on the utility of the datasets, specifically for Machine Learning tasks.

Keywords: Fingerprinting · Relational databases · Data utility

1 Introduction

An increased interest in data collection, sharing and analysis has lead to the emergence of data economies, where various stakeholders gather and store data, and others consume this data to create additional value. Data is thus on the one hand a valuable asset to its owner, and therefore any type of unauthorised distribution or usage of data by a third party, violating the owner's rights and rights of the authorised buyers, needs to be prevented. In some cases, it might be required to prove ownership of the data. On the other hand, the collected data often concerns individuals. It can either be data directly containing information about individuals, such as contact or residence information, or data about the behaviour of individuals, e.g. interaction with online resources, shopping preferences. For these situations, data leakages should be detectable, respectively attributable, i.e. it should be possible to trace the initial (authorised) receiver of a certain data set. Such a mechanism can on the one hand help in litigation cases, but on the other hand can also be a preventive measure that deters malicious behaviour, at least for some potential adversaries.

© IFIP International Federation for Information Processing 2019
Published by Springer Nature Switzerland AG 2019
A. Holzinger et al. (Eds.): CD-MAKE 2019, LNCS 11713, pp. 209–228, 2019.
https://doi.org/10.1007/978-3-030-29726-8_14

Fingerprinting techniques, which can be seen as a personalised version of generic watermarks applied to a digital object, can be utilised as a mechanism enabling ownership attribution. They generally embed a pattern in the data, i.e. they distort the original data set to a certain extent. A good fingerprint should (i) be recognisable by the original owner of the data, (ii) not be detectable (and consequently, removable) by recipients of the data, (iii) be robust to intentional or unintentional modifications of the data, such as creating a subset, and (iv) should not lower the utility of the data too much.

The assumption in a fingerprinting scenario is that every recipient (e.g. a buyer) of the data has her own fingerprint attributed, therefore every copy that is fingerprinted and distributed by the owner is different from each other. By detecting the fingerprint within the dataset, the owner is able to detect the exact buyer of that instance of dataset.

Fingerprinting therefore usually relies on two steps: fingerprint insertion and fingerprint detection. In the first step, the fingerprint of a recipient is embedded into the dataset. Fingerprint detection then strives for detecting the fingerprint in a suspicious dataset in order to connect it with the recipient who distributed the dataset without authorisation (or is at least the first step in the chain from which the leakage originated). Fingerprint detection could be disrupted by (i) malicious attempts of the recipient to remove the fingerprint from the data, or (ii) by benign changes in the dataset, such as an well-intended sub-setting of the data, if only the subset is of relevance for a certain operation.

In this paper, we compare a number of popular fingerprinting algorithms for the above mentioned properties. We evaluate the robustness of the fingerprinting techniques towards various types of attacks by an adversary intending to disable the fingerprint. We then evaluate the effects of the fingerprint on the utility of the data by comparing the effectiveness of various machine learning models trained on both the original and the fingerprinted data sets.

The remainder of this paper is organised as follows. Section 2 discusses related work and introduces the fingerprinting schemes that we analyse. In Sect. 3, we describe our experiment setup and the data sets employed and, while we discuss the robustness towards attacks and the data utility aspects in our evaluation in Sect. 4. Finally, we provide conclusions and an outlook on future work in Sect. 5.

2   Related Work

Fingerprinting is, in the literature, often discussed as an extension of *watermarking*. Watermarking is an information hiding technique that allows identifying the source of digital objects by embedding secret owner-specific information into the dataset. Fingerprinting extends the functionality of watermark by providing the identification of the source of unauthorised data leakage. Fingerprint combines thus secret owner-specific and recipient-specific information embedded in a specific release of a digital object.

The concepts of fingerprinting and watermarking digital data firstly appear in domains of multimedia data and have been extensively studied over last two decades [6,7,16]. Most of these techniques were initially developed for images [15], and later extended to other modalities such as video [9] and audio [3].

Approaches for applying a watermarking scheme in other domains such as text and software have been studied as well. Techniques for watermarking text data typically exploit properties of text formatting and semantics. Watermarks are often introduced by altering the spacing between words and lines of text [14]. Other techniques rely on natural language processing and rephrasing some sentences in the text [2], thereby noticeably modifying the content, especially if more than one copy of the (differently fingerprinted) object is available.

Regarding **relational databases**, which is the focus of this work, most of the current state-of-the-art fingerprinting methods extend the watermarking technique proposed by Agrawal [1]. As mentioned above, the technique in principle contains two algorithms: watermark insertion and watermark detection.

The insertion step marks certain numerical attributes such that the least significant bits (LSBs) are altered. Thus this technique assumes that the dataset contains one or more numerical attributes. The number of LSBs available for marking is a trade-off between the robustness and imperceptibility of the mark. The insertion uses a cryptographic pseudo-random sequence generator $\mathcal{G}$, seeded by a secret key known only to the owner of the database and concatenated with the primary key attribute value of each tuple from a database. The numbers generated determine the bits to be marked, as well as the mark itself. It is computationally unfeasible to predict the next number generated by $\mathcal{G}$, thus unfeasible to guess the marking pattern without the knowledge of the owner's private key.

The detection calculates the same sequence as in the insertion algorithm, thus identifying which bits within the database should have been marked, and counts how many of them match the bits from a specific database. If the number of matches is "large", defined by a parameter called *significance level*, the database owner can suspect a leakage. The authors analysed the robustness of this technique against the number of malicious attacks: subset attacks, bit-flipping attacks, mix-and-match attack and false claim of ownership.

Li [12] extends this watermarking technique into a fingerprinting technique, by embedding different bit-strings – *fingerprints* in different releases of the data. The owner generates a fingerprint from her secret key and the recipient's identifier, using a cryptographic hash function. This way, storing a recipient-to-fingerprint pair, and entailing security management for this database, is not required. The insertion step is similar to [1], additionally embedding the generated fingerprint by an XOR function applied on the mark (called *mask*) and a selected fingerprint bit. Also the detection step is similar to [1] – it locates the bits that should have been altered and compares the matching of the extracted fingerprint with recipients fingerprints, with a τ as a parameter related to the assurance of the detection process.

In [13] a block-oriented fingerprinting scheme, inspired by a fingerprinting scheme for images from [8], is presented. In the insertion step, the LSBs of numerical values are combined into a two-dimensional matrix and separated into blocks of size $\beta \times \beta$. All blocks receive a fingerprint, the position within the block being randomly selected. The fingerprint is produced in the same manner as in [12], using the owner's secret key and the recipients's identifier as seed. If the fingerprint is shorter than the number of blocks, it might be embedded multiple times.

The detection step first tries to restore the database to be examined by filling in the original values in case of data deletion. The expected location of the fingerprint bit is computed as in the insertion step, and the bit is recorded. As the fingerprint is embedded multiple times in the dataset, if most of the detected values for a single fingerprint bit are found, the detected fingerprint is said to be found, otherwise it is regarded as not found.

The *Watermill* scheme [5, 11] further considers constraints of data alteration and treating fingerprinting as an optimisation problem. By using a declarative language the usability constraints that the fingerprinted dataset must meet are specified. One of two proposed fingerprinting strategies consists of translating the weight-independent constraints into an integer linear program (ILP) and using ILP solver to solve it. The second fingerprinting strategy is *pairing heuristics* for larger datasets where using ILP solver might not be efficient.

2.1   Fingerprinting Categorical Data

All of the previously mentioned fingerprinting techniques have one restriction in common – they are applicable only on numerical attributes since they are all bit-resetting techniques. Few solutions have been proposed for categorical data. One approach is the watermarking technique presented in [17,18], which, similar to the AK scheme, uses a pseudo-random sequence generator to choose tuples for marking, and marks categorical data by changing the values to another, also pseudorandomly chosen, value from the attribute domain. One of the requirements for the technique is the presence of the primary key in the dataset, which is together with owner's secret key used as a seed for pseudo-random sequence generator. In case of multiple categorical attributes in the dataset, the technique consists of several marking iterations, one categorical attribute at a time, where in each iteration the marking pattern of some attribute is additionally controlled by adding combination of other attributes' values to the seed of pseudo-random number generator. This method prevents the attribute removal attack, but (i) increases the complexity of the marking technique, (ii) is not suitable for database relations that need frequent updates and (iii) marks are possibly overlapping because a single attribute is marked several times. The authors do not mention possibility of extending this technique to fingerprinting technique, but claim robustness against serious attacks.

Another approach is a fingerprinting technique that incorporates the k-anonymity property into the fingerprinted data [10]. k-anonymity [19] strives to modify a dataset so that at least k data samples (individuals) become indiscernible, when considering quasi-identifying attributes. This is commonly achieved by generalising values in the dataset to a broader meaning. There are generally multiple solutions of achieving the same level of k by choosing different attributes to modify. The idea in the proposed scheme is therefore to utilise these multiple, equivalent versions of the dataset as one fingerprinted version for each recipient.

K-anonymity is applied on both categorical data and numerical, therefore this fingerprinting approach can, unlike the previous schemes, operate on categorical data in the process. However, there are also several limitations: (i) the number of available fingerprints is inherently limited to the number of different equivalent versions of achieving k-anonymity, (ii) the fingerprinted copies are generally rather different from each other, and thus certain attacks might be more feasible, (iii) the utility of the differently fingerprinted (anonymised) datasets can vary significantly, and (iv), the fingerprint can not be computed alone by the recipients identifier, but rather, a mapping of fingerprint and recipients needs to be stored, with all associated security risks.

We therefore do not consider this approach in this paper. Instead, we employ a rather simple modification of the above schemes for numerical data. We first convert the categorical data to an integer representation, by simply assigning increasing integer values to each unique categorical value (a process sometimes referred to as *label encoding* in data mining settings). We can then proceed to simply applying the fingerprinting scheme by modifying the LSBs of this numerical representation. After the modification is done, we convert the label-encoded variable back to the corresponding categorical value. This process works fine as long as the number of distinct values is a multiple of 2, and thus all modified numerical values have a corresponding categorical value. For other cases, we consider passing the modified value through a *modulo* function before the transformation to a categorical value. This ensures syntactical correct values in the dataset, but introduces potential issues with detecting the fingerprint, where a different numeric value might be expected than the one resulting from the modulo function. We will study the effects of these on the data utility as well as on the robustness of the fingerprint in our evaluation.

3   Experiment Setup

In this section, we describe the datasets used in our experiment, as well as the approach for the robustness and utility evaluation.

3.1   Datasets

For the empirical evaluation, we selected two publicly available datasets. The first dataset is the so-called *Forest Cover Type* dataset, obtained from the UCI Machine Learning repository[1]. The dataset contains measurements related to the forest cover originally obtained from US Geological Survey (USGS) and US Forest Service (USFS) data. This dataset consists of 581,012 instances, each describing a Forest Cover Type by 54 attributes, which are Integer or Binary values. The output variable to be predicted is one of seven different cover types. As binary variables can be easily treated as numerical/integer types, this dataset can thus be considered to contain numerical values only. The dataset is chosen

---

[1] https://archive.ics.uci.edu/ml/datasets/covertype.

due to its desired properties of containing multiple integer-valued attributes; further, this dataset is often used for experiments in watermarking and fingerprinting literature [1,12]. For the purpose of fingerprint insertion, one extra attribute *id* is added to serve as the primary key, since the chosen fingerprinting techniques require the presence of a primary key for fingerprint embedding. 44 out of the 54 attributes of the dataset contain binary values – to minimise the impact of the distortion introduced by the fingerprint, we use the remaining 10 integer-valued attributes for embedding.

The second dataset is the *Adult* dataset, obtained as well from the UCI Machine Learning repository[2]. This dataset contains 15 attributes in 30,162 samples (after removing samples containing missing values), where the attributes are both numerical and categorical (five continuous numerical and ten categorical). This dataset will thus be used for evaluating the effect of the simple fingerprinting technique for categorical data, as mentioned in Sect. 2.1. This dataset contains five categorical attributes that have a number of distinct values that is not a power of two, which is potentially problematic for our fingerprinting scheme because the marking algorithm may produce values out of the domain of categorical attribute. The algorithm in that case applies modulo function as an error correction step and may erase the mark.

3.2 Robustness Analysis

Fingerprinting schemes should be robust against different attacks that aim at preventing the correct detection of the fingerprint. Modifying, deleting and adding values to the fingerprinted data, which can be both benign updates and malicious attacks, can modify or erase the fingerprint. A robust fingerprinted scheme should make it difficult for an attacker to erase the fingerprint, to modify it in the way that an innocent recipient is indicted as a culprit, or to modify unmarked data such that a valid fingerprint is detected.

We will analyse robustness against different attacks using robustness measures proposed in [12].

- **Misattribution false hit** (fh^A): The probability of detecting an incorrect (but valid) fingerprint from fingerprinted data, i.e. a fingerprint of a different recipient.
- **False negative** (fn): The probability of not detecting the valid fingerprint from fingerprinted data.
- **False miss** (fm): The probability of failing to detect an embedded fingerprint correctly. The *false miss rate* is the sum of the false negative and misattribution false hit rates, i.e. $fm = fh^A + fn$.
- **Misdiagnosis false hit** (fh^D): The probability of detecting a valid fingerprint from data that has not been fingerprinted. This measure differs from the others as it does not measure the success of a malicious attack or benign updates on the dataset. In contrast to the ability of the detection algorithm to

---

[2] https://archive.ics.uci.edu/ml/datasets/adult.

detect the correct fingerprint from the pirated (and fingerprinted) data, the fingerprinting scheme may also, purely by chance, extract a valid fingerprint from unmarked data.

We will experimentally perform the following attacks to the fingerprinted data sets:

- **Subset attack.** In the attempt to erase the fingerprint from the dataset, the attacker may release only a subset of tuples of a fingerprinted dataset. In our attack model, we assume the attacker selects each tuple independently with probability p to include it in the pirated dataset. We also assume no other updates on dataset are applied and no other attacks performed. As each fingerprint might be embedded multiple times in a dataset, a subset attack therefore succeeds when all embedded bits for at least one fingerprint bit are deleted.
- **Superset attack.** In this attack, additional tuples to the fingerprinted data are added. This attack considers only addition of new tuples, while the original set of tuples remains unchanged. The sources of the additional tuples can be various, such as related datasets with similar attributes, artificial tuples with some semantic meaning, tuples generated from the dataset itself – or the values can be completely random. This attack can only be applied on fingerprinting schemes whose algorithms do function without the access to the original dataset (e.g. AK scheme). Otherwise it is trivial to compare the distributed dataset to the original and remove the tuples that are added by an attacker. In other cases, defending against such an attack can be helped by syntactical examination of the dataset – completely randomly generated tuples might be easy to spot. Also semantic information on the database can serve as a preliminary step in deletion of the superfluous tuples.
- **Bit-flipping attack.** The attacks mentioned above to not alter the values of the original tuples – however, an attacker may change these values in attempt to destroy the fingerprint. In a bit-flipping attack, some bits are selected and flipped. The choice of the bits is assumed random, as the attacker in our threat model is defined as having no knowledge about the fingerprint insertion scheme.
- **Additive attack.** In the additive attack [1], the attacker tries to claim the ownership of a dataset by inserting an additional fingerprint in the dataset he received. The competing ownership claims can be resolved if there exists at least one bit that both the owner and the attacker have marked, each with a different value. The way to resolve the ownership claim competition is to determine which owner's marks win, i.e. which mark has overwritten the other. The winning owner's mark was inserted later, therefore his claim of ownership is false. In case there is no overwritten mark, one approach for dealing with the false claims of ownership could be to ask both the owner and the attacker to produce the original dataset, i.e. the dataset before it was fingerprinted, and to demonstrate the presence of the fingerprint in each other's original datasets. The real owner will be able to demonstrate the presence of her fingerprint in attacker's original unlike the attacker in the owner's original.

3.3   Utility Analysis

Besides the robustness, the effect of embedding fingerprints on the data utility is
of interest. Fingerprinting datasets entails introducing distortions to the values,
which might have a negative impact on the utility of the data, similarly as it is
the case when data sensitisation methods are applied [4]. The utility of a fingerprinted dataset, for researchers, economists or other data analysts, can thus
be measured by the extent to which it preserves aggregate and statistical information. A *utility metric* quantifies the utility of a modified dataset. In general,
utility can be measured by two approaches. One approach is to utilise one or
more quantitative measures of information loss (see [4] for an overview). As these
measures do not necessarily reflect the final utility of a *machine learning model*,
a second approach is to measure the effects of the fingerprinting on the quality
of the analysis based on the data. In this paper, we employ both approaches.

For the measures on the data itself, we analyse the mean and variance of
attributes, resp. the changes of those statistical moments introduced by the
fingerprinting. We first discuss the expected behaviour on the example of the
AK scheme, while the estimation is generally similar for the other schemes.

The procedure of embedding the fingerprint generally is controlled by the
parameter γ, the number of attributes v, and the number of least significant bits
ξ. In a dataset with η tuples, on average η/γ tuples are selected for marking,
and within each of those tuples, a single bit of a single attribute is selected for
marking. As the mark value is calculated as XOR of the fingerprint bit and
pseudorandomly selected mask bit, the bit value will match the original value
on average half of the times and therefore not lead to a change. Thus, a value
of a tuple i will be selected and changed with probability $P\{L_i = 1\} = \frac{1}{2\gamma v}$.
The changes in the attributes after fingerprinting, i.e. the errors introduced,
are $\{\Delta_1, \Delta_2, ..., \Delta_\eta\}$, i.i.d. random variables. Each $\Delta_i, 1 \leq i \leq \eta$, is defined
as $\Delta_i = L_i S_i 2^{U_i}$, where $S_i \in \{-1, 1\}$, depending whether the perturbed value
is smaller or greater than the original value, both with probability 0.5, and
$U_i \in \{0, 1, ..., \xi - 1\}$ is the uniformly distributed variable representing position
of the marked bit.

The expected **mean** value of the changed attribute values is

$$\bar{x}' = (1/\eta) \sum_{i=1}^{\eta} x_i + \overline{\Delta} = (1/\eta) \sum_{i=1}^{\eta} x_i + (1/\eta) \sum_{i=1}^{\eta} \Delta_i$$

It can be shown that the expected mean error $\overline{\Delta}$ of a single attribute value is

$$E[\Delta_i] = \frac{1}{2} L_i 2^{U_i} - \frac{1}{2} L_i 2^{U_i} = 0, \forall i : 1 \leq i \leq \eta,$$

thus the expected error in attribute mean value after embedding the fingerprint is 0.

The expected variance of the perturbed attribute values is

$$V_x' = \frac{1}{\eta} \sum_{i=1}^{\eta} [(x_i + \Delta_i) - (\bar{x} + \overline{\Delta})]^2.$$

where the error in variance can be shown to be

$$\frac{1}{\eta}\sum_{i=1}^{\eta}(\Delta_i - \overline{\Delta})^2 + 2 * \frac{1}{\eta}\sum_{i=1}^{\eta}(x_i - \overline{x})(\Delta_i - \overline{\Delta}).$$

The expected error in computing the variance is thus given by

$$E[V_\Delta] \approx \frac{2^{2\xi}}{6\gamma v\xi}.$$

Also, we will employ the second approach, by directly using the fingerprinted dataset as an input to the machine learning model building, and evaluate the quality of the result. We approached the building of a classification model by applying several machine learning algorithms, namely k-nearest Neighbours (k-NN), Logistic Regression, and Random Forests. All classifiers are implemented in the Python sklearn package[3]. We present the resulting accuracy and F1-measure scores in the tables in Sect. 4.

4   Evaluation

4.1   Robustness Evaluation

Misdiagnosis False Hit. We briefly derive an expected value for this error for the AK scheme. Assume that the detection algorithm from the unmarked data extracts a potential fingerprint $f = (f_0, ..., f_{L-1})$, i.e. some bit string of length L. Furthermore, assuming that a single fingerprint bit f_i is extracted from the dataset multiple times, it is decided to be a single value (0 or 1) if that value is extracted more than $\tau\omega_i$, where ω_i is the number of times f_i is extracted. Due to the use of pseudo-random mask bits in this scheme, each time f_i is extracted, it will be extracted as 0 or 1 with a probability of 0.5, which is modelled as an independent Bernoulli trial. Once when the detection algorithm is done processing the dataset, the probability of the value of one fingerprint bit f_i of the extracted potential fingerprint f being 0 is $B(\lfloor\tau\omega_i\rfloor; \omega_i, 0.5)$, and the same probability stands for f_i being 1. Therefore, the algorithm detects the potential fingerprint with the probability $\prod_{i=0}^{L-1} 2B(\lfloor\tau\omega_i\rfloor; \omega_i, 0.5)$. The probability that the extracted fingerprint is matching one of the N valid ones equals to choosing N bit strings out of 2^L possible ones: $N/2^L$. Now the overall misdiagnosis false hit rate is

$$fh^D = \frac{N}{2^L} \prod_{i=0}^{L-1} 2B(\lfloor\tau\omega_i\rfloor; \omega_i, 0.5)$$

The misdiagnosis false hit rate is exponentially dependant on the length of the fingerprint L. The rate can be reduced by increasing L. Table 1 shows the misdiagnosis false hit rate under different values of L and $\omega_i \approx \{100, 50\} : \forall i \in \{0, ..., L-1\}$, where $N = 100$ and $\tau = 0.5$ are fixed values. We can see that for $L \gg log(N)$ we can almost completely avoid the misdiagnosis false hit ($fh^D \simeq 0$), becoming thus an important influence on the fingerprint size to be chose.

---

[3] https://scikit-learn.org/stable/ (specifically, we used version 0.20.3).

Table 1. Misdiagnosis false hit rate for exemplary fingerprint sizes

L	8	16	32	64	128
$fh^D(\omega_i = 100)$	0.7208	0.0052	2.70×10^{-7}	7.30×10^{-16}	5.31×10^{-33}
$fh^D(\omega_i = 50)$	0.9151	0.0084	7.01×10^{-7}	4.92×10^{-15}	2.42×10^{-31}

Subset Attack. For the AK Scheme, assuming that each fingerprint bit f_i is embedded ω_i times, the probability that all embedded bits for f_i are deleted is $(1 - p)^{\omega_i}$. The probability that no valid fingerprint will be detected from the dataset is then

$$fm = 1 - \prod_{i=0}^{L-1} (1 - (1 - p)^{\omega_i}).$$

We show empirically the success of a subset attack, with an attack performed on the Forest Cover Type dataset (where $\eta = 581,012$ and $v = 10$), using different parameter settings. The experimental results, for $L = 96$ and $\xi = 4$, are shown in Table 2, where every experiment is run 500 times. We can see from Table 2 that the results roughly match the theoretical expectation. The best rate of success have those attacks where the most of the tuples are deleted ($>95\%$), and the percentage of fingerprinted tuples is low (γ is high). Therefore, we can argue that the AK scheme is robust against subset attacks.

It has to be considered that as few as 1% of the tuples in this example is approximately 5,810 tuples, which for the attacker might still be an acceptable amount of tuples to release without authorisation, and to perform the successful subset attack if γ is set high enough ($\gamma \geq 25$). In those cases where p' is large, γ should be set to the smaller value, since the probability for a successful subset attack decreases when γ decreases for the same p'.

Table 2. Experimental results of subset attack success against the AK scheme, on the Forest Cover Type dataset

	$p' = 70\%$	$p' = 80\%$	$p' = 90\%$	$p' = 95\%$	$p' = 99\%$
$\gamma = 6$	0	0	0	0	0.004
$\gamma = 12$	0	0	0	0	0.5
$\gamma = 25$	0	0	0	0	1.0
$\gamma = 50$	0	0	0.002	0.194	1.0
$\gamma = 100$	0	0	0.20	0.9975	1.0

For e.g. the *block scheme* algorithm, it is crucial to have the same number of tuples and attributes, and their right sequence, in the suspicious database to be able to detect a valid fingerprint. When the attacker removes tuples, the detection scheme first has to replace these with the corresponding ones from the original dataset. In general, for this scheme the number of tuples to be removed is much smaller – with half of the dataset still available, the success rate for large values of γ reaches values comparable to the best chance presented for the AK scheme. Theoretical success of the subset attack against the block scheme is shown in Table 3.

Table 3. The probability of a successful subset attack in block scheme

	$p' = 30\%$	$p' = 40\%$	$p' = 45\%$	$p' = 50\%$
$\beta = 5$	0	0	0	1.0
$\beta = 10$	0	0	0.001	1.0
$\beta = 15$	0	6.8233×10^{-7}	0.2320	1.0
$\beta = 20$	0	9.7949×10^{-4}	0.8301	1.0
$\beta = 30$	2.0832×10^{-7}	0.2151	0.9998	1.0

The *extended AK scheme* for categorical data described in Sect. 2.1 differs from original AK scheme in an additional step in the fingerprinting embedding for categorical values. As mentioned before, we trade the strength of detection algorithm for fingerprinting categorical data successfully, as the additional operations in the fingerprint insertion phase cause errors in the detection phase that cannot be avoided. Having errors in unaffected fingerprinting scheme increases also the vulnerability of the scheme to attacks. To show this, we conducted experiments are on Adult dataset, which contains categorical data. We measure the success of a subset attack on the extended AK scheme over 500 runs and parameters set as follows: $L = 80$, $\xi = 1$, $\tau = 0.5$, $\gamma = \{3, 6, 12, 25, 50, 100\}$ and $p' = \{0.30, 0.60, 0.80, 0.90, 0.95, 0.99\}$, where p' represents the percentage of tuples that are deleted. The results are shown in Table 4.

Table 4. Experimental results of subset attack success, on the Adult dataset

	$p' = 30\%$	$p' = 60\%$	$p' = 80\%$	$p' = 90\%$	$p' = 95\%$	$p' = 99\%$
$\gamma = 3$	0.0	0.0	0.0	0.004	0.22	1.0
$\gamma = 6$	0.08	0.18	0.20	0.354	0.954	1.0
$\gamma = 12$	0.078	0.0	0.212	0.97	1.0	1.0
$\gamma = 25$	0.012	0.284	0.99	1.0	1.0	1.0
$\gamma = 50$	0.346	1.0	1.0	1.0	1.0	1.0
$\gamma = 100$	0.976	1.0	1.0	1.0	1.0	1.0

Even though the detection algorithm is able to detect the correct fingerprint from the full set of tuples, the errors introduced by the modulo operation are enhancing the success of the attack. For a comparison, the results attack success results when no error correction step has been applied, are given in Table 5. In this experiment, the fingerprint is embedded only in numerical values of the Adult dataset, otherwise using the same scheme. If an error correction step is being applied, the attack success rate is generally higher. Only for small values of γ, and if not a large portion of tuples are deleted, the scheme is robust to subset attacks.

Table 5. Experimental results of subset attack success for the case where fingerprint is marking only numerical values, on the Adult dataset

	$p' = 30\%$	$p' = 60\%$	$p' = 80\%$	$p' = 90\%$	$p' = 95\%$	$p' = 99\%$
$\gamma = 3$	0.0	0.0	0.0	0.0	0.07	1.0
$\gamma = 6$	0.0	0.0	0.0	0.0	0.11	0.98
$\gamma = 12$	0.0	0.0	0.16	0.97	1.0	1.0
$\gamma = 25$	0.0	0.11	0.98	1.0	1.0	1.0
$\gamma = 50$	0.15	0.98	1.0	1.0	1.0	1.0
$\gamma = 100$	0.97	1.0	1.0	1.0	1.0	1.0

Bit-Flipping Attack. As an example, for the Block scheme, we assume that the attacker examines every bit available for fingerprinting independently and selects it for flipping with probability p. Let us approximate the number of times that each fingerprint bit is embedded in the data to ω. For the detection algorithm to fail to recover the correct fingerprint bit, at least $(1 - \tau)\omega$ embedded bits corresponding to the single fingerprint bit f_i must be changed, i.e. more than $\omega - \lceil \tau\omega \rceil + 1$ bits must be changed. The probability that one fingerprint bit is destroyed is $B(\omega - \lceil \tau\omega \rceil + 1; \omega, p)$. The probability that the entire fingerprint will be detected incorrectly is therefore

$$fm = 1 - (1 - B(\omega - \lceil \tau\omega \rceil + 1; \omega, p)^L).$$

We run experiments on the Forest dataset both for Block scheme and AK scheme. Table 6 shows the obtained empirical results for the success of the bit-flipping attack on the block scheme where each experiment is run 100 times, while Table 7 shows the results for the AK scheme.

Table 6. Experimental results of the bit-flipping attack on the Block scheme, for the Forest Cover Type data

	p = 30%	p = 40%	p = 45%	p = 50%
$\beta = 5$	0	0	0.50	1.0
$\beta = 10$	0	0.50	0.50	1.0
$\beta = 15$	0	0.50	0.92	1.0
$\beta = 20$	0.08	0.50	1.0	1.0

Table 7. Experimental results of the bit-flipping attack on the AK scheme, for the Forest Cover Type Data

	$p = 20\%$	$p = 30\%$	$p = 40\%$	$p = 45\%$
$\gamma = 6$	0	0	0.50	0.56
$\gamma = 12$	0	0	0.50	1.0
$\gamma = 25$	0	0	0.54	1.0
$\gamma = 50$	0	0.50	0.72	1.0
$\gamma = 100$	0	0.36	1.0	1.0

We can observe that the number of bits to be flipped needs to be rather high - more than 30% of the bits available for fingerprinting, to achieve an attack with a certain guarantee of success. Such a large modification is expected to render the utility of the dataset obtain rather low. Choosing smaller β for the Block scheme or γ for the AK scheme contributes to better robustness against bit-flipping attack.

Additive Attack. We consider a scenario where the attacker tries to claim the ownership of the dataset by inserting an additional fingerprint in the received dataset. The competing ownership claims can be resolved if there exists at least one bit that both the owner and the attacker have marked, each with a different value. In that case it is possible to decide which mark appeared later, "on top of the other". In all of the considered techniques it is justified to conclude that the odds of finding such conflicting bits are low, unfortunately for the owner.

Let us take AK Scheme as an example. Suppose that the data fingerprinted by the owner is marked ω times with parameters γ, v and ξ and that the attacker performs the fingerprinting insertion algorithm with parameters γ', v' and ξ'. Under the usual probabilistic model of AK scheme's bit-marking process, the probability that a specified bit marked by original fingerprint is also marked by the attacker is the product of probabilities that the tuple containing the bit is chosen for marking $(1/\gamma')$, that the attribute containing the bit is also chosen for marking $(1/v')$ and that the specified bit is chosen $(1/\xi')$. The probability that the attacker's mark is different from the original mark is $1/2$, so that the overall probability that the specified bit is a conflict bit is $1/(2\gamma'v'\xi')$. The tuples are marked independently of each other, therefore the probability that the attack is successful, i.e. no conflicting bits are found, is

$$P\{success|\omega\} = (1 - \frac{1}{2\gamma'v'\xi'})^{\omega}.$$

For example, let the dataset have 500,000 tuples and let $\omega = 1000$. Assume that attacker wants to increase his chances of success, i.e. minimise the likelihood to overwrite an existing fingerprinted bit, thus she sets $\gamma' = 10,000$ (a rather large value, considering this means that only $1/10,000$ tuples will be marked), $v' = 10$ and $\xi' = 5$, then $P\{success|\omega\} = (1 - 10^{-6})^{1000} \approx 0.999$.

4.2 Utility

Utility Measured on the Data. For the utility evaluation on the data directly, we discuss the results of applying the AK scheme on the Forest Cover Type dataset. We choose a set of values for the parameters, specifically $\gamma = \{12, 25, 50, 100\}$, and $\xi = \{4, 8\}$. Table 8 contains recorded changes in the variance introduced by fingerprinting for each of the attributes and parameter setting. These measured values support the analysis previously made on errors in mean and variance of the attribute values in Sect. 3.3.

The error in the mean in all of the cases of this experiment was zero or very close to zero, thus only the error in the variance is presented in the table. The largest changes are, as expected, occurring when γ is small and ξ is big, i.e. in the cases where more tuples are selected and more bits of a value are available for marking. The errors in variance between cases with the same γ value and different xi vary noticeable, implying that the imperceptibility of the fingerprint is sensitive to the number of LSBs available for marking. The magnitude of the unperturbed values of the variances in general does not affect the relative error of the perturbed counterparts. The only exception is the attribute "HD-Roadways" with large original values for both mean and variance.

Table 8. Change in variance introduced by the AK fingerprinting scheme, on the Forest Cover Type dataset

| | | γ | 100 | | 50 | | 25 | | 12 | |
| | | ξ | 4 | 8 | 4 | 8 | 4 | 8 | 4 | 8 |
Attribute	Mean	Variance								
Elevation	2,959	78,391	0	+1	0	+1	+1	+5	+1	+9
Aspect	156	12,525	0	+1	0	+1	+1	+5	0	+8
Slope	14	56	0	+1	0	+3	0	+5	0	+11
HD-Hydrology	269	45,177	0	+1	0	+1	0	+2	+1	+2
VD-Hydrology	46	3,398	0	+1	0	+2	0	+4	0	+9
HD-Roadways	2,350	2,431,276	0	+10	0	+10	−1	+5	+2	+37
Hillshade-9am	212	717	0	+1	0	+2	0	+4	0	+9
Hillshade-noon	223	391	0	+1	0	+2	0	+4	0	+10
Hillshade-3pm	143	1,465	0	+1	0	+2	0	+4	0	+8
HD-Fire-Points	1,980	1,753,493	0	−2	0	+5	0	+8	+1	+30

Table 9 shows that for the Block scheme, there is also an impact on the mean values, even though still a rather marginal one. However, for the variance, the changes in values are now much more pronounced than for the AK scheme, especially when setting higher values for ξ. While some changes in variance occur in attributes that have a rather high variance, and therefore constitute only a small relative change, for attributes like *Hillshade-3pm* or especially *Hillshade-noon*, the differences are also relatively large, with an increase of 11% and 51% percent, respectively.

Table 9. Change in mean and variance introduced by fingerprinting with the Block scheme, on the Forest Cover Type dataset

		β	30		25		15		10		β	30		25		15		10	
		ξ	4	8	4	8	4	8	4	8	ξ	4	8	4	8	4	8	4	8
Attribute	Mean										Variance								
Elevation	2,959										78,391	0	+13	+1	+15	+1	+48	+1	+178
Aspect	156										12,525	0	+7	0	+12	0	+35	0	+127
Slope	14							+1			56	0	+12	0	+18	0	+48	0	0
HD-Hydrology	269										45,177	0	+6	+1	+4	+1	+13	+2	0
VD-Hydrology	46					+1		+1		+1	3,398	0	+10	0	+15	0	+38	0	+87
HD-Roadways	2,350										2,431,276	0	+3	0	+3	0	+44	−2	0
Hillshade-9am	212										717	0	+11	0	+15	0	+41	0	+8
Hillshade-noon	223									−2	391	0	+11	0	+16	0	+45	0	+200
Hillshade-3pm	143					−1		−1		−1	1,465	0	0	0	+13	0	+35	0	+160
HD-Fire-Points	1,980										1,753,493	0	0	0	−4	0	+54	0	+68

The fingerprinting scheme that deals with categorical data requires a different type of measure for data utility since mean and variance are not applicable in this case. One possible measure is the number of changes introduced by marking the data.

Table 10 shows the utility effects on the Adult dataset (which contains 30,162 tuples) introduced by the extended AK scheme for fingerprinting categorical data. The utility of numerical attributes is still measured by mean and variance, where the difference in the mean is negligible (it does not exceed 0.02 and is therefore excluded from the table). The change in variance introduced by errors for numerical attributes is also rather small, as it was the case with previously presented schemes. For each categorical attribute we count how many changes in values are introduced by the fingerprint. The Number of values that change in a single categorical attribute is approximately $30,162/(2\gamma v)$. For the presented set of parameters, the introduced total number of changes is <4% of the total number of tuples in the dataset. Due to the random nature of fingerprint insertion process, the distributions of attributes are not significantly affected.

Utility on a Machine Learning Task. In this section, we evaluate the utility of the fingerprinted data sets by comparing the effectiveness of a machine learning model on correctly predicting the target class of the datasets. As we are interested only in the changes in effectiveness as compared to the original dataset, the following results report the difference in the effectiveness scores F1 and classification accuracy (on a scale of $[0, 100]\%$).

On the Adult data set, we can conclude that the differences observed when using the Logistic Regression classifier (see Table 11 are rather minute, and would

Table 10. Change in variance and value-flips introduced by fingerprinting with the extended AK scheme, on the Forest Cover Type dataset

Attribute	γ ξ Variance	50 2	4	25 2	4	12 2	4	6 2	4
Age	173	0	0	0	0	0	0	0	+0.05
Capital Gain	54,853,968	−1	−3	−5	−11	−23	−56	−31	−67
Capital Loss	163,457	0	−1	0	−1	−1	−2	−2	−5
Hours per Week	144	0	0	0	0	0	+0.2	0	+0.3
	Value changes								
Workclass		26	19	45	45	81	90	165	165
Education		26	18	49	43	83	84	172	173
Marital Status		24	24	46	44	101	87	207	189
Occupation		23	20	44	47	75	73	148	135
Relationship		22	22	29	41	81	89	175	189
Race		19	20	47	51	87	91	160	174
Sex		12	5	19	13	39	25	77	46
Native country		19	21	45	30	94	78	173	164

Table 11. Effect on F1 score and classification accuracy with Logistic Regression, on the Adult dataset

	$\xi = 1$ F1	Accuracy	$\xi = 2$ F1	Accuracy	$\xi = 4$ F1	Accuracy	$\xi = 6$ F1	Accuracy
$\gamma = 50$	−0.15%	−0.07%	−0.02%	−0.01%	−0.07%	−0.03%	−0.03%	−0.02%
$\gamma = 25$	−0.25%	−0.14%	−0.13%	−0.06%	−0.10%	−0.06%	−0.14%	−0.06%
$\gamma = 12$	−0.46%	−0.22%	−0.27%	−0.12%	−0.12%	−0.08%	−0.39%	−0.15%
$\gamma = 6$	−0.68%	−0.38%	−0.41%	−0.22%	−0.46%	−0.19%	−0.80%	−0.33%
$\gamma = 3$	−2.12%	−1.01%	−1.08%	−0.52%	−0.75%	−0.32%	−1.33%	−0.62%

not constitute a noticeable degradation of effectiveness. The trend is the same also for other classifiers, as can be seen in Table 12 for k-NN, and Table 13 for Decision Trees, as well as with Random Forests and Gradient Boosting, which are not depicted here for brevity. In a few rare cases for the k-NN Classifier and Decision Tree Classifier the classification results obtained even improved, though by the same rather marginal order of magnitude as the observed decline.

For the Forest Cover Type dataset, the results are provided in Table 14 for Decision Trees, Table 15 for Random Forests, and Table 16 for Logistic Regression. Similar to the Adult dataset, we can note that there are very small effects on the classification accuracy and F1 score.

Table 12. Effect on F1 score and classification accuracy with KNN, on the Adult dataset

	$\xi = 1$		$\xi = 2$		$\xi = 4$		$\xi = 6$	
	F1	Accuracy	F1	Accuracy	F1	Accuracy	F1	Accuracy
$\gamma = 50$	+0.05%	+0.03%	−0.10%	−0.05%	−0.06%	−0.02%	−0.02%	+0.01%
$\gamma = 25$	−0.10%	−0.05%	+0.05%	+0.02%	+0.07%	+0.03%	−0.02%	+0.03%
$\gamma = 12$	−0.32%	−0.19%	−0.10%	−0.06%	+0.02%	+0.03%	−0.20%	−0.04%
$\gamma = 6$	−0.70%	−0.42%	−0.50%	−0.22%	−0.36%	−0.15%	−0.60%	−0.21%
$\gamma = 3$	−1.79%	−1.02%	−0.70%	−0.36%	−0.61%	−0.22%	−0.81%	−0.32%

Table 13. Effect on F1 score and classification accuracy with Decision Tree, on the Adult dataset

	$\xi = 1$		$\xi = 2$		$\xi = 4$		$\xi = 6$	
	F1	Accuracy	F1	Accuracy	F1	Accuracy	F1	Accuracy
$\gamma = 50$	+0.02%	−0.08%	+0.72%	−0.04%	+0.43%	−0.03%	−0.01%	−0.07%
$\gamma = 25$	−0.05%	−0.25%	+0.32%	−0.05%	+0.49%	−0.16%	+0.36%	−0.22%
$\gamma = 12$	−0.83%	−0.36%	−0.16%	−0.05%	+0.49%	−0.12%	−0.24%	−0.04%
$\gamma = 6$	−0.93%	−0.58%	−0.34%	−0.28%	+0.30%	−0.14%	−0.93%	−0.41%
$\gamma = 3$	−2.09%	−1.04%	−0.30%	−0.64%	−0.54%	−0.39%	+0.19%	−0.54%

In experiments with both datasets the classification accuracy and F1 score generally slightly decrease for smaller γ, i.e. by introducing more error, which is expected. However, bigger errors introduced by fingerprinting did not significantly affect the performance of any of the classifiers. This property meets the requirement of a fingerprinting scheme to be imperceptible by the users and to keep the utility of the data on the reasonable level.

Table 14. Effect on F1 score and classification accuracy with Decision Trees, on the Forest Cover Type dataset

	$\xi = 2$		$\xi = 4$		$\xi = 6$	
	F1	Accuracy	F1	Accuracy	F1	Accuracy
$\gamma = 100$	0.0%	+0.01%	+0.17%	+0.01%	+0.16%	+0.01%
$\gamma = 50$	0.0%	+0.01%	0.0%	0.0%	0.0%	+0.01%
$\gamma = 25$	−0.0%	+0.01%	+1.15%	+0.31%	+1.17%	+0.32%
$\gamma = 12$	−0.01%	−0.01%	−0.01%	0.0%	−0.01%	−0.12%
$\gamma = 6$	−0.01%	0.0%	−0.04%	−0.01%	−0.49%	−0.18%

Table 15. Effect on F1 score and classification accuracy with Random Forests, on the Forest Cover Type dataset

	$\xi = 2$		$\xi = 4$		$\xi = 6$	
	F1	Accuracy	F1	Accuracy	F1	Accuracy
$\gamma = 100$	+0.02%	−0.03%	+0.04%	−0.05%	+0.04%	+0.02%
$\gamma = 50$	+0.08%	0.0%	+0.04%	+0.6%	+0.03%	+0.04%
$\gamma = 25$	+0.09%	+0.02%	−0.09%	−0.03%	−0.05%	−0.03%
$\gamma = 12$	−0.01%	−0.0%	+0.04%	+0.03%	−0.03%	−0.05%
$\gamma = 6$	−0.06%	−0.11%	−0.01%	−0.03%	−0.0%	−0.01%

Table 16. Effect on F1 score and classification accuracy with Logistic Regression, on the Forest Cover Type dataset

	$\xi = 2$		$\xi = 4$		$\xi = 6$	
	F1	Accuracy	F1	Accuracy	F1	Accuracy
$\gamma = 100$	0.0%	0.0%	+0.01%	0.0%	−0.01%	+0.01%
$\gamma = 50$	0.02%	0.0%	+0.01%	0.0%	−0.01%	+0.01%
$\gamma = 25$	0.0%	0.0%	0.01%	0.01%	−0.05%	+0.02%
$\gamma = 12$	0.0%	0.0%	−0.02%	0.0%	−0.11%	+0.02%
$\gamma = 6$	0.0%	0.0%	−0.03%	0.0%	−0.14%	+0.03%

5   Conclusions and Future Work

In this paper, we compared a number of previously published methods for finger-printing relational databases with structured data. We then tested the robust-ness of the schemes against various types of attacks, such as sub-setting or bit-flipping. We further analysed empirically, on two benchmark datasets, how the perturbation from the fingerprint embedding affects the data utility. We followed two approaches, on the one hand computing effects directly measurable on the data, such as mean or variance, and on the other hand by measuring the effects of the fingerprint on a specific machine learning target, by comparing the achievable results on classification effectiveness. We could observe that for the selected schemes, parameters and datasets, the effects on utility of the data on the machine learning task were rather small, which is an encouraging result from a security perspective.

Table 17 illustrates the impact of common parameters on the robustness against attacks respectively on the data utility - the number of marks ω, the number of LSBs available for marking ξ, the detection threshold τ, the length of a fingerprint L, and number of recipients N. When increasing the values of these parameters, an upwards arrow denotes an increase in robustness/utility, and a downwards arrow a decrease.

Parameter ω increases the robustness against each of the presented attacks, but decreases the utility of the data, leaving the owner of the dataset the decision of how much error is it acceptable to introduce as a trade-off for the robustness. Some other parameters rather have a conflicting effect on different robustness aspects. For instance, increasing the detection threshold τ, the technique loses its robustness against subset attack, bit-flipping attack and additive attack, but on the other hand gains robustness against misdiagnosis false hit. L shows the similar effect, except that it does not have an impact on the additive attack.

Table 17. Impact of parameters on robustness against attacks resp. on data utility

↑	ω	ξ	τ	L	N
Misdiagnosis false hit	↑		↑	↑	↓
Subset Attack	↑		↓	↓	
Bit-flipping Attack	↑	↑	↓	↓	
Additive Attack	↑	↓	↓		
Utility	↓	↓			

Future work will specifically deal in more detail with approaches for fingerprinting categorical data, as this aspect has not been studied extensively in the literature so far, while categorical data (e.g. in the form of binary categories) is present in several datasets, benchmark and from real world applications. We also want to extend the analysis to other datasets, to verify that the conclusions drawn in this paper are generally valid and can be used to effectively influence the choice of parameters to obtain a secure fingerprint against the decrease in data utility.

Acknowledgments. This work was partially funded by the EU Horizon 2020 research and innovation programme under grant agreement No. 732907.

References

1. Agrawal, R., Haas, P.J., Kiernan, J.: Watermarking relational data: framework, algorithms and analysis. VLDB J. — Int. J. Very Large Data Bases **12**(2), 157–169 (2003)
2. Atallah, M., et al.: Natural language watermarking: design, analysis, and a proof-of-concept implementation. In: Moskowitz, I.S. (ed.) IH 2001. LNCS, vol. 2137, pp. 185–200. Springer, Heidelberg (2001). https://doi.org/10.1007/3-540-45496-9_14
3. Boney, L., Tewfik, A.H., Hamdy, K.N.: Digital watermarks for audio signals. In: Proceedings of the Third IEEE International Conference on Multimedia Computing and Systems, pp. 473–480. IEEE (1996)
4. Chen, B.C., Kifer, D., LeFevre, K., Machanavajjhala, A.: Privacy-preserving data publishing. Found. Trends Databases **2**(1–2), 1–167 (2009). https://doi.org/10.1561/1900000008

5. Constantin, C., Gross-Amblard, D., Guerrouani, M.: Watermill: an optimized fin-gerprinting system for highly constrained data. In: Proceedings of the 7th Work-shop on Multimedia and Security, pp. 143–155. ACM (2005)

6. Cox, I.J., Kilian, J., Leighton, F.T., Shamoon, T.: Secure spread spectrum water-marking for multimedia. IEEE Trans. Image Process. **6**(12), 1673–1687 (1997)

7. Cox, I.J., Miller, M.L., Bloom, J.A., Honsinger, C.: Digital Watermarking, vol. 53. Springer, Heidelberg (2002)

8. Das, T.K., Maitra, S.: A robust block oriented watermarking scheme in spatial domain. In: Deng, R., Bao, F., Zhou, J., Qing, S. (eds.) ICICS 2002. LNCS, vol. 2513, pp. 184–196. Springer, Heidelberg (2002). https://doi.org/10.1007/3-540-36159-6_16

9. Hartung, F., Girod, B.: Watermarking of uncompressed and compressed video. Sig. Process. **66**(3), 283–301 (1998)

10. Kieseberg, P., Schrittwieser, S., Mulazzani, M., Echizen, I., Weippl, E.: An algo-rithm for collusion-resistant anonymization and fingerprinting of sensitive micro-data. Electron. Markets **24**(2), 113–124 (2014). https://doi.org/10.1007/s12525-014-0154-x

11. Lafaye, J., Gross-Amblard, D., Constantin, C., Guerrouani, M.: Watermill: an opti-mized fingerprinting system for databases under constraints. IEEE Trans. Knowl. Data Eng. **20**(4), 532–546 (2008)

12. Li, Y., Swarup, V., Jajodia, S.: Fingerprinting relational databases: schemes and specialties. IEEE Trans. Dependable Secure Comput. **2**(1), 34–45 (2005)

13. Liu, S., Wang, S., Deng, R.H., Shao, W.: A block oriented fingerprinting scheme in relational database. In: Park, C., Chee, S. (eds.) ICISC 2004. LNCS, vol. 3506, pp. 455–466. Springer, Heidelberg (2005). https://doi.org/10.1007/11496618_33

14. Maxemchuk, N.F.: Electronic document distribution. AT&T Tech. J. **73**(5), 73–80 (1994)

15. O'Ruanaidh, J., Dowling, W., Boland, F.: Watermarking digital images for copy-right protection. IEEE Proc. Vis. Image Sig. Process. **143**(4), 250–256 (1996)

16. Petitcolas, F.A., Katzenbeisser, S.: Information Hiding Techniques for Steganogra-phy and Digital Watermarking (Artech House Computer Security Series). Artech House (2000)

17. Sion, R.: Proving ownership over categorical data. In: Proceedings. 20th Interna-tional Conference on Data Engineering, pp. 584–595. IEEE (2004)

18. Sion, R., Atallah, M., Prabhakar, S.: Rights protection for categorical data. IEEE Trans. Knowl. Data Eng. **17**(7), 912–926 (2005)

19. Sweeney, L.: K-anonymity: A model for protecting privacy. Int. J. Uncertainty Fuzziness Knowl. Based Syst. **10**(5), 557–570 (2002). https://doi.org/10.1142/S0218488502001648

Differentially Private Obfuscation of Facial Images

William L. Croft[✉], Jörg-Rüdiger Sack, and Wei Shi

Carleton University, Ottawa, ON, Canada
leecroft@cmail.carleton.ca

Abstract. The pervasiveness of camera technology in every-day life begets a modern reality in which images of individuals are routinely captured on a daily basis. Although this has enabled many benefits, it also infringes on personal privacy. To mitigate the loss of privacy, researchers have investigated methods of facial obfuscation in images. A promising direction has been the work in the k-same family of methods which employ the concept of k-anonymity from database privacy. However, there are a number of deficiencies of k-anonymity which carry over to the k-same methods, detracting from their usefulness in practice. In this paper, we first outline several of these deficiencies and discuss their implications in the context of facial obfuscation. We then develop the first framework to apply the formal privacy guarantee of differential privacy to facial obfuscation in generative machine learning models for images. Next, we discuss the theoretical improvements in the privacy guarantee which make this approach more appropriate for practical usage. Our approach provides a provable privacy guarantee which is not susceptible to the outlined deficiencies of k-same obfuscation and produces photo-realistic obfuscated output. Finally, while our approach provides a stronger privacy guarantee, we demonstrate through experimental comparisons that it can achieve comparable utility to k-same approaches in the context of preservation of demographic information in the images. The preservation of such information is of particular importance for enabling effective data mining on the obfuscated images.

Keywords: Privacy protection · Facial obfuscation ·
Differential privacy · Neural networks

1 Introduction

With the ever expanding presence of devices used to capture photos and video, visual privacy has become increasingly important. Images and video frames containing faces are routinely captured, e.g., through cameras, closed-circuit television systems [5], visual sensor networks [40] and a host of other devices and

The authors gratefully acknowledge the financial support from the Natural Sciences and Engineering Research Council of Canada (NSERC) under Grants No. RGPIN-2015-05390, No. RGPIN-2016-06253 and No. CGSD2-503941-2017.

A. Holzinger et al. (Eds.): CD-MAKE 2019, LNCS 11713, pp. 229–249, 2019.
https://doi.org/10.1007/978-3-030-29726-8_15

methods. These systems have many benefits including mitigation of crime [5,40], improved care in assisted-living [34], and useful services such as Google Street View [17]. However, despite the benefits of the legitimate applications, the potential for infringement on personal privacy must be taken seriously.

Although many systems require only visual monitoring of behaviour, identities are often captured as well [40]. In some areas, the degree of public surveillance is reaching levels where it becomes possible to profile and track much of the population [34]. In cases where visual information is disseminated to the public, such as with Google Street View, it is imperative to hide the identities of individuals before the images are published [15]. Failure to sufficiently protect privacy may allow undesirable inferences to be drawn about individuals or enable malicious activities such as voyeurism or stalking. Users of mobile devices have also expressed strong aversion to the collection of images from their mobile devices via the applications they use [31]. Even in scenarios where users willingly share images to online platforms, they have expressed concerns over who is able to view their images [22]. In a similar context, the privacy of individuals captured in the backgrounds of images uploaded to such platforms should be taken into consideration. Rich visual information from these sources combined with the great advances in machine learning approaches to facial recognition (e.g., VGGFace [35]) make the exploitation of unprotected visual data a relatively easy task. While such machine learning algorithms are no doubt beneficial in many contexts, it is essential for approaches of privacy protection to be resistant to them.

To protect the privacy of individuals, methods for hiding identity via manipulation of the data can be employed. Many methods focus on the face as it is often the most identifiable piece of information. Trivially, the face could be covered by a uniformly coloured rectangle. This destroys all information about the face, guaranteeing that it can no longer be exploited to reveal an identity. However, this also destroys a great deal of utility. In scenarios where images are shared in online platforms, users have expressed a strong aversion to the use of such rectangles with respect to the visual quality and information content of the images [28]. Less severe methods of obfuscation present trade-offs between the level of privacy attainable and the utility of the data. Preservation of utility is especially important for machine learning and data mining. Visual data can be used to learn about customers in retail environments [29] and to detect anomalous or illegal events [39]. It is therefore essential for a good method of obfuscation to preserve as much of the non-sensitive information as possible.

A number of research directions have been explored for the obfuscation of visual identity in images, e.g., pixelization, blurring, etc. [34]. While many approaches lack a formal privacy guarantee, the k-same [33] family of approaches has gained a great deal of traction, largely thanks to its guarantee that for a chosen privacy parameter k, obfuscated individuals are indistinguishable within a group of k potential true identities. While this privacy guarantee is appealing, it suffers from susceptibilities (e.g., composition attacks [16]) carried over from the disclosure control method of k-anonymity on which it is based. In this paper, we outline these susceptibilities in the context of privacy in images and propose an alternative, based on differential privacy, which addresses these susceptibilities.

1.1 Contributions and Paper Outline

Our contributions in this work are as follows:

- We examine susceptibilities of k-same obfuscation to composition attacks and background knowledge. We demonstrate how the privacy guarantee can be violated and discuss the implications this has on privacy in images.
- To address the deficiencies of k-same obfuscation carried over from k-anonymity, we propose as an alternative, the formal guarantee of differential privacy. We develop the first framework to apply differential privacy for the obfuscation of facial identity in images via generative machine learning models.
- We conduct a series of experiments to compare the quality of differential privacy to k-same obfuscation on two well-known datasets. The results of our experiments suggest that differential privacy offers a comparable level of utility in the obfuscated images to k-same obfuscation even though the privacy guarantee is improved.

We provide a review of existing work on the obfuscation of facial images in Sect. 2. We then cover the deficiencies of k-same in Sect. 3 and lay out a framework for differentially private obfuscation of images in Sect. 4. Finally, we describe our experimental comparisons and their results in Sect. 5.

2 Literature Review

Perhaps the most well-known and earliest studied alterations to images for the prevention of human recognition of faces are pixelization [21] and blurring [20]. Pixelization decreases the information conveyed in an image by dividing the image into a grid of cells and setting all pixels within each cell to a common pixel intensity. Blurring involves the addition of, typically Gaussian, noise to the image. While these methods have been successful at foiling human recognition, they have been shown to be highly ineffective against machine recognition [33].

Other ad hoc methods of privacy protection involving variations on blurring [26], warping [24], morphing [23] and face swapping [3] have been studied, however, the methods which have gained the most momentum are those which offer a formal guarantee of privacy. This trend has been reinforced by the legal and legislative demands in the broader context of the release of sensitive data [4,38]. To this end, k-same approaches have been quite successful. These approaches use an adaptation of k-anonymity [36], a concept from the field of database privacy which guarantees that an anonymized database record is linkable to at least k possible identities. The first adaptation of this concept to image obfuscation worked by aligning a set of input images on their facial features, partitioning the set into clusters of k or more similar images, and then averaging the pixels within each cluster to produce an averaged face which would replace each of the original faces in the cluster [33]. By releasing only the averaged faces, it could be guaranteed that neither human nor machine recognition could do better than identifying the cluster of identities which produced the image, thus limiting the probability of successful re-identification by an upper bound of $\frac{1}{k}$.

One issue with the original k-same averaging of pixels was poor visual quality due to inexact alignment of facial features, leading to superimposed features. The k-same-m [18] approach improved upon this by using an active appearance model (AAM) [8] to obfuscate faces. AAMs are generative machine learning models for the approximation of visual representations of a particular class of objects (e.g., human faces). A model is trained on a set of images in order to learn about visual patterns and minimize differences with respect to shape and texture between the original images and the generated output of the model. The k-same-m approach first trains an AAM and then performs the clustering and averaging process within the parameter space of the model representations of faces to be obfuscated, thus eliminating the issue of superimposed features.

More recently, generative neural networks (GNNs) have been applied for k-same obfuscation [7,32]. GNNs are machine learning models which have shown great success in the generation of visual representations of input class labels [9]. A GNN passes the input labels through a sequence of convolutional layers, transforming them into features of finer granularity at each layer until reaching a pixel-space output. A training process adjusts weights used by filters in each convolutional layer in order to learn feature representations which minimize a loss function measuring the quality of the output. When trained on a set of images using identities as class labels, a GNN is able to produce a visual approximation of an identity based on an input class vector. By providing input vectors in which k identities are specified, the GNN produces k-anonymous output.

Efforts have also been devoted to the preservation of utility in the obfuscated images. The k-same-select approach [19] proposed partitioning the input images into classes based on the information to be preserved (e.g., male and female identities) before clustering such that the images within each cluster would share the same class, thus preserving this information in the averaged version. This idea has been extended to the k-same-m model by training a different AAM for each combination over the demographic attributes of age, gender and race [10]. By using the appropriately trained AAM for obfuscation, the attributes for which it was trained can be preserved in the output. In the context of GNNs, preservation of information has been considered by designing the network architecture to allow for multiple input vectors over different types of classes [9]. This has been applied to produce obfuscated images with specific facial expressions [32].

We note that differentially private obfuscation of images has been studied in the context of noise applied to pixel intensities [13]. While achieving a strong privacy guarantee, this leads to poor visual quality in the output as the obfuscated images no longer resemble the original class of the object (e.g., a human face). To the best of our knowledge, our work is the first to study differential privacy applied to generative models for the obfuscation of facial images.

3 Weaknesses in Existing Facial Obfuscation

Given the importance of preserving privacy in images, a good method of obfuscation must assert a meaningful guarantee about the level of privacy it provides. Without such a guarantee, it is impossible to formally assess the effectiveness of the obfuscation. Empirical results may help to gain intuition on which approaches appear promising, however, without a formal guarantee to back up the results, it is impossible to assert that privacy will remain protected in untested scenarios against unknown attacks. For this reason, we focus our attention only on methods of obfuscation which offer a formal privacy guarantee.

The necessity of this restriction is underscored by the concept of parrot attacks [33]. A parrot attack uses a neural network to classify identities using labeled instances of obfuscated images as the training set. Having learned about patterns in the obfuscation during training, the network is made much more effective at defeating the obfuscation. Despite pixelization being reasonably effective against human recognition and even naive machine recognition, it can be completely defeated by a parrot attack. This formed a strong basis for the need of a formal privacy guarantee such as that provided by the k-same family.

The k-same approaches employ a privacy guarantee derived from k-anonymity [36] which asserts that the original identity for any obfuscated image is indistinguishable from at least $k - 1$ others. This guarantee is a result of the obfuscation process which draws upon clusters of k or more images to produce averaged instances as replacements for all images in each cluster. This makes it impossible for any software to achieve a better probability of re-identification than $\frac{1}{k}$.

However, the k-same guarantee relies on assumptions about the nature of the attack. In this section, we discuss these assumptions. We show why they are often unrealistic in practice, making the guarantee weaker than it appears to be.

3.1 Background Information

A well-known deficiency of k-anonymity is its susceptibility to attacks which employ background information [2]. This refers to cases where the attacker uses prior knowledge about the sensitive information to draw inferences which violate the privacy guarantee. This concept carries directly over to the k-same privacy guarantee. If, via prior knowledge, the attacker knows with certainty that some of the k individuals could not be in the obfuscated image, they can discount them from the set of k identities. An attacker could come by this knowledge in a number of ways: personal knowledge about friends and family, information scrapped from other data sources such as social media, etc. The simple combination of knowledge about the time at which an photo was taken and the approximate locations of some of the k individuals at that time can be enough to derive a proper subset of the k individuals which violates the privacy guarantee.

Contextual information in an image can often enable these types of inferences. Using signs, architecture or landscapes in an image, an attacker might recognize the location or employ software to determine it. Knowledge about locations that individuals frequent may greatly increase the probability of some possibilities over others. Similarly, if some of the k identities are known to live in different cities than where the photo was taken or worse yet, different continents, these identities become much less probable. Other cues such as accessories or clothing on obfuscated individuals may also greatly impact the probabilities accorded to the k possible identities. Since the privacy guarantee asserts that each of the k identities are equally probable, this is also in violation of the guarantee.

We note that the original k-same paper does acknowledge this vulnerability to contextual information and asserts that the privacy guarantee applies strictly to the information contained within the face, not to the image as a whole [33]. While this important distinction allows for the privacy guarantee to be upheld, it is a major restriction on the practical applicability of the k-same guarantee. Most contexts in which facial obfuscation is applied will be rich with contextual information, making the privacy guarantee much less meaningful.

3.2 Composition Attacks

Another deficiency of k-anonymity is a susceptibility to composition attacks [16]. This is a class of attacks which exploit information from multiple, potentially uncoordinated, obfuscated releases to violate the privacy guarantee. A simple instance of this is the intersection attack. An attacker first identifies the clusters in which a particular individual exists from two different releases. If the releases were uncoordinated, the clusters likely differ, allowing the attacker to take their intersection to achieve a new set with a cardinality less than k.

This attack again carries directly over to the k-same approach. Consider a scenario where an individual takes a photo which they wish to upload to social media. Privacy protection might be applied to the individual or perhaps to bystanders who were captured in the background of the photo. Should the individual decide to upload the same photo to two or more social media platforms, the issue of uncoordinated obfuscation immediately arises. An attacker needs only scrape these platforms for similar photos to apply an intersection attack.

Intersection attacks may even be effective for multiple releases from the same organization if care is not taken. For example, an individual may take consecutive photos and then upload all of them. Algorithms for k-same determine clusters based on the similarity of faces but many factors beyond facial identity (e.g., pose, angle and lighting) could impact similarity. It is therefore not unlikely that multiple images of the same individual will result in different clusters. Sequences of images uploaded in this way would be an ideal target for intersection attacks.

Most k-same approaches require each individual to appear only once in the gallery of images to be obfuscated. This prevents intersection attacks for releases from the same organization but does not protect against uncoordinated releases across multiple organizations. Furthermore, enforcing this restriction may be very challenging in practice. While the primary subject in a photo might be determined based on the account used to upload the photo, other individuals in the photo cannot be correctly identified 100% of the time. Face recognition software has not yet reached this level of accuracy. Without manual labeling, such a policy cannot be enforced. Beyond this, the restriction of one image per identity is very severe and does not match typical use cases for image sharing.

3.3   Other Difficulties

We discuss here two other difficulties that arise when using k-same obfuscation in practice. Although these difficulties do not violate the privacy guarantee, they hinder meaningful applications for k-same obfuscation in some contexts.

The first problem arises from the requirement of an input gallery of images. This may be appropriate for scenarios where batches of images are obfuscated but it is awkward to apply to cases where images are sporadically uploaded (e.g., in social media platforms). One might consider the use of a preloaded static gallery or even a dynamic gallery that gets updated as new images are uploaded. This, however, is not a good solution since identities can then participate in more than one cluster. Furthermore, if an attacker records information about identities known to be in the gallery, those identities can be discounted when an image is uploaded for a new identity. An alternative solution could rely on buffering uploaded images to form a gallery which can eventually be used to release a batch of obfuscated images. However, this necessitates a trade-off between the size of the gallery (and thus the quality of the output) and ability to deliver a timely service. In an era where users expect images to be uploaded instantly, this is not likely to be a manageable trade-off. The release of multiple batches also increases the chances of enabling composition attacks.

The second problem relates to the preservation of utility in the obfuscated output. Approaches which partition the gallery according to classes to be preserved (e.g., combinations of age and gender) place an even greater strain on the input gallery requirement. Working separately with the subset of images from each class greatly reduces the number of images available for clustering. Such an approach is not scalable for large numbers of classes that would be needed for finely grained attention to utility. In the worst case, some classes may be outliers in the overall distribution and could lack sufficient images to form a cluster. These classes would have to be merged with others in order to achieve the k-same guarantee, thus failing to achieve the desired granularity of classes.

4   Differential Privacy for Generative Models

Due to the deficiencies of the k-same privacy guarantee in practical applications of facial obfuscation, we argue that a more robust privacy guarantee is required. Following the advances in the field of database privacy, we consider the potential of differential privacy to provide a stronger privacy guarantee. In this section, we first review basic theory of differential privacy. We then adapt the privacy guarantee to fit the context of generative machine learning models for images and we formalize a framework to apply differential privacy to facial images. We discuss how the derived privacy guarantee addresses the issues identified with the k-same approach. Finally, we apply our framework to implement differentially private facial obfuscation using an AAM and a GNN.

4.1 Differential Privacy for Databases

A privacy guarantee which offers an absolute bound on re-identification risk necessitates restrictive assumptions about the attacker. This is due to the fact that it is impossible to prevent an attacker from learning about the sensitive information through means other than the obfuscated release [11]. Differential privacy recognizes this difficulty and instead adopts a privacy guarantee which limits the increase in an attacker's knowledge about the sensitive information. In the context of databases, the goal is to release aggregate information about the database while preventing that information from being exploited to derive sensitive details about the individual records. Differential privacy functions by using a *randomization mechanism* to add controlled noise to database query responses in order to release useful responses while achieving a desired level of indistinguishability between potential configurations of the database contents.

Two databases are considered to be *adjacent* if they differ by a single record. Informally, the privacy guarantee enforces that any pair of adjacent databases must be bounded within a multiplicative factor of e^ϵ (where ϵ is the *privacy parameter*) in their probabilities of producing the same noisy query response. This is often interpreted as a ratio of e^ϵ between these probabilities. With a sufficiently small ratio, similar databases have similar probability distributions over their noisy query responses, causing them to behave similarly with respect to the noisy query responses they produce. This limits the usefulness of the noisy responses as a means to distinguish between potential configurations of the database. The privacy guarantee [12] in Formula 1 formally states this requirement in terms of any pair of adjacent databases $D_1, D_2 \in \mathbb{D}$, where $\mathbb{D}$ is the set of valid database configurations, and a randomization mechanism $K : \mathbb{D} \to \mathbb{R}^n$, where $n \in \mathbb{Z}^+$.

$$\Pr\left(K\left(D_1\right) = R\right) \le e^\epsilon \Pr\left(K\left(D_2\right) = R\right) \qquad \forall R \in \mathbb{R}^n. \tag{1}$$

To achieve this privacy guarantee, the mechanism K must take into account the value of ϵ and the *query sensitivity*. The sensitivity ΔF of a query $f : \mathbb{D} \to \mathbb{R}^n$ is defined as the maximum possible L_1 distance between the query responses for any pair of adjacent databases. The guarantee can be achieved by adding to the query response a vector of n continuous random variables, each drawn independently from a Laplace distribution with $\frac{\Delta F}{\epsilon}$ as its scaling parameter [12]. The exponential decay of probability density in the Laplace distribution benefits the utility of the mechanism by limiting the expected perturbation of the query responses. Through the selection of an appropriate value for ϵ, a data custodian can control how much information is revealed about the contents of the database.

4.2 Framework for Generative Models

We now consider how differential privacy can be applied to generative models for images. A generative model can represent images of instances from specific classes (e.g., human faces) using a numeric representation which abstracts from pixel intensities. Our goal is to protect the privacy of individuals in images by modifying these numeric representations to prevent facial identification while maintaining utility and visual quality. Differential privacy is ideal for this purpose as it provides a robust guarantee against the accuracy of the inferences an attacker can make about the original data. The application of noise to the numeric representation of the model allows for the generation of photo-realistic instances of novel human faces. This avoids the significant degradation in visual quality which results from the addition of noise to pixel intensities.

When moving from the domain of databases to that of generative model representations, the concepts of adjacency and query sensitivity can no longer be applied for the configuration of a mechanism. In place of a database where each record is an individual, we have a numeric representation of a single individual. To protect sensitive data in this form, one can apply a generalization of differential privacy to arbitrary *secrets* [6], where a secret is any numeric representation of data. In our case, the secret is the generative model representation of an individual. This generalization substitutes the notion of adjacency between databases with distance between secrets. By controlling noise according to an appropriate distance metric, the privacy guarantee is adapted to ensure that similar secrets are highly indistinguishable while very different secrets remain distinguishable. For a pair of databases, the distance between them is the number of records by which they differ. For other types of secrets, the distance metric must be carefully chosen in order to provide an appropriate privacy guarantee.

The notion of distance between secrets is appropriate for the representation of images within a generative model. Any model which employs a numeric representation of images allows for the calculation of distance between images. While the exact representation of an image differs from model to model, they can generally mapped to a vector of fixed length with little difficulty. We provide details on how this concept can be applied to both AAMs and GNNs in Sect. 4.4. To develop a general framework here, we consider the representation of an image to be a vector $X \in \mathbb{R}^n$ and the randomization mechanism to be a function $K : \mathbb{R}^n \to \mathbb{R}^n$ used to produce an obfuscated instance of the image. Although the differential privacy generalization only deals explicitly with one and two-dimensional secrets [6], its generalization to an n-dimensional vector is straightforward. We therefore adapt the privacy guarantee to suit this purpose in Formula 2, using a distance function $d : \mathbb{R}^n \times \mathbb{R}^n \to \mathbb{R}$.

$$\Pr\left(K\left(X_1\right) = R\right) \leq e^{\epsilon d(X_1, X_2)} \Pr\left(K\left(X_2\right) = R\right) \qquad \forall X_1, X_2, R \in \mathbb{R}^n. \qquad (2)$$

Comparing this to Formula (1), the databases D_1 and D_2 have been replaced by secrets X_1 and X_2 and the distance function now appears in the exponent of the multiplicative factor e^ϵ. The distance between any pair of secrets acts as a coefficient to ϵ when interpreting the ratio of their probabilities. Intuitively, the meaning is that the more similar a pair of images are to each other, the harder is it to determine which of them led to a given obfuscated instance. This hampers the accuracy with which attempts at re-identification can be made. To achieve this guarantee, we must first determine an appropriate distance metric to measure the distinguishability of the numeric representations of images.

A natural choice for the distance metric is L_1 distance, however, we must be wary of the meaning of each element in the vectors. Should certain elements have differently sized ranges, they should be obfuscated using different magnitudes of noise. If one element has a much larger range than the others, the addition of noise configured to the smaller range would do little to prevent an inference of high accuracy on the original value of the element. We therefore apply normalization such that the distance between any pair of elements in the ith position of a pair of vectors falls within the range $[0, 1]$. Letting $R_i = [i_{min}, i_{max}]$ be the range of elements in the ith position of a model representation vector, we define a normalized, element-wise distance metric as follows:

$$d_e(x_1, x_2) = \frac{|x_1 - x_2|}{i_{max} - i_{min}} \qquad \forall x_1, x_2 \in R_i. \tag{3}$$

A distance metric for vectors defined as the sum of the element-wise distances for each position would be appropriate for images represented by the same model. However, a more useful framework would allow for reasoning about the level of privacy across different models. Ideally, the meaning of a privacy parameter ϵ applied to one model should have a similar meaning for a different model. For this, we require another normalization to account for models having vectors of different lengths. We therefore define the distance metric for vectors as follows:

$$d(X_1, X_2) = \frac{\sum\limits_{i=1}^{n} d_e(X_{1i}, X_{2i})}{n} \qquad \forall X_1, X_2 \in \mathbb{R}^n. \tag{4}$$

By using this distance metric in combination with Formula 2, we obtain a meaningful privacy guarantee for the model representations of images. Although this type of metric is not novel, its use in this context is. We must therefore address how to configure a mechanism to satisfy this instantiation of the privacy guarantee. This leads to our main result in the development of a framework for the application of differential privacy to generative models for images.

Theorem 1. *Any image $X \in \mathbb{R}^n$ can be protected by ϵ-differential privacy through the addition of a vector $(Y_1, ..., Y_n) \in \mathbb{R}^n$ where each Y_i is a random variable independently drawn from a Laplace distribution using a scaling parameter $\sigma_i = \frac{n(i_{max} - i_{min})}{\epsilon}$.*

Proof. We must satisfy the privacy guarantee (Formula 2) using our proposed distance metric (Formula 4). The form this privacy guarantee takes is our starting point in Formula 5. Through manipulation of this inequality and the substitution of mechanism probabilities with a Laplace distribution, we prove that the selection of an appropriate scaling parameter for each instance of the Laplace distribution allows for the privacy guarantee to be satisfied.

$$\prod_{i=1}^{n} \Pr\left(K\left(X_{1i}\right) = R_i\right) \leq e^{\frac{\epsilon \sum_{i=1}^{n} d_e(X_{1i}, X_{2i})}{n}} \prod_{i=1}^{n} \Pr\left(K\left(X_{2i}\right) = R_i\right) \forall X_1, X_2, R \in \mathbb{R}^n. \quad (5)$$

$$\prod_{i=1}^{n} \Pr\left(K\left(X_{1i}\right) = R_i\right) \leq \prod_{i=1}^{n} e^{\frac{\epsilon d_e(X_{1i}, X_{2i})}{n}} \prod_{i=1}^{n} \Pr\left(K\left(X_{2i}\right) = R_i\right) \forall X_1, X_2, R \in \mathbb{R}^n. \quad (6)$$

$$\prod_{i=1}^{n} \frac{e^{-\frac{|X_{1i}, R_i|}{\sigma}}}{2\sigma} \leq \prod_{i=1}^{n} e^{\frac{\epsilon d_e(X_{1i}, X_{2i})}{n}} \prod_{i=1}^{n} \frac{e^{-\frac{|X_{2i}, R_i|}{\sigma}}}{2\sigma} \quad \forall X_1, X_2, R \in \mathbb{R}^n. \quad (7)$$

$$\prod_{i=1}^{n} e^{\frac{|X_{2i}, R_i| - |X_{1i}, R_i|}{\sigma}} \leq \prod_{i=1}^{n} e^{\frac{|X_{2i} - X_{2i}|}{\sigma}} \leq \prod_{i=1}^{n} e^{\frac{\epsilon d_e(X_{1i}, X_{2i})}{n}} \quad \forall X_1, X_2, R \in \mathbb{R}^n. \quad (8)$$

$$\prod_{i=1}^{n} e^{\frac{\epsilon d_e(X_{1i}, X_{2i})}{n}} = \prod_{i=1}^{n} e^{\frac{\epsilon |X_{2i} - X_{2i}|}{n(i_{max} - i_{min})}} \quad \forall X_1, X_2. \quad (9)$$

From Formula 9, it becomes clear that the inequality holds when using an independent Laplace distribution for each pair of elements X_{1i}, X_{2i}, substituting the scaling parameter σ with a corresponding value $\sigma_i = \frac{n(i_{max} - i_{min})}{\epsilon}$. □

Using the generalization of differential privacy, the notion of query sensitivity is implicitly captured in the distance metric. Since the distance metric of Formula (4) has a range of $[0, 1]$, the ratio of probabilities for a pair of maximally dissimilar images to produce the same obfuscated output is e^{ϵ}. This is akin to the meaning of the privacy guarantee for a pair of databases which differ on every record. In order to select an appropriate value of ϵ, a data custodian must keep in mind that similar images will have a very small distance between them, requiring much larger values of ϵ to provide a reasonable ratio. In Sect. 5, we demonstrate the implications of the choice of ϵ on the levels of privacy and utility.

4.3   Benefits of Differentially Private Facial Obfuscation

We now describe the improvements we obtain from the use of differential privacy for each of the problems identified in Sect. 3.

Background Information. By removing dependence of the attack model on an absolute level of re-identification risk, we are able to reason about the level of privacy in the presence of attackers with background knowledge. If the location in a photo is identified as a particular city, no facial obfuscation can prevent the inference that individuals living in the identified city have a higher probability of being the obfuscated identity than individuals living elsewhere. Yet, the differential privacy guarantee continues to hold as the background information does not impact the conditional probability distribution used by the randomization mechanism. Since the privacy guarantee concerns only the change in the attacker's knowledge when presented with the obfuscated data (e.g., the face), it is unaffected by other sources of information the attacker may gain access to.

Composition Attacks. Another very important property of differential privacy is its resilience to composition attacks. The composition theorem [12] states that for two differentially private releases using privacy parameters ϵ_1 and ϵ_2 respectively, the privacy guarantee holds for a privacy parameter $\epsilon = \epsilon_1 + \epsilon_2$. Thus, even in the case of uncoordinated releases, we still have a valid privacy guarantee. Furthermore, this removes the restriction on the same individual appearing only once in the release of obfuscated images.

Input Image Gallery. Differentially private image obfuscation has no need for a gallery of images in order to perform obfuscation. Since noise is added on a per-image basis, there is no computation of clusters required. Given a trained model, obfuscation of a single image or a batch of images can be performed with ease. This makes the obfuscation process much more versatile.

4.4   Implementation Details

AAMs and GNNs have the very useful property of producing photo-realistic images. We now describe how our framework can be applied to these models. Provided that the addition of noise is properly controlled, the output will be a photo-realistic image of any newly created identity.

The AAM representation of an image consists of a shape vector and a texture vector. We take the concatenation of these vectors as the overall model vector. It is important to note that this gives a representation of the identity which is strictly contained within the contour of the face, leaving features such as hair and ears as contextual information which is untouched by the obfuscation. This is not ideal for the goal of hiding an identity since this contextual information can greatly facilitate inferences about the identity. Although in theory an AAM could be designed to incorporate the hair and ears, we are unaware of any research in which this has been done. Thus, although we include AAMs in our experimental comparisons, we recommend the use of GNNs instead.

For GNNs, we consider architectures which take one or more class vectors as input and employ up-convolution to transform the input into a visual representation in pixel space [9]. By considering each identity to be a different class, an input vector can specify the individual to be generated. The identity class vector is an obvious choice as the model vector to be obfuscated. However, this leads to some form of interpolation between the identities. To apply a finer degree of modification to the identity, we propose the application of obfuscation at the second layer of the network. Typically, the second layer applies convolution to the class vector and transforms it into a vector of numeric representations of high level facial features. By applying obfuscation to these features instead, we can achieve a richer variety in the potential modifications to the face. We therefore apply obfuscation to the output of the first convolutional layer of the network and pass the obfuscated feature vector on as the input to the next layer of the network. A sample architecture is shown in Fig. 1.

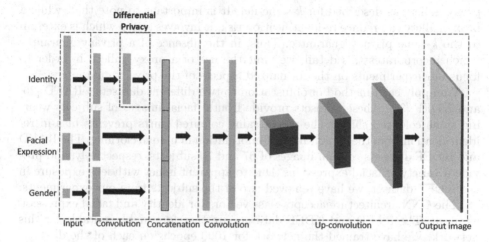

Fig. 1. Visualization of the layer architecture in an up-convolutional neural network using differential privacy. Noise is applied to the output of the second identity layer. The numbers and shapes of the convolutional layers shown here are not exact and represent only the general structure of such a network.

Information about the range of each model vector element can be used as a means to preserve the visual quality of the obfuscated output. Noisy elements which have gone too far beyond the valid range may lead to visual artifacts or distortions in the output image. To prevent this, we snap any out-of-bounds noisy value back to the nearest valid value. Since differential privacy is resistant to any form of post-processing [12] and the ranges of the elements are non-sensitive information, this step cannot violate the privacy guarantee.

5   Experiments

In this section, we run an experimental comparison between our proposed implementation of Sect. 4.4 and k-same implementations following the designs of k-same-m [18] and k-same-net [32]. We employ these experiments to gain insight into the relative performances of differential privacy and k-same obfuscation in the context of a trade-off between re-identification risk and utility.

5.1   Experimental Design and Results

We have implemented AAM obfuscation using AAM-API [37] for model training and generation of output. For GNN obfuscation, we have built on top of the

DeconvFaces [14] network which implements the concept of up-convolution for the generation of images of input classes [9]. For both models, we have applied differential privacy as described in Sects. 4.2 and 4.4. For the k-same implementation in the AAM, we have followed the process of k-same-m [18]. For the GNN, we have followed the approach of k-same-net [32]. In both cases, we have implemented clustering as described for k-same-m. This deviates from the use of a proxy gallery as described for k-same-net. It is important to note that, while a proxy gallery can reduce re-identification risk, it involves a step which is external to the k-same privacy guarantee. Thus, in the absence of a privacy guarantee which incorporates this detail, we omit the use of a proxy gallery in order to focus our experiments on the formalized aspects of the privacy guarantees.

We apply each method of obfuscation to two different datasets - RAFD [25] and KDEF [30]. These datasets provide frontal facial images of subjects wearing same coloured shirts. The use of same coloured shirts prevents bias in re-identification from the exploitation of information in unique clothing. The RAFD and KDEF datasets contain images of 67 and 70 subjects, respectively, and provide a variety of facial expressions. Due to apparent issues with lens exposure in the KDEF dataset, we have removed two of the subjects from our experiments.

The GNN architecture accepts class vectors for identity and facial expression as input. The RAFD and KDEF datasets are therefore highly suitable for this network. We have trained the network for 1000 epochs on each of the datasets to obtain models capable of reproducing these identities. An example of obfuscated output is shown in Fig. 2. The AAM has the advantage of being able to approximate previously unseen identities. To use this in our experiments, we have trained a model for each dataset using the other dataset (e.g., RAFD as training data for KDEF) with pre-processing to adjust the colour saturation of the training data in order to better match the target data. Since the training requires annotations of facial landmarks, we have employed OpenFace [1] to compute high accuracy approximations of the landmarks.

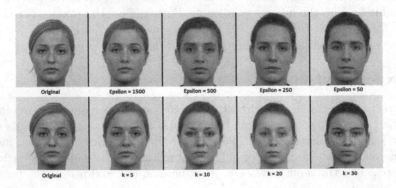

Fig. 2. Obfuscation via the GNN on the RAFD dataset. The top row employs differential privacy and the bottom row employs k-same obfuscation.

To measure re-identification risk, we have employed VGGFace D [35], a deep convolutional neural network which has been shown to achieve excellent facial identity classification accuracy. This simulates how an attacker might leverage machine learning models to launch an attack on obfuscated images. We have trained a separate model for each dataset, using the neutral and sad expressions for each identity for validation and the remaining expressions for training. To improve the robustness of the models, we have also augmented the datasets by creating two additional versions of each image - one with increased contrast and one with decreased contrast.

In our experiments, we generate obfuscated images having a neutral facial expression. We measure re-identification risk based on the accuracy of the top 1 guesses of the VGGFace network. Given that differential privacy is a stochastic process, for each combination of a privacy parameter and an identity to be protected, we have generated 10 obfuscated instances over which we take the average of the re-identification risk. We measure overall re-identification risk for a given privacy parameter as the average risk over all individuals in the dataset. Since the k-same approaches are deterministic, we produce only a single output image per identity and then take the average re-identification risk over the whole dataset (Fig. 3). In contrast to the typical ϵ values applied to differentially private mechanisms for databases, the values used in our experiments may appear unusually high. The larger magnitude is simply a side-effect of the normalization for the model vector, resulting in the interpretation of ϵ on a different scale.

To compare the methods of obfuscation in terms of utility, we have focused on gender classification in the obfuscated output. As forms of demographic classification may be desirable for data mining purposes, we consider high classification accuracy to reflect good utility. To this end, we employ a convolutional neural network for the classification of gender in facial images [27]. Since we wish to compare differential privacy to k-same obfuscation, we plot the data as a function of identity classification error in order to abstract away from the proprietary privacy parameters (Fig. 4). Given the poor obfuscation achieved by the AAMs, we omit them from this comparison. To highlight the ability of GNNs to incorporate properties relevant to image utility into the network architecture, we have also created a modified version of the architecture which preserves gender in the obfuscated output. To do so, we have created an input layer having two classes which specify the gender in the image. By training a model with gender labels, it learns to separate features relevant to gender from those relevant to identity. This enables us to focus obfuscation only on the features relevant to identity while leaving the gender feature vector untouched. An example of gender-preserving obfuscation is shown in Fig. 5.

5.2   Discussion

We first consider the results on re-identification risk. It is immediately notable that the AAMs are ineffective at privacy protection, even under severe privacy settings. This is due to the contextual information outside of the facial contour such as the hair, ears and neck. In an alternative setting, the obfuscated face

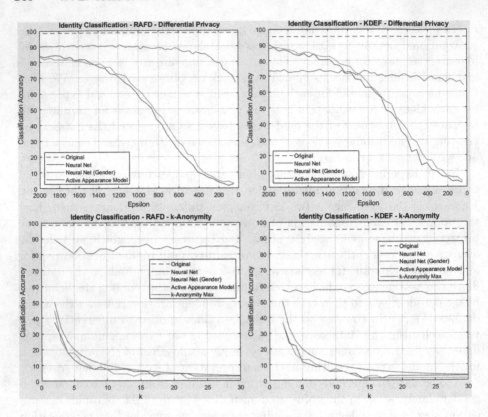

Fig. 3. Identity classification accuracy for the methods of obfuscation

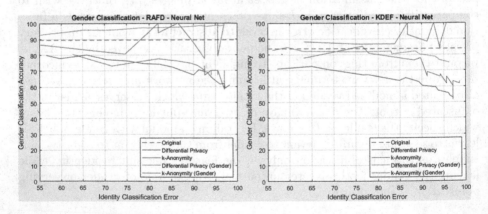

Fig. 4. Gender classification accuracy for the methods of obfuscation

could be released on its own (e.g., against a black background) to prevent leaking this contextual information. While this would greatly improve the privacy protection, it would omit a great deal of useful information in the image and would lead to an output which is not very visually pleasing. We expect that in most practical situations, it is desirable to release a full version of the image. We therefore recommend against the use of AAMs for obfuscation, at least in a form which does not obfuscate the full head.

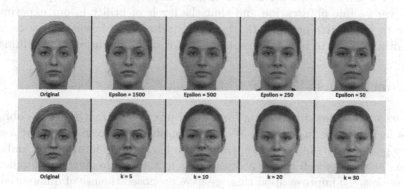

Fig. 5. Gender-preserving obfuscation via the GNN on the RAFD dataset. The top row employs differential privacy and the bottom row employs k-same obfuscation.

The k-same AAM results clearly demonstrate the violation of the k-same privacy guarantee since the re-identification risk is well above the theoretical maximum. We again note that if we only consider information within the contour of the face, the guarantee is not violated, however, such a guarantee is not useful in most practical situations. With differential privacy, we maintain a meaningful privacy guarantee for images which contain such information. Although our experiments do not illustrate the susceptibility of k-same obfuscation to composition attacks, we plan to demonstrate this in future work.

In the gender classification comparisons, we see that the basic models for differential privacy and k-same obfuscation suffer a degradation in classification accuracy at high levels of privacy protection. Comparing the gender-preserved models to their basic counterparts, we see a large improvement in the classification accuracy, suggesting that this is an effective approach for the preservation of specific properties in the obfuscated output. In some cases, the classification accuracy of the obfuscated images has surpassed that of the original data. This is a result of the explicit specification of gender labels in the network input which can lead to obfuscated identities that more prominently display these features.

The overall comparison between the utility preserved in differential privacy and k-same obfuscation appears to be inconclusive in these results. Some experiments show better results for differential privacy while other experiments show better results for k-same obfuscation. The k-same results are also more difficult to assess given the sporadic nature of the plots. This is likely due to changes in clusters between each level of obfuscation which can greatly impact classification accuracy, both for identity and for gender. It is clear that the utility is also data-dependent given the variations in the results seen on the two datasets. Notably, many subjects in the KDEF dataset, including some males, have long hair whereas all subjects in the RAFD dataset have short hair. The males with long hair in KDEF may have contributed to the lower gender classification accuracy.

Given that differential privacy offers great improvements in the privacy guarantee over k-same obfuscation and that the levels of utility in our experiments appear to be similar between the two approaches of obfuscation, we consider differential privacy to be a preferable choice for the obfuscation of facial images.

6   Conclusions

We have studied how to obtain a formalized privacy guarantee for the obfuscation of facial images in practice. We have identified shortcomings of the k-same privacy guarantee including susceptibilities to background information and composition attacks as well as the awkwardness in the requirement for a gallery of input images. To improve upon this, we have proposed the use of differential privacy in the context of obfuscation applied to generative models for images. We have developed a framework which provides a meaningful privacy guarantee for such models and we have derived the configuration of Laplace mechanism which can achieve this privacy guarantee. Our approach preserves the privacy guarantee in the presence of attackers with background information, provides resistance to composition attacks and removes the requirement for a gallery of input images. We have implemented both our proposed framework as well as k-same obfuscation in order to run experimental comparisons. Through our comparisons, we have shown that this application of differential privacy can achieve comparable utility to k-same obfuscation. We conclude that the key improvements in the privacy guarantee combined with comparable levels of utility make differential privacy a much more appropriate choice for the obfuscation of facial images.

References

1. Amos, B., Ludwiczuk, B., Satyanarayanan, M.: OpenFace: a general-purpose face recognition library with mobile applications. Technical report, CMU-CS-16-118, CMU School of Computer Science (2016)
2. Basu, A., Nakamura, T., Hidano, S., Kiyomoto, S.: k-anonymity: risks and the reality. In: 2015 IEEE Trustcom/BigDataSE/ISPA, vol. 1, pp. 983–989 (2015). https://doi.org/10.1109/Trustcom.2015.473
3. Bitouk, D., Kumar, N., Dhillon, S., Belhumeur, P., Nayar, S.K.: Face swapping: automatically replacing faces in photographs. ACM Trans. Graph. **27**(3), 39:1–39:8 (2008). https://doi.org/10.1145/1360612.1360638
4. Brand, A., Lal, J.A.: European best practice for quality assurance, provision and use of genome-based information and technologies. Drug Metab. Drug Interact. **27**, 177–182 (2012). https://doi.org/10.1515/dmdi-2012-0026

5. Cavailaro, A.: Privacy in video surveillance [In the Spotlight]. IEEE Signal Process. Mag. **24**(2), 166–168 (2007). https://doi.org/10.1109/MSP.2007.323270
6. Chatzikokolakis, K., Andrés, M.E., Bordenabe, N.E., Palamidessi, C.: Broadening the scope of differential privacy using metrics. In: Privacy Enhancing Technologies, pp. 82–102 (2013). https://doi.org/10.1007/978-3-642-39077-7_5
7. Chi, H., Hu, Y.H.: Face de-identification using facial identity preserving features. In: 2015 IEEE Global Conference on Signal and Information Processing (Global-SIP), pp. 586–590 (2015). https://doi.org/10.1109/GlobalSIP.2015.7418263
8. Cootes, T.F., Edwards, G.J., Taylor, C.J.: Active appearance models. IEEE Trans. Pattern Anal. Mach. Intell. **23**(6), 681–685 (2001). https://doi.org/10.1109/34. 927467
9. Dosovitskiy, A., Springenberg, J.T., Tatarchenko, M., Brox, T.: Learning to generate chairs, tables and cars with convolutional networks. IEEE Trans. Pattern Anal. Mach. Intell. **39**(4), 692–705 (2017). https://doi.org/10.1109/TPAMI.2016. 2567384
10. Du, L., Yi, M., Blasch, E., Ling, H.: GARP-face: balancing privacy protection and utility preservation in face de-identification. In: IEEE International Joint Conference on Biometrics, pp. 1–8 (2014). https://doi.org/10.1109/BTAS.2014.6996249
11. Dwork, C.: Differential privacy. In: Bugliesi, M., Preneel, B., Sassone, V., Wegener, I. (eds.) ICALP 2006. LNCS, vol. 4052, pp. 1–12. Springer, Heidelberg (2006). https://doi.org/10.1007/11787006_1
12. Dwork, C., Roth, A.: The algorithmic foundations of differential privacy. Found. Trends Theor. Comput. Sci. **9**(3–4), 211–407 (2014). https://doi.org/10.1561/0400000042
13. Fan, L.: Image pixelization with differential privacy. In: Kerschbaum, F., Paraboschi, S. (eds.) DBSec 2018. LNCS, vol. 10980, pp. 148–162. Springer, Cham (2018). https://doi.org/10.1007/978-3-319-95729-6_10
14. Flynn, M.: Generating faces with deconvolution networks (2016). https://github.com/zo7/deconvfaces
15. Frome, A., et al.: Large-scale privacy protection in google street view. In: 2009 IEEE 12th International Conference on Computer Vision, pp. 2373–2380 (2009). https://doi.org/10.1109/ICCV.2009.5459413
16. Ganta, S.R., Kasiviswanathan, S.P., Smith, A.: Composition attacks and auxiliary information in data privacy. In: Proceedings of the 14th ACM SIGKDD International Conference on Knowledge Discovery and Data Mining, pp. 265–273 (2008). https://doi.org/10.1145/1401890.1401926
17. Google: Google Maps. https://www.google.be/maps. Accessed 27 Feb 2019
18. Gross, R., Sweeney, L., de la Torre, F., Baker, S.: Model-based face de-identification. In: 2006 Conference on Computer Vision and Pattern Recognition Workshop (CVPRW 2006), pp. 161–161 (2006). https://doi.org/10.1109/CVPRW. 2006.125
19. Gross, R., Airoldi, E., Malin, B., Sweeney, L.: Integrating utility into face de-identification. In: Privacy Enhancing Technologies, pp. 227–242 (2006). https://doi.org/10.1007/11767831_15
20. Harmon, L.: The recognition of faces. Sci. Am. **229**(5), 71–82 (1973)
21. Harmon, L., Julesz, B.: Masking in visual recognition: effects of two-dimensional filtered noise. Science **180**(4091), 1194–1197 (1973). https://doi.org/10.1126/science. 180.4091.1194
22. Hu, X., et al.: How people share digital images in social networks: a questionnaire-based study of privacy decisions and access control. Multimed. Tools Appl. **77**(14), 18163–18185 (2018). https://doi.org/10.1007/s11042-017-4402-x

23. Korshunov, P., Ebrahimi, T.: Using face morphing to protect privacy. In: 2013 10th IEEE International Conference on Advanced Video and Signal Based Surveillance, pp. 208–213 (2013). https://doi.org/10.1109/AVSS.2013.6636641

24. Korshunov, P., Ebrahimi, T.: Using warping for privacy protection in video surveillance. In: 2013 18th International Conference on Digital Signal Processing (DSP), pp. 1–6 (2013). https://doi.org/10.1109/ICDSP.2013.6622791

25. Langner, O., Dotsch, R., Bijlstra, G., Wigboldus, D., Hawk, S., van Knippenberg, A.: Presentation and validation of the radboud faces database. Cogn. Emot. **24**(8), 1377–1388 (2010). https://doi.org/10.1080/02699930903485076

26. Letournel, G., Bugeau, A., Ta, V., Domenger, J.: Face de-identification with expressions preservation. In: 2015 IEEE International Conference on Image Processing (ICIP), pp. 4366–4370 (2015). https://doi.org/10.1109/ICIP.2015.7351631

27. Levi, G., Hassner, T.: Age and gender classification using convolutional neural networks. In: IEEE Conference on Computer Vision and Pattern Recognition Workshops (CVPRW), pp. 34–42 (2015). https://doi.org/10.1109/CVPRW.2015.7301352

28. Li, Y., Vishwamitra, N., Knijnenburg, B.P., Hu, H., Caine, K.: Effectiveness and Users' Experience of Obfuscation As a Privacy-Enhancing Technology for Sharing Photos. Proc. ACM Hum.-Comput. Interact. 1, 1–24 (2017). https://doi.org/10.1145/3134702

29. Liu, X., Krahnstoever, N., Yu, T., Tu, P.: What are customers looking at? In: 2007 IEEE Conference on Advanced Video and Signal Based Surveillance, pp. 405–410 (2007). https://doi.org/10.1109/AVSS.2007.4425345

30. Lundqvist, D., Flykt, A., Öhman, A.: The Karolinska Directed Emotional Faces – KDEF (1998). ISBN 91-630-7164-9

31. Martin, K., Shilton, K.: Putting mobile application privacy in context: an empirical study of user privacy expectations for mobile devices. Inf. Soc. **32**(3), 200–216 (2016). https://doi.org/10.1080/01972243.2016.1153012

32. Meden, B., Emersic, Z., Struc, V., Peer, P.: k-same-net: neural-network-based face deidentification. In: 2017 International Conference and Workshop on Bioinspired Intelligence (IWOBI), pp. 1–7 (2017). https://doi.org/10.1109/IWOBI.2017.7985521

33. Newton, E.M., Sweeney, L., Malin, B.: Preserving privacy by de-identifying face images. IEEE Trans. Knowl. Data Eng. **17**(2), 232–243 (2005). https://doi.org/10.1109/TKDE.2005.32

34. Padilla-López, J.R., Chaaraoui, A.A., Flórez-Revuelta, F.: Visual privacy protection methods. Expert Syst. Appl. **42**(9), 4177–4195 (2015). https://doi.org/10.1016/j.eswa.2015.01.041

35. Parkhi, O.M., Vedaldi, A., Zisserman, A.: Deep face recognition. In: British Machine Vision Conference (2015)

36. Samarati, P., Sweeney, L.: Protecting Privacy when Disclosing Information: k-Anonymity and its Enforcement through Generalization and Suppression. Technical report (1998). http://www.csl.sri.com/papers/sritr-98-04/

37. Stegmann, M.B.: The AAM-API (2003). http://www.imm.dtu.dk/~aam/aamapi/, platform: MS Windows

38. U.S. Department of Health & Human Services: Guidance Regarding Methods for De-identification of Protected Health Information in Accordance with the Health Insurance Portability and Accountability Act (HIPAA) Privacy Rule (2015). https://www.hhs.gov/hipaa/for-professionals/privacy/special-topics/de-identification/index.html. Accessed 09 Feb 2018

39. Venetianer, P.L., Zhang, Z., Scanlon, A., Hu, Y., Lipton, A.J.: Video verification of point of sale transactions. In: 2007 IEEE Conference on Advanced Video and Signal Based Surveillance, pp. 411–416 (2007). https://doi.org/10.1109/AVSS.2007.4425346
40. Winkler, T., Rinner, B.: Security and privacy protection in visual sensor networks: a survey. ACM Comput. Surv. **47**(1), 1–42 (2014). https://doi.org/10.1145/2545883

Insights into Learning Competence Through Probabilistic Graphical Models

Anna Saranti[1,2], Behnam Taraghi[1,2], Martin Ebner[1,2],
and Andreas Holzinger[2,3(✉)]

[1] Department Educational Technology,
Graz University of Technology, Münzgrabenstrasse 36/I, 8010 Graz, Austria
s0473056@sbox.tugraz.at
{b.taraghi,martin.ebner}@tugraz.at
[2] Institute of Interactive Systems and Data Science, Graz University of Technology,
Inffeldgasse 16c/I, 8010 Graz, Austria
[3] Institute for Medical Informatics, Statistics and Documentation,
Medical University Graz, Auenbruggerplatz 2, 8036 Graz, Austria
andreas.holzinger@medunigraz.at

Abstract. One-digit multiplication problems is one of the major fields in learning mathematics at the level of primary school that has been studied over and over. However, the majority of related work is focusing on descriptive statistics on data from multiple surveys. The goal of our research is to gain insights into multiplication misconceptions by applying machine learning techniques. To reach this goal, we trained a probabilistic graphical model of the students' misconceptions from data of an application for learning multiplication. The use of this model facilitates the exploration of insights into human learning competence and the personalization of tutoring according to individual learner's knowledge states. The detection of all relevant causal factors of the erroneous students answers as well as their corresponding relative weight is a valuable insight for teachers. Furthermore, the similarity between different multiplication problems - according to the students behavior - is quantified and used for their grouping into clusters. Overall, the proposed model facilitates real-time learning insights that lead to more informed decisions.

Keywords: Bayesian networks · Probabilistic graphical models · Learning analytics

The authors declare that there are no conflicts of interest and no ethical issues, no particular funding was achieved.

1   Introduction

1.1   Previous Work

The field of learning analytics is of increasing importance for educational research [16,19,45]. Moreover, it aims to assist the learning process by providing teachers a deeper insight into learning processes and learning results. Teachers play an essential role because they are responsible for intervening in a pedagogical adequate manner. Recently, the field makes heavy use of statistical machine learning [41]. Whilst educational data mining targets on automating learning activities, learning analytics supports educators in their daily routine.

The so called "1 × 1 trainer"[1] is a learning analytics application developed by the department Educational Technology of Graz University of Technology, Austria. It uses the benefits of both fields, learning analytics as well as educational data mining [17,18,43].

The application poses exercises to students from the multiplication table with one decimal digit operands. The algorithm of the "1 × 1 trainer" adapts the sequence of given questions subject to the students answers individually, in order to improve individual learning progress. However, technically, this needs to react adaptively to the changes of the learning progress of a student. In that way, it would support each student according to the user's distinct learning progress over the whole learning period. This underlying personalized adaptive learning algorithm shall discover weak mathematical knowledge of single students and alerts teachers just in time to adequately intervene.

The work is based on previous research that used data gathered by the "1 × 1 trainer". Firstly, different mathematical questions were roughly classified according to the learners' answers. Questions were considered to be more difficult than others when students required more attempts to answer them [48,51]. Some specific questions could be identified as difficult for the majority of the users. The next step was to analyze the explanation of the errors made. Therefore, error types were assigned to falsely answered questions, which correspond to the innate cognitive and conceptual learning shortcomings of the users [50]. The "relative difficulty" of those questions - 2 × 3 seems to be simpler than 7 × 8 - played no role in identifying the error types. More explanation on error types follows in Sect. 2.

1.2   Bayesian Student Models

There is a plethora of learning applications that use probabilistic graphical models (also called Bayesian models/networks [39]) to model student's knowledge. These models have started making an impact in the research of causal learning and inference generally, but there are good arguments that even children's causal learning could also be modeled in this way [10].

Most applications belong to the category of intelligent tutoring systems (ITS) [13] or adaptive educational systems (AES) [4]. The main goal of an ITS is to

---

[1] https://schule.learninglab.tugraz.at/einmaleins/, Last accessed 18 June 2019.

provide personalized recommendations according to the different learning styles, whereas AES adapt the learning content as well as its sequence according to the student's profile [42]. As explained in the literature review by Chrysafiadi & Virvou [11], there are two classes of intelligent tutoring systems: Systems that make diagnosis with the student's knowledge, misconceptions, learning style or cognitive state, and systems that plan a personalized strategy using diagnosis for each learner individually. Student modelling is considered a subproblem of user modelling which is of central importance to ITS since otherwise each student is treated the same [37].

Primarily, we are interested in Bayesian networks because of their ability to model uncertainty [36] and, at the same time, to support a decision making process. The user modelling goals of a Bayesian network for knowledge modelling is mainly to have an adaptive estimation of the knowledge itself, since it may increase or decrease during the learning process [4]. Since scalar models and fuzzy logic approaches [14] have lower precision, structural models are built with the assumption that the knowledge is composed mainly by independent parts. On the other hand, bug/perturbation models [11] represent errors and misconceptions of the student. In this case, the Bayesian network is used to find the error that most probably caused the observable behavior (also called evidence) [36] which is called credit/blame assignment problem [38]. Bayesian networks can model the assumption that a wrongly answered question having two potential causes is most probably caused by the one that is more prevalent, according to the data provided so far. Sometimes, random slips or typos are included in the model and do not rely on assumptions as for example: A wrongly answered question does not necessarily mean that the student does not know a concept completely, or a correctly answered one wasn't a guess. The structure in both cases constitutes the qualitative model; its definition uses domain knowledge and (optionally) data. The parametrization is learned from the data during a training phase and constitutes the quantitative model.

The reason for the creation of the model is in some cases to assist the teachers of large classes that suffer from a high dropout rate [52]. A model recognizes the student's knowledge faster and more accurate [36] which is primarily beneficial when the class has a large number of students. In other cases that are summed up in [5], the goal is to provide a personalized optimal sequence of the learning material or even to sequence the curriculum according to the student's individual needs. And yet, further cases [46,47] show that the learning application that is based on the model provides long-term learning effects as opposed to traditional methods. This was studied by a post-test that was made several weeks after the learning sessions.

The issue of defining the prior beliefs, which consist the starting parameterization, is often coupled with user clustering; demographics, longitudinal data [47], pre-tests [12,25], defining the prior beliefs as well as the starting groupings [4] with respect to the learners. In other cases, the teacher sets the prior beliefs from his/her experience [37] or a uniform prior is used [12,38]. Another common characteristic is the definition of hidden structural elements that represent unobservable entities, which must be estimated from the observed ones. The design of the structure must take correct assumptions into account, based on a solid theoretical background, otherwise the model will not work correctly [36].

In the work of Millán et al. [33], the researchers draw a parallel between medical diagnosis systems and student's knowledge diagnosis [34]. Actually, this is an important comparison as the development of clinical reasoning and decision-making skills is very similar [3].

The student answers a set of questions that can only be answered correctly, when several concepts are known. In this case the knowledge of the concepts is the cause of the answer. The noise in the process, for instance when a student knows the concept but answers wrongly and vice versa a correct guess, is also modelled. The initialization of the model parameters is made by teachers-in-the-loop; afterwards the parameters are learned from the data. The model is used to efficiently determine those concepts the student knows less and the deductive proposal of the next question.

The "eTeacher" is a web-based education system [21,42] that recognizes the learning style of a student according to the performance in exams as well as email, chat and messaging usage. The number of different learning strategies and their characteristics is the "domain knowledge" defining the structure of the Bayesian network. The initialization of the parameters uses in some cases uniform priors and in others priors defined by experts. After that initial phase, the parameters are continuously learned from the behaviour of the students. After identifying the learning style, a recommendation engine proposes different ways to learn the same material to each student according to his or her learning style.

"ANDES" is an ITS developed by Conati et al. [13], which mainly focuses on knowledge tracing but also on recognition of the learning plan of the user. The students solve Newtonian physics problems with different possible solution paths that define the Bayesian network's structure. Since each action may belong to different solution paths and the user does not provide its reasoning explicitly, the credit assignment problem is to find and quantify the most likely solution an action belongs to. This triggers personalized help instructions and hints in two cases: when a wrong answer is given or when the model predicts that the answer might be wrong. The parameters of the network change in an online manner while the student is solving the problems. Firstly, the evaluation was made by simulating students that have different knowledge profiles and measuring the accuracy of the predictions made by the model. In a second step, a post-test was carried out to compare real students having used "ANDES" to students who have not. Regression analysis was used to recognize the correlation between the use of the program and the learning gain [6].

Specifically for mathematical problems there are several approaches that specialize in dealing with decimals misconceptions. In the work of Stacey et al. [46,47] the misconceptions that define the structure of the model are provided by two main factors: the domain knowledge and data of a Decimal Conception Test (DCT) that students had to go through. Wrongly answered questions provided by the students depend on their misconceptions. The researchers defined the distinct misconception by computing which of them has the highest probability according to the data. Although the model drives different question sequencing strategies, some of the misconceptions were not correctly recognized. Therefore, the researchers decided that the teacher and not the system should provide instructions.

Also, the research work of Goguadze concentrates on the modelling of decimals misconceptions [24,25]. The "AdaptErrEx" project selected the most frequently occurring misconceptions and ordered them a taxonomy (higher and lower level misconceptions), which is reflected in the dependencies of the Bayesian network. As the previous application, a wrong answer may be caused by different misconceptions. The prior beliefs are defined by a pre-test; the researchers assert that sufficient training data diminish the role of the prior in the computation of the posterior. This prior defines the typical/average student and then each user's parameters can be updated and individualized accordingly. One aspect that has not been considered in this model yet, is the difficulty of each question: easy questions will more likely be answered correctly than difficult ones, even if there is a high probability of misconception.

Several student modelling models track the progress of knowledge through time with Dynamic Bayesian Networks (DBN). The knowledge of the learner at each time point can be considered to be dependent on the knowledge and (optionally) the observed result of the interaction at the previous time point [35]. The project "LISTEN" [9] represents the hidden knowledge state of the student at each time point. The observable entities are the tutor interventions and the student's performance which are used to infer the knowledge state. In the work of Käser et al. [28] there is an overview and comparison of Bayesian Knowledge Tracing (BKT), which is a technique for student modelling using a Hidden Markov model (HMM) modelling and DBN for various learning topics, such as number representation, mathematical operations, physics, algebra and spelling. A HMM is a special case of a DBN, which, according to the researchers, cannot represent dependencies that would lead to hierarchies of skills; in these case DBNs create more adequate models.

All above described applications have a Bayesian network of the students model at its architecture core. There are a number of other components that either support the teacher or the student. One of them, for example, is the visualization of the model in the "VisMod" application [53], which is displayed in (among other things) color and arrow intensity instead of number-filled tables. This increases the readability of the model and enhances the tutor's understanding. Gamification elements can also be found in "Maths Garden" [29], an application that lets users gain and loose points and coins depending on answering correctly or wrongly. A coaching component that provides feedback and hints to refresh the memory can be found in "ANDES" 's architecture [6]. An overview about the design and architectural elements of intelligent tutoring systems that have a Bayesian network as user model is provided in the work of Gamboa et al. [20].

A detailed overview about intelligent techniques other than Bayesian networks, such as recommender systems for the computation of the learning path as well as clustering and classification for learner profiles that are used in e-learning systems, is provided in [32]. Specifically in [4], the demand for the most appropriate activity proposed - neither too easy nor too difficult - can only be fulfilled, if the used model is both accurate and adaptive.

1.3 Research Question

The main objective of this research work is to answer the research question, whether Bayesian networks can quantify the defined misconceptions of one-digit multiplication problems. In order to answer this question the "1×1 trainer" application is taken as the underlying data provider. The application focuses on the recognition of the current learning status. However learning aware applications maintain an adaptive learning model that represents the knowledge of the learner/user with regard to the learned topic. The application is expected to support individual learning needs and abilities as well as considering common characteristics in the learning process of different persons. The progress of the learning model itself will be used to transform the learning application into an adaptive one; that may change the content and sequence of assessments constantly to improve the learning process and to maximize the learning efforts.

The "1×1 trainer" application has a current overall report that is accessible to teachers. It contains information about the actual number as well as the proportion of correct and wrong answers of each posed question. It uses color encoding that helps distinguishing four sets of questions with similar proportion of correct and wrong answers. The implementation of the Bayesian model provides further insights to detailed cognitive information that enriches the information content of the current report. Furthermore, the new report can concentrate on individualized learning status, considering the causes of the wrong answers and can be updated in real time after each action of the student.

1.4 Outline

This research work proposes a Bayesian model for the learning competence of students using the "1×1 trainer" application. The first step is to specify the error types that are relevant for this research; their detailed description is made in Sect. 2. Data analysis (specifically descriptive statistics) is used to guide the necessary assumptions about the modelled entities and their independences. Based on this information, the structure of the model and its parametrization is defined. The personalized model of each student and the method by which it adapts its parameters to new data is described in Sect. 3. The usage of the model and the insights that are provided to the teachers in the form of an enhanced report are explained in Sect. 4. Finally, a conclusion about future research and improvement possibilities is in Sect. 5.

2 Error Types of One-Digit Multiplication and Descriptive Statistics

2.1 Error Types of One-Digit Multiplication Problems

The bug library [11] of the proposed learning competence model contains six error types: operand, intrusion, consistency, off-by, add/sub/div, and pattern. Any false answer that does not belong to one of those six categories is assigned to the unclassified category. The description of the error types is explained in detail in [49]; a brief description follows here:

1. Operand error: It occurs, when the student mistakes at least one operand for one of its neighbours [7]. In the implementation only a neighbourhood of overall absolute distance of 2 from the correct operands was considered. One example is the answer 48 to the question 7×8 since the user may mistakenly multiply 6×8. Research shows that this is the most frequently occurred error, but it occurs with a different proportion in each posed question [8].
2. Operand intrusion (abbreviated intrusion) error: It happens, when the decades digit and/or the unit digit of the result equals one of the two operands of the posed question, for example $7 \times 8 = 78$. It is argued by [7] that the two operands of the multiplication question are perceived as one number by the student (the first operand corresponding to the decades digit and the second to the unit digit).
3. Consistency: The student's answer has either the unit digit or the decade digit of the correct answer [15, 44]. For example, the answer 46 to the question 7×8 indicates that the unit digit is correct, but the decades digit is false.
4. Off-by-± 1, Off-by-± 2: It occurs, when the answer of the student deviates from the correct one by ± 1 or ± 2, for example, when the answer of the question $5 * 8$ is one of the following: $\{38, 39, 41, 42\}$.
5. Add/Sub/Div: The student confuses the operation itself and performs for example an addition instead of a multiplication; in that case the answer to 7×8 is 15.
6. Pattern: The student mistakes the order of the digits of the result, for example, question 7×8 provides the answer 65 (the decades digit and the unit digit are permuted).
7. Unclassified: Any answer that can not be matched to one of the above error types.

All questions that have a correct answer with value smaller than 10 do not have consistency error. These are: $1 \times 1, 1 \times 2, 2 \times 1, 1 \times 3, 3 \times 1, 1 \times 4, 4 \times 1, 1 \times 5, 5 \times 1, 1 \times 6, 6 \times 1, 1 \times 7, 7 \times 1, 1 \times 8, 8 \times 1, 1 \times 9, 9 \times 1, 2 \times 2, 2 \times 3, 2 \times 4, 3 \times 2, 3 \times 3, 4 \times 2$.

One of the main reasons to use a probabilistic graphical model, is the fact that a specific false answer can be classified to multiple error types. The identification of the most probable error type causing a wrong answer is called credit assignment. The Table 1 shows the possible false answers for the question 7×8. One can see that for example the answer 72 could occur because of an operand or an intrusion error.

Table 1. Answers for question 7×8 listed by error types

Error type	Answers
operand	$40, 42, 48, 49, 54, 63, 64, 72$
intrusion	$18, 28, 38, 48, 58, 68, 71, 72, 73, 74, 75, 76, 77, 78, 79, 88, 98$
consistency	$16, 26, 36, 46, 51, 52, 53, 54, 55, 57, 58, 59, 66, 76, 86, 96$
off-by-± 1/off-by-± 2	$54, 55, 57, 58$
add/sub/div	$1, 15$
pattern	65
unclassified	$4, 5, 6, 7, 8, 9, 10, 11, 12, 19, 20, 21, 22, 23, 24, 25, 27, 29, 30,$ $31, 32, 33, 34, 35, 37, 39, 41, 43, 44, 45, 47, 50, 60, 61, 62, 67,$ $69, 80, 81, 82, 83, 84, 85, 87, 89, 90, 91, 92, 93, 94, 95, 97, 99$
correct	56

2.2 Data of the "1 × 1 Trainer"

The data that were used for building the model were provided from the "1 × 1 trainer" application. The application is for both students and teachers. For this work it was also used for a preliminary categorization of learners. Users of this application are confronted with multiplication questions with both multiplicands being one-digit integer numbers. The possible questions range from 1×1 up to 10×9 (a total of 90 questions) and are posed in a pre-specified order. The application does not provide any means of help or hints to the students so far; the only feedback users get, is whether their answer is correct or not. It is expected that by repeated use of the application the students will learn and get better through exercise. But there is no individualisation that takes care of the personal needs of the learning style and knowledge level of the users. Furthermore, personal information such as age, gender, demographics, and educational level were not collected.

The data were cleaned in the preprocessing phase. The answers that did not lie in the interval $[0 - 100)$ were considered invalid and were removed. Overall there were 1179720 question-answer pairs with 1164786 valid. The number of unique users that gave at least one valid answer is 9058. The file covers eight columns providing the user ID, session ID, platform ID, date and time of the answer, as well as the reaction time of the student. Along with the posed question and the provided answer the ID of the result type as one of {R, WR, W, WWR, WW, WWWR, WWW, WWWW}, whereas W means "wrong" and R "right" is stored. The detailed description of the result type is shown in [49] and is basically a way to quantify the relative difficulty of each question by keeping the recent history of the user's answering behaviour for each question. This information and the reaction time was not used in the model.

2.3 Data Analysis and Descriptive Statistics

To help designing the probabilistic graphical model, some analysis steps were necessary to be carried out with the data. The analysis and descriptive statistics provides insights about the overall similarities and differences between the students [22].

Firstly, not every user has answered the same number of questions. The vast majority of the users $(98, 6\%)$ have ≤ 1000 valid answered questions. For the training of the model, the prior must have an equal amount of answers for each question. This does not take the sequence of posed questions into consideration.

Secondly, for each question the proportion of wrong answers was computed and depicted with a heatmap, whereas the x-axis is the first operand and the y-axis the second (see Fig. 1). As it turned out, the most difficult question is 6×8 with 26.8% of wrong answers given. It must be advised to remember that not all questions are posed the same number of times, because of the algorithm that chooses the question sequence. Therefore the belief about the relative difficulty of the questions has not an equal confidence for all the questions.

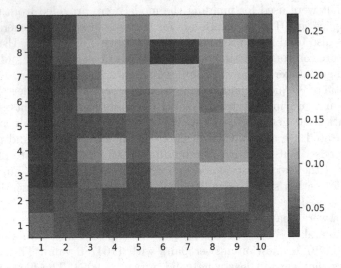

Fig. 1. Relative difficulty of the questions measured by the proportion of wrong answers. The x-axis is the first operand and the y-axis the second.

3 Probabilistic Graphical Model of Learning Competence

The use of a learning-aware data-driven application cannot assume that the user's learning competence remains unchanged. Simple statistical descriptions are not practical in representing a continuous change and do not effectively capture the differences between the learning process of the users. Furthermore, the purpose is to choose intelligent actions (also called "actionable information" [1]) based on the data and this is not possible simply by one rigid and non-adaptive analysis of the data.

The choice of a probabilistic graphical model has several benefits. Firstly, it allows the representation of conditional dependencies (and independencies correspondingly) in the graphical representation of the model of the data. Those are assumed to be the same for all users and stay stable over the course of application usage. Secondly, its parameters (that can be thought of as a configuration or instance) are adaptive and change with each new data sample that is observed. They may be a temporary snapshot description that characterizes the learning competence but unlike the statistics there is an effective way to adapt those and not recompute them from scratch each time the model confronts new data. Thirdly, they've already been extensively used for decision problems [1, 30] which are the forefronts of reinforcement learning algorithms.

3.1   Introduction to Probabilistic Graphical Models

Probabilistic Graphical Models are representations of joint probability distributions over random variables that have probabilistic relationships expressed through a graph. The random variables involved can be discrete which have categorical values or continuous with real values. The set of possible values that a random variable can take - sometimes also referred as the possible outcomes of the experiment described by the random variable - is its domain. The random variables can be either visible or hidden. The visible ones have outcomes that can be directly observed and their values are contained in the dataset. The hidden variables are defined by human experts using the domain knowledge of the problem, but their outcomes are not directly accessible. They usually represent latent causes of visible random variables and can improve the accuracy and the interpretability of the model [31].

To specify the dependencies of the variables in general, one needs to specify their direction, type and intensity. This is made with the use of graphs which provide the terminology and theory for understanding and reasoning about Probabilistic Graphical Models. The nodes (also called vertices) of the graph represent the random variables and the edges their dependencies which can be directed or undirected. Undirected models - also called Markov networks - on the other hand represent symmetric probabilistic interactions where there is no dependency with direction, only factors that represent the degree of the strength of the connection. In case where the dependencies are directed, the graph must be a directed acyclic graph (DAG), otherwise circular reasoning would be possible. These two categories are used in different applications.

3.2   Model Structure

Domain knowledge about the already described error types that are encountered in one-digit multiplication, as described in Sect. 2, was used to define the model. This is in accordance with the data-driven approach of model construction [31] where the structure of the model is specified by the designer and the parameters are learned from the data.

A question is either answered correctly or faulty. The student can make one of the following errors: Operand, intrusion, consistency, off-by-±1 and off-by-±2, pattern, confusion with addition, subtraction, division or an unclassified error (meaning none of the above). Therefore a multinomial random variable called **Learning State$_q$** - individual for each question q was chosen to represent the proportion of each of these misconceptions of the user, when he or she is answering a one-digit multiplication question. The variable follows the categorical distribution; in this case the **Learning State$_q$** has eight possible outcomes and the domain of this random variable is Val (**Learning State$_q$**) = {operand, intrusion, consistency, pattern, confusion, unclassified, correct} (meaning that 1 is the operand error, 2 the intrusion error and so on). The **Learning State$_q$** of a specific user can be described for example 5% operand error, 4% consistency error and 91% correct answering (the rest possible outcomes have 0%). This parametrization must be learned from the data.

In the previous section it is shown that a specific faulty answer may be classified to more than one error types. Although in reality the model does not assume that more than one error type created a particular answer, the model cannot know a priori which error type was more prevalent and played the decisive role in choosing the wrong answer. The **Learning State$_q$** is hidden and the percent of each error type is expected to be learned by the provided answers. Thereby, a dominant error type (for a specific user) can be still discovered and weaken the belief that multiple error types played a role for a specific faulty answer. In Sect. 4 the inference of the most probable error type (credit assignment problem) of a specific wrong answer will be made after the learning of the parameters is completed.

The proportion of correct and false answers is different for each question. Even though each question is not posed the same number of times and the belief about the possibility of correctly answering each question is different, this was also taken into account. That means that the probability of answering correctly is not. Therefore, there are 90 random variables called **Correctness$_{1 \times 1}$** to **Correctness$_{10 \times 9}$** (abbreviated by **Correctness$_q$**) that have each two possible outcomes. Therefore the Bernoulli distribution was chosen, which is equivalent to a categorical distribution with a domain of two values.

Each question has a distinct random variable, named accordingly as **Answers$_{1 \times 1}$** to **Answers$_{10 \times 9}$** (abbreviated as **Answers$_q$**), which is a child of the **Learning State$_q$** random variable. The arrows from the **Learning State$_q$** to its children reflect the dependency of the answer to a question from the misconception or correct understanding of the user.

The conditional independence property of each Learning Competence model is expressed by the following equation:

$$\text{Answes}_q \perp \text{Correctness}_q | \text{Learning State}_q \tag{1}$$

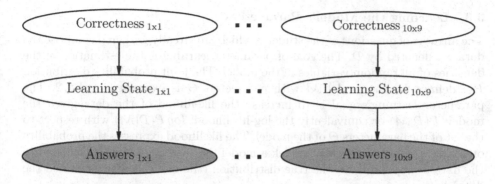

Fig. 2. The structure of all Probabilistic Graphical models for Learning Competence. The shaded **Answes$_q$** nodes are the ones that are observed, whereas the **Correctness$_q$**, **Learning State$_q$** random variables remain unobserved.

The joint probability distribution for each question **q** has the following factorization:

$$P(\textbf{Correctness}_q, \textbf{Learning State}_q, \textbf{Answers}_q) =$$
$$P(\textbf{Correctness}_q)\, P(\textbf{Learning State}_q | \textbf{Correctness}_q) \qquad (2)$$
$$P(\textbf{Answers}_q | \textbf{Learning State}_q)$$

Each error type can only produce a specific subset of answers, so the others will have zero probability of occurring given this particular error type. Every row of the conditional probability tables of the **Answers$_q$** random variables has values that sum to one and the last row has only one entry with probability 1.0 at the column with the correct answer and 0.0 everywhere else. Figure 2 depicts the described structure of Learning Competence models.

The model needs to express the following procedure: First knowing if the question is answered correctly; this is provided by the **Correctness$_q$** random variable. If this is true then there are no more steps to follow. In the case where the answer is false, there must have been an error which belongs to the hidden **Learning State$_q$**. One of the possible answers of this error, as seen and quantified by **Answers$_q$** will be the actual answer of the user. The possibility of guessing the answer is a valid one, but there is no way to get that kind of evidence in this application. The probability of continuously guessing the correct answer is very low and students that continuously provide random answers need to be discovered by the inconsistency of their model.

The model reflects our belief about the overall learning competence of the user. Its structure is considered to be the same for all users, but the conditional probability values (entries in the conditional probability tables) will differ for each individual user. Nevertheless the model can also reveal similarities between the users, meaning at this stage models that have similar parameter values.

3.3   Learning the Model's Parameters

The answers of questions of the students which are already gathered comprise the data set denoted by $\mathcal{D}$. The goal of parameter learning is the estimation of the densities of all random variables in the model. The joint probability distribution $P_{\mathcal{M}}$ defined by the model $\mathcal{M}$ with parameters Θ is expressed by Eq. 2. The parameter learning's goal is to increase the likelihood of the data given the model: $P(\mathcal{D}|\mathcal{M})$ or equivalently the log-likelihood: $log\,P(\mathcal{D}|\mathcal{M})$ with respect to the set of the parameters Θ of the model. The likelihood expresses the probability of the data given a particular model; a model that assigns a higher likelihood to the data $\mathcal{D}$ approximates the true distribution (the one that has generated the data) better.

The algorithm that is used to estimate the parameters in cases where some of the variables are hidden is expectation-maximization (EM). Since the latent variables **Correctness$_q$** and **Learning State$_q$** are not observed, the direct maximization of the likelihood is not possible. The EM algorithm initializes all model's parameters randomly and iteratively increases the likelihood by stepwise maximizing the expected value of the currently estimated parameters [2]. If the likelihood's increase or the parameters' change is not significant compared to the previous iteration, then the algorithm can be stopped. The procedure of updating the log-likelihood in this manner is shown to guarantee convergence to a stationary point, which can be a local minimum, local maximum or saddle point. Fortunately, by initializing the iterations from different starting parameters and injecting small changes to the parameters, the local minima and saddle points can be avoided [31].

The models are simple enough; therefore the EM-algorithm has a straightforward analytical solution. The available data were divided into a training and test set, with a dataset containing data from users that have answered all the questions at least one time (The number of users that have answered all questions exactly once is 2218). The models parameters are computed by the EM-algorithm on the training data and it iterates 4 times. After 4 iterations the likelihood of the training set increases, but the likelihood of the test set decreases which consists an indication of overfitting. The diagram in Fig. 3 describes the main computational blocks of this process.

Figure 4 depicts the learned parameters for the Learning Competence model of question 8×5 as an example.

4   Insights

After the model of a particular student is learned - by using the informed prior as starting point and as evidence the answers he or she has given so far. The better and more accurate the model captures the learning competence of the student, the better the performance of the predictions of the answers will be. In some probabilistic modelling frameworks such as Figaro[2] the parameter learning part is made by the offline component and the probability queries by the online component.

---

[2] https://www.cra.com/work/case-studies/figaro, Last accessed 18 June 2019.

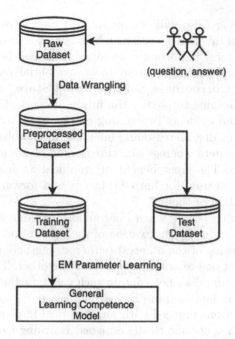

Fig. 3. Computational blocks diagram of data preprocessing, splitting into training and test set until the computation of the learned model.

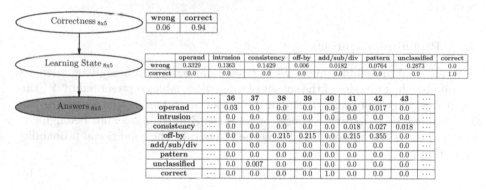

Correctness $_{8x5}$

wrong	correct
0.06	0.94

Learning State $_{8x5}$

	operand	intrusion	consistency	off-by	add/sub/div	pattern	unclassified	correct
wrong	0.3329	0.1363	0.1429	0.006	0.0182	0.0764	0.2873	0.0
correct	0.0	0.0	0.0	0.0	0.0	0.0	0.0	1.0

Answers $_{8x5}$

	...	36	37	38	39	40	41	42	43	...
operand	...	0.03	0.0	0.0	0.0	0.0	0.0	0.017	0.0	...
intrusion	...	0.0	0.0	0.0	0.0	0.0	0.0	0.0	0.0	...
consistency	...	0.0	0.0	0.0	0.0	0.0	0.018	0.027	0.018	...
off-by	...	0.0	0.0	0.215	0.215	0.0	0.215	0.355	0.0	...
add/sub/div	...	0.0	0.0	0.0	0.0	0.0	0.0	0.0	0.0	...
pattern	...	0.0	0.0	0.0	0.0	0.0	0.0	0.0	0.0	...
unclassified	...	0.0	0.007	0.0	0.0	0.0	0.0	0.0	0.0	...
correct	...	0.0	0.0	0.0	0.0	1.0	0.0	0.0	0.0	...

Fig. 4. Learned parameters of Learning Competence model for the question 8 × 5.

There are three types of reasoning one can make with probabilistic graphical models: causal, evidential and explaining away. Causal reasoning (also called prediction) consists of statements that start with the knowledge of the causes as evidence and provide information about the effects. In our model this would be possible if the **Correctness$_q$** and **Learning State$_q$** were known: the computation of the answer to a posed question would be accurately determinable. The direction of causal reasoning in directed graphical models goes from parent to child variables ("downstream") in general and is used to predict future events.

Evidential reasoning (also called explanation) on the other hand has the opposite direction and involves situations where effects lead to the specification of causes. This is the most important reasoning in our case because the answers of the students provide the information to do evidential reasoning and learn the hidden variables **Correctness$_q$** and **Learning State$_q$** which in turn can be used for causal reasoning to predict the future answers of each student. The difference in causal and evidential reasoning can be understood by considering the direction of time; evidential reasoning infers the past probability distribution from the current set of data whereas causal reasoning makes a prediction for the future given the data. The great benefit of graphical models over statistics is that the same model is used for both backward and forward reasoning (with respect to the perception of time).

Intercausal reasoning occurs when one random variable depends on two or more parents. In this case, the observation of the value of one parent influences the belief about the value of the other(s) (either strengthen or weaken). In this situation it is said that one reason explains away the other. The Learning Competence model's structure does not contain such cases; further discussion about this reasoning type can be found in [2,27,31].

The upcoming sections proceed with an analytical implementation of probabilistic queries which is specific to the designed Learning Competence models. Personalized insights computed by the latent explanations of wrong answers of each student are made possible by exact and efficient inference as described in Sect. 4.2.

4.1 Probability Queries

A conditional probability query $P(\boldsymbol{Y}|\boldsymbol{E} = e)$ - also called probabilistic inference - computes the posterior of the subset of random variables represented by $\boldsymbol{Y}$ (target of query) given observations e of the subset of evidence variables denoted by $\boldsymbol{E}$ (of course there may be a subset of variables $\boldsymbol{Z}$ in the model not belonging to either of these two subsets). By using the Bayes rule, the conditional probability is written as:

$$P(\boldsymbol{Y}|\boldsymbol{E} = e) = \frac{P(\boldsymbol{Y}, e)}{P(e)} \tag{3}$$

The MAP query, which is also called most probable explanation (MPE) [31], [40] is a query that maximizes the posterior of the joint distribution of a subset of random variables $\boldsymbol{Y}$:

$$\text{MAP}(\boldsymbol{Y}|\boldsymbol{E} = e) = \arg\max_y P(y, e) \tag{4}$$

In the case of MAP Query the whole set of random variables is $\mathcal{X} = \{\boldsymbol{Y}, \boldsymbol{E}\}$. In other words the MPE, after observing (clamping) a subset of variables, it computes the most likely values of the rest of them jointly.

A slightly different query is the marginal MAP which is written as follows:

$$\text{Marginal MAP}(\boldsymbol{Y}|\boldsymbol{E}=e) = \arg\max_{y} P(y|e) =$$

$$\arg\max_{\boldsymbol{Y}} \sum_{\boldsymbol{Z}} P(\boldsymbol{Y},\boldsymbol{Z}|\boldsymbol{E}=e) \qquad (5)$$

which directly follows from the fact that $\mathcal{X} = \{\boldsymbol{Y},\boldsymbol{E},\boldsymbol{Z}\}$.

The computation of the query result can be made with the variable elimination algorithm which is described in the following Sect. 4.2; in this case the exact value of Eq. 3 is computed by dividing $P(y,e) = \sum_{w} P(y,e,w)$ and $P(e) = \sum_{y} P(y,e)$. Alternatively, the normalization of a vector containing all $P(y^{k},e)$ (where y^{k} are all possible outcomes of the variables $\boldsymbol{Y}$) so that it has sum that equals to one, provides also the desired result. For more complex Bayesian networks approximate inference algorithms are applied since exact inference is NP-hard [30], but even those can be in the worst-case also NP-hard [31]. The Learning Competence models are simple and the variable elimination algorithm is fast enough.

4.2   Variable Elimination in the Learning Competence Model

The probability query that is of high relevance for the teachers is the probability of error types regarded as causes of a specific wrong answer. The sum of products expression in Eq. 6 computes the distribution of the **Learning State**$_q$ by means of the joint distribution $P(\mathbf{C_q},\mathbf{LS_q},\mathbf{A_q})$:

$$P(\mathbf{LS_q}) = \sum_{\mathbf{C_q},\mathbf{A_q}} P(\mathbf{C_q},\mathbf{LS_q},\mathbf{A_q}) = \sum_{\mathbf{C_q}} P(\mathbf{C_q})P(\mathbf{LS_q}|\mathbf{C_q}) \sum_{\mathbf{A_q}} P(\mathbf{A_q}|\mathbf{LS_q}) \quad (6)$$

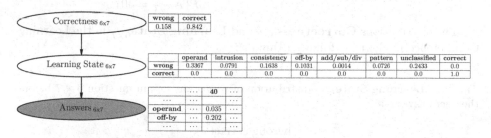

Fig. 5. Parameters of Learning Competence model of question 6×7 that are relevant to the computation of the MAP query when the answer is 40

The first step of the Variable Elimination algorithm, in case it is applied where an evidence exists, is to compute the unnormalized joint distribution $P(\mathbf{C_{6\times7}},\mathbf{LS_{6\times7}},\mathbf{A_{6\times7}}=40)$. For example, the faulty answer 40 for the question 6×7 eliminates all cases for which the answer is not equal to 40; it can belong only to two potential error types: consistency and off-by. The remaining rows of the joint distribution - those containing the unnormalized proportion unequal to 0 are listed in Table 2. The computations use the corresponding parameters of the Learning Competence Model of question 6×7 depicted in Fig. 5.

Table 2. Unnormalized joint distribution $P(\mathbf{C}_{6 \times 7}, \mathbf{LS}_{6 \times 7}, \mathbf{A}_{6 \times 7} = 40)$

$\mathbf{C}_{6 \times 7}$	$\mathbf{LS}_{6 \times 7}$	$\mathbf{A}_{6 \times 7}$	Unnormalized proportions
wrong	operand	40	$0.158 \cdot 0.336 \cdot 0.035 = 1.85 \cdot 10^{-3}$
wrong	off-by	40	$0.158 \cdot 0.103 \cdot 0.202 = 3.28 \cdot 10^{-3}$

The sum of the unnormalized proportions, $1.85 \cdot 10^{-3} + 3.28 \cdot 10^{-3} = 5.14 \cdot 10^{-3}$ (which is the value of $P(\mathbf{A}_{6 \times 7} = 40)$), can be used to compute the normalized probabilities of the causes of answer 40 as depicted in Table 3.

Table 3. Normalized joint distribution $P(\mathbf{C}_{6 \times 7}, \mathbf{LS}_{6 \times 7}, \mathbf{A}_{6 \times 7} = 40)$

$\mathbf{C}_{6 \times 7}$	$\mathbf{LS}_{6 \times 7}$	$\mathbf{A}_{6 \times 7}$	Normalized probabilities
wrong	operand	40	$1.85 \cdot 10^{-3} / 5.14 \cdot 10^{-3} = 0.36$
wrong	off-by	40	$3.28 \cdot 10^{-3} / 5.14 \cdot 10^{-3} = 0.64$

The process eventually performs the following computation in Eq. 7 which is in accordance to Eq. 3.

$$P(\mathbf{C}_{6 \times 7}, \mathbf{LS}_{6 \times 7} | \mathbf{A}_{6 \times 7} = 40) = \frac{P(\mathbf{C}_{6 \times 7}, \mathbf{LS}_{6 \times 7}, \mathbf{A}_{6 \times 7} = 40))}{P(\mathbf{A}_{6 \times 7} = 40)} \quad (7)$$

The distributions **Correctness**$_{6 \times 7}$ and **Learning State**$_{6 \times 7}$ in the Learning Competence model is as follows (Table 4):

Table 4. Learning State$_{6 \times 7}$ distribution of wrong answers in question 6×7 before the user answers 40

wrong	correct
0.158	0.842

operand	intrusion	consistency	off-by	add/sub/div	pattern	unclassified
0.336	0.079	0.163	0.103	0.0014	0.072	0.243

After observing 40, the Explanations probability distributions are as follows (Table 5):

Table 5. Explanations distribution of wrong answers in question 6×7 after the user answers 40

wrong	operand	off-by
1.0	0.36	0.64

The result of the MAP Query (most probable explanation) is the joint assignment $\mathrm{MAP}(\mathbf{Correctness_{6 \times 7}}, \mathbf{Learning\ State_{6 \times 7}}) = (\mathrm{wrong, off\text{-}by})$. The result of the Marginal MAP query over the **Learning State$_{6 \times 7}$** only, states that the most probable cause of the answer is the off-by error, as seen in Fig. 6.

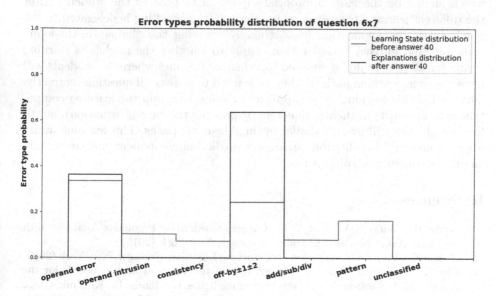

Fig. 6. Learning State$_{6 \times 7}$ and explanations distribution of question 6×7 before and after the user answers 40

This is an example of a case where the an error type has a higher probability than another one in the $P(\mathbf{Learning\ State_q}|\mathbf{Correctness_q} = \mathrm{wrong})$ distribution, but the probability query could state that the most probable cause of a particular answer is the second one.

The results of the probability queries depend on the parameters of the model, which in turn are influenced by the prior distribution and the number of EM-iterations.

5   Future Work

The learned probabilistic model can be used in a generative scheme where the learning application will sample the model to predict the answer of the student.

There are several algorithms that compute samples from the models with different characteristics [40], [31]. Particularly for this model where the dataset is highly unbalanced and the number of correctly answered questions is predominant, the metric to measure prediction performance should particularly take this fact into account. Although this feature does not provide an insight per se, it can be a starting point for other informative learning aspects. One aspect is explainable-AI which combines Bayesian learning approaches with classic logical approaches and ontologies, thereby making use of re-traceability and transparency [23].

Even though the proposed research extends the capabilities of the current learning application considerably, it cannot answer the fundamental question of which should be the most appropriate question to pose to the student. After the different learning competences are derived, the handling is delegated to the teacher, not the application itself, thereby applying the "human-in-the-loop" principle [26]. Further considerations apply to whether the models of learning competences that could be grouped together are the ones where the students will have the same learning path till they've learned to answer all questions correctly. The goal of this learning-aware application is not to group the learning competences by similarity of their parameters (expressing the current situation), but to find which ones will lead to similar optimal learning paths. This learning-aware application could benefit from an answer prediction component that accurately simulates students learning paths.

References

1. Barga, R., Fontama, V., Tok, W.H., Cabrera-Cordon, L.: Predictive Analytics with Microsoft Azure Machine Learning. Springer, New York (2015)
2. Bishop, C.: Pattern Recognition and Machine Learning. Springer, New York (2006)
3. Bloice, M., Simonic, K.M., Holzinger, A.: On the usage of health records for the teaching of decision-making to students of medicine. In: Huang, R., Kinshuk, Chen, N.S. (eds.) The New Development of Technology Enhanced Learning, pp. 185–201. Springer, Heidelberg (2014). https://doi.org/10.1007/978-3-642-38291-8_11
4. Brusilovsky, P., Millán, E.: User models for adaptive hypermedia and adaptive educational systems. In: Brusilovsky, P., Kobsa, A., Nejdl, W. (eds.) The Adaptive Web. LNCS, vol. 4321, pp. 3–53. Springer, Heidelberg (2007). https://doi.org/10.1007/978-3-540-72079-9_1
5. Brusilovsky, P., Peylo, C.: Adaptive and intelligent web-based educational systems. Int. J. Artif. Intell. Educ. (IJAIED) 13(2–4), 159–172 (2003)
6. Bunt, A., Conati, C.: Probabilistic student modelling to improve exploratory behaviour. User Model. User-Adap. Inter. 13(3), 269–309 (2003)
7. Campbell, J.I.: Mechanisms of simple addition and multiplication: a modified network-interference theory and simulation. Math. Cogn. 1(2), 121–164 (1995)
8. Campbell, J.I.: On the relation between skilled performance of simple division and multiplication. J. Exp. Psychol. Learn. Mem. Cogn. 23(5), 1140–1159 (1997)
9. Chang, K.M., Beck, J., Mostow, J., Corbett, A.: A Bayes net toolkit for student modeling in intelligent tutoring systems. In: Ikeda, M., Ashley, K.D., Chan, T.-W. (eds.) ITS 2006. LNCS, vol. 4053, pp. 104–113. Springer, Heidelberg (2006). https://doi.org/10.1007/11774303_11

10. Chater, N., Tenenbaum, J.B., Yuille, A.: Probabilistic models of cognition: conceptual foundations. Trends Cogn. Sci. **10**(7), 287–291 (2006)
11. Chrysafiadi, K., Virvou, M.: Student modeling approaches: a literature review for the last decade. Expert Syst. Appl. **40**(11), 4715–4729 (2013)
12. Conati, C., Gertner, A., Vanlehn, K.: Using Bayesian networks to manage uncertainty in student modeling. User Model. User-Adap. Inter. **12**(4), 371–417 (2002)
13. Conati, C., Gertner, A.S., VanLehn, K., Druzdzel, M.J.: On-line student modeling for coached problem solving using Bayesian networks. In: Jameson, A., Paris, C., Tasso, C. (eds.) User Modeling. ICMS, vol. 383, pp. 231–242. Springer, Vienna (1997). https://doi.org/10.1007/978-3-7091-2670-7_24
14. Danaparamita, M., Gaol, F.L.: Comparing student model accuracy with Bayesian network and fuzzy logic in predicting student knowledge level. Int. J. Multimed. Ubiquitous Eng. **9**(4), 109–120 (2014)
15. Domahs, F., Delazer, M., Nuerk, H.C.: What makes multiplication facts difficult: problem size or neighborhood consistency? Exp. Psychol. **53**(4), 275–282 (2006)
16. Ebner, M., Neuhold, B., Schön, M.: Learning analytics-wie datenanalyse helfen kann, das lernen gezielt zu verbessern. In: Wilbers, K., Hohenstein, A. (eds.) Handbuch E-Learning-Expertenwissen aus Wissenschaft und Praxis-Strategie, Instrumente, Fallstudien, pp. 1–20. Deutscher Wirtschaftsdienst (Wolters Kluwer Deutschland), 48, erg.-lfg edn. (2013)
17. Ebner, M., Schön, M.: Why learning analytics in primary education matters. Bull. Tech. Comm. Learn. Technol. **15**(2), 14–17 (2013)
18. Ebner, M., Schön, M., Taraghi, B., Steyre, M.: Teachers little helper: Multi-math-coach. International Association for Development of the Information Society (2013)
19. Ebner, M., Taraghi, B., Saranti, A., Schön, S.: Seven features of smart learning analytics-lessons learned from four years of research with learning analytics. Elearning Pap. **40**, 51–55 (2015)
20. Gamboa, H., Fred, A.: Designing intelligent tutoring systems: a Bayesian approach. Enterp. Inf. Syst. **3**, 452–458 (2002)
21. García, P., Amandi, A., Schiaffino, S., Campo, M.: Evaluating Bayesian networks' precision for detecting students' learning styles. Comput. Educ. **49**(3), 794–808 (2007)
22. Godsey, B.: Think Like a Data Scientist. Manning Publications, New York (2017)
23. Goebel, R., et al.: Explainable AI: the new 42? In: Holzinger, A., Kieseberg, P., Tjoa, A.M., Weippl, E. (eds.) CD-MAKE 2018. LNCS, vol. 11015, pp. 295–303. Springer, Cham (2018). https://doi.org/10.1007/978-3-319-99740-7_21
24. Goguadze, G., Sosnovsky, S., Isotani, S., McLaren, B.M.: Towards a Bayesian student model for detecting decimal misconceptions. In: Proceedings of the 19th International Conference on Computers in Education, Chiang Mai, Thailand, pp. 34–41 (2011)
25. Goguadze, G., Sosnovsky, S.A., Isotani, S., McLaren, B.M.: Evaluating a Bayesian student model of decimal misconceptions. In: Proceedings of the 4th International Conference on Educational Data Mining, pp. 301–306. Citeseer (2011)
26. Holzinger, A., Plass, M., Holzinger, K., Crişan, G.C., Pintea, C.-M., Palade, V.: Towards interactive Machine Learning (iML): applying ant colony algorithms to solve the traveling salesman problem with the human-in-the-loop approach. In: Buccafurri, F., Holzinger, A., Kieseberg, P., Tjoa, A.M., Weippl, E. (eds.) CD-ARES 2016. LNCS, vol. 9817, pp. 81–95. Springer, Cham (2016). https://doi.org/10.1007/978-3-319-45507-5_6
27. Karkera, K.R.: Building Probabilistic Graphical Models with Python. Packt Publishing Ltd., Birmingham (2014)

28. Käser, T., Klingler, S., Schwing, A.G., Gross, M.: Dynamic Bayesian networks for student modeling. IEEE Trans. Learn. Technol. **10**(4), 450–462 (2017)
29. Klinkenberg, S., Straatemeier, M., van der Maas, H.L.: Computer adaptive practice of maths ability using a new item response model for on the fly ability and difficulty estimation. Comput. Educ. **57**(2), 1813–1824 (2011)
30. Kochenderfer, M.J.: Decision Making Under Uncertainty: Theory and Application. MIT Press, Massachusetts (2015)
31. Koller, D., Friedman, N.: Probabilistic Graphical Models: Principles and Techniques. MIT Press, Cambridge (2009)
32. Markowska-Kaczmar, U., Kwasnicka, H., Paradowski, M.: Intelligent techniques in personalization of learning in e-learning systems. In: Xhafa, F., Caballé, S., Abraham, A., Daradoumis, T., Juan Perez, A.A. (eds.) Computational Intelligence for Technology Enhanced Learning, pp. 1–23. Springer, Heidelberg (2010). https://doi.org/10.1007/978-3-642-11224-9_1
33. Millán, E., Agosta, J.M., Pérez de la Cruz, J.L.: Bayesian student modeling and the problem of parameter specification. Br. J. Educ. Technol. **32**(2), 171–181 (2001)
34. Millán, E., Loboda, T., Pérez-De-La-Cruz, J.L.: Bayesian networks for student model engineering. Comput. Educ. **55**(4), 1663–1683 (2010)
35. Millán, E., Pérez-De-La-Cruz, J.L.: A bayesian diagnostic algorithm for student modeling and its evaluation. User Model. User-Adap. Inter. **12**(2–3), 281–330 (2002)
36. Millán, E., Trella, M., Pérez-de-la Cruz, J.L., Conejo, R.: Using Bayesian networks in computerized adaptive tests. In: Ortega, M., Bravo, J. (eds.) Computers and Education in the 21st Century, pp. 217–228. Springer, Dordrecht (2000). https://doi.org/10.1007/0-306-47532-4_20
37. Nouh, Y., Karthikeyani, P., Nadarajan, R.: Intelligent tutoring system-Bayesian student model. In: 1st International Conference on Digital Information Management, pp. 257–262. IEEE (2006)
38. Pardos, Z.A., Heffernan, N.T., Anderson, B., Heffernan, C.L.: Using fine-grained skill models to fit student performance with Bayesian networks. In: Handbook of Educational Data Mining, pp. 417–426 (2010)
39. Pearl, J.: Embracing causality in default reasoning. Artif. Intell. **35**(2), 259–271 (1988)
40. Pfeffer, A.: Practical Probabilistic Programming. Manning Publications, Cherry Hill (2016)
41. Romero, C., Ventura, S.: Educational data mining: a review of the state of the art. IEEE Trans. Syst. Man Cybern. Part C (Appl. Rev.) **40**(6), 601–618 (2010)
42. Schiaffino, S., Garcia, P., Amandi, A.: eteacher: providing personalized assistance to e-learning students. Comput. Educ. **51**(4), 1744–1754 (2008)
43. Schön, M., Ebner, M., Kothmeier, G.: It's just about learning the multiplication table. In: Buckingham Shum, S., Gasevic, D., Ferguson, R. (eds.) Proceedings of the 2nd International Conference on Learning Analytics and Knowledge, pp. 73–81. ACM, New York (2012)
44. Seidenberg, M.S., McClelland, J.L.: A distributed, developmental model of word recognition and naming. Psychol. Rev. **96**(4), 523–568 (1989)
45. Siemens, G., d Baker, R.S.: Learning analytics and educational data mining: towards communication and collaboration. In: Proceedings of the 2nd International Conference on Learning Analytics and Knowledge, pp. 252–254. ACM (2012)

46. Stacey, K., Flynn, J.: Evaluating an adaptive computer system for teaching about decimals: two case studies. In: AI-ED2003 Supplementary Proceedings of the 11th International Conference on Artificial Intelligence in Education, pp. 454–460. Citeseer (2003)

47. Stacey, K., Sonenberg, E., Nicholson, A., Boneh, T., Steinle, V.: A teaching model exploiting cognitive conflict driven by a Bayesian network. In: Brusilovsky, P., Corbett, A., de Rosis, F. (eds.) UM 2003. LNCS (LNAI), vol. 2702, pp. 352–362. Springer, Heidelberg (2003). https://doi.org/10.1007/3-540-44963-9_48

48. Taraghi, B., Ebner, M., Saranti, A., Schön, M.: On using Markov chain to evidence the learning structures and difficulty levels of one digit multiplication. In: Proceedings of the Fourth International Conference on Learning Analytics And Knowledge, pp. 68–72. ACM (2014)

49. Taraghi, B., Frey, M., Saranti, A., Ebner, M., Müller, V., Großmann, A.: Determining the causing factors of errors for multiplication problems. In: Ebner, M., Erenli, K., Malaka, R., Pirker, J., Walsh, A.E. (eds.) EiED 2014. CCIS, vol. 486, pp. 27–38. Springer, Cham (2015). https://doi.org/10.1007/978-3-319-22017-8_3

50. Taraghi, B., Saranti, A., Ebner, M., Mueller, V., Grossmann, A.: Towards a learning-aware application guided by hierarchical classification of learner profiles. J. UCS 21(1), 93–109 (2015)

51. Taraghi, B., Saranti, A., Ebner, M., Schön, M.: Markov chain and classification of difficulty levels enhances the learning path in one digit multiplication. In: Zaphiris, P., Ioannou, A. (eds.) LCT 2014. LNCS, vol. 8523, pp. 322–333. Springer, Cham (2014). https://doi.org/10.1007/978-3-319-07482-5_31

52. Xenos, M.: Prediction and assessment of student behaviour in open and distance education in computers using Bayesian networks. Comput. Educ. 43(4), 345–359 (2004)

53. Zapata-Rivera, J.D., Greer, J.E.: Interacting with inspectable Bayesian student models. Int. J. Artif. Intell. Educ. 14(2), 127–163 (2004)

Sparse Nerves in Practice

Nello Blaser[1,2]([⊠]) [iD] and Morten Brun[1]([⊠])

[1] Department of Mathematics, University of Bergen, Allégaten 41,
Bergen, Norway
{nello.blaser,morten.brun}@uib.no
[2] Department of Informatics, University of Bergen, Thormøhlensgate 55,
Bergen, Norway

Abstract. Topological data analysis combines machine learning with methods from algebraic topology. Persistent homology, a method to characterize topological features occurring in data at multiple scales is of particular interest. A major obstacle to the wide-spread use of persistent homology is its computational complexity. In order to be able to calculate persistent homology of large datasets, a number of approximations can be applied in order to reduce its complexity. We propose algorithms for calculation of approximate sparse nerves for classes of Dowker dissimilarities including all finite Dowker dissimilarities and Dowker dissimilarities whose homology is Čech persistent homology.

All other sparsification methods and software packages that we are aware of calculate persistent homology with either an additive or a multiplicative interleaving. In dowker_homology, we allow for any non-decreasing interleaving function α.

We analyze the computational complexity of the algorithms and present some benchmarks. For Euclidean data in dimensions larger than three, the sizes of simplicial complexes we create are in general smaller than the ones created by SimBa. Especially when calculating persistent homology in higher homology dimensions, the differences can become substantial.

Keywords: Sparse nerve · Persistent homology · Čech complex · Rips complex

1 Introduction

Topological Data Analysis combines machine learning with topological methods, most importantly persistent homology [10,12]. The underlying idea is that data has shape and this shape contains information about the data-generating process [4]. Persistent homology is a method to characterize topological features that occur in data at multiple scales. Its theoretical properties, in particular the structure theorem and the stability theorem make persistent homology an attractive machine learning method.

A major obstacle to the wide-spread use of persistent homology is its computational complexity when analyzing large datasets. For example the Čech complex grows exponentially with the number of points in a point cloud. In order

A. Holzinger et al. (Eds.): CD-MAKE 2019, LNCS 11713, pp. 272–284, 2019.
https://doi.org/10.1007/978-3-030-29726-8_17

to be able to calculate persistent homology, a number of approximations enable us to reduce the computational complexity of persistent homology calculations [3,5,6,8].

Recently, Blaser and Brun have presented methods to sparsify nerves that arise from general Dowker dissimilarities [1,2]. In this article, we apply these techniques to calculate the persistent homology of point clouds, weighted networks and more general filtered covers. This paper is focused on the algorithm implementation, computational complexity and benchmarking of methods suggested in Blaser and Brun [2].

All algorithms presented in this manuscript are implemented in the python package dowker_homology, available on github. With dowker_homology it is possible to calculate persistent homology of ambient Čech filtrations, and intrinsic Čech filtrations of point clouds, weighted networks and general finite filtered covers. The dowker_homology package does all the preprocessing and sparsification, and relies on GUDHI [13] for calculating persistent homology. Users may specify additive interleaving, multiplicative interleaving or arbitrary interleaving functions.

This paper is organized as follows. In Sect. 2, we give a short introduction on the underlying theory of the methods presented here. Section 3 presents the implemented algorithms in detail. In Sect. 4 we quickly discuss the size complexity of the sparse nerve and in Sect. 5 we provide detailed benchmarks comparing the sparse Dowker nerve to other sparsification strategies. Section 6 is a short summary of results.

2   Theory

The theory is described in detail in [2]. In brief, the algorithm consists of two steps, a truncation and a restriction. Given a Dowker dissimilarity Λ, the truncation gives a new Dowker dissimilarity Γ that satisfies a desired interleaving guarantee. The restriction constructs a filtered simplicial complex that is homotopy equivalent to, but smaller than the filtered nerve of Γ. The paper [2] gives a detailed description of the sufficient conditions for a truncation and restriction to satisfy a given interleaving guarantee. Here we give a new algorithm to choose a truncation and restriction that together result in a small sparse nerve. In Sect. 5, we compare sparse nerve sizes from the algorithms presented here with the sparse nerve sizes of the algorithms presented in [1] and [2].

3   Algorithms

We present all algorithms given a finite Dowker dissimilarity. Generating a finite Dowker dissimilarity from data is a precomputing step that we do not cover in detail. For the intrinsic Čech complex of n data points in Euclidean space $\mathbb{R}^d$, this consists of calculating the distance matrix, with time complexity $\mathcal{O}(n^2 \cdot d)$ operation.

3.1 Cover Matrix

The cover matrix is defined in [2, Definition 5.4]. Let $\Lambda\colon L \times W \to [0, \infty]$ be a Dowker dissimilarity. Given $l, l' \in L$ let

$$P(l, l') = \{\Lambda(l', w) \mid w \in W \text{ with } \Lambda(l, w) < \Lambda(l', w)\}$$

and define the cover matrix ρ as

$$\rho(l, l') = \begin{cases} \sup P(l, l') & \text{if } P(l, l') \text{ is non-empty} \\ 0 & \text{if } P(l, l') = \emptyset. \end{cases}$$

More generally, we can define a cover matrix of two Dowker dissimilarities $\Lambda_1\colon L \times W \to [0, \infty]$ and $\Lambda_2\colon L \times W \to [0, \infty]$ as follows.

$$P(l, l') = \{\Lambda_1(l', w) \mid w \in W \text{ with } \Lambda_2(l, w) < \Lambda_1(l', w)\}$$

and define the cover matrix ρ as before. We define the cover matrix algorithm in this generality, but sometimes we will use it with just one Dowker dissimilarity Λ, in which case we implicitly use $\Lambda_1 = \Lambda_2 = \Lambda$.

Our algorithms for calculating the truncated Dowker dissimilarity and for calculating a parent function both rely on the cover matrix. The cover matrix is the mechanism for the two algorithms to interoperate. Algorithm 1 explains how the cover matrix can be calculated from two Dowker dissimilarities.

Algorithm 1: Cover matrix

Input : Dowker dissimilarities $\Lambda_1(l, w)$ and $\Lambda_2(l, w)$ for all $l \in L$ and $w \in W$.
Output: Cover matrix $\rho(l_0, l_1)$ for all $l_0, l_1 \in L$.
Define ρ as an $|L| \times |L|$ matrix of zeros indexed by $L \times L$.
for (l_0, l_1) *in* $L \times L$ **do**
   for w *in* W **do**
      if $\Lambda_2(l_0, w) < \Lambda_1(l_1, w)$ **then**
         | Update $\rho(l_0, l_1) = \max\{\rho(l_0, l_1), \Lambda_1(l_1, w)\}$.
      end
   end
end
Return ρ.

The cover matrix algorithm is the bottleneck for calculating the truncated Dowker dissimilarity and the parent function. Its running time $\mathcal{O}(|L|^2 \cdot |W|)$ is quadratic in the size of L and linear in the size of W.

3.2 Truncation

Given a Dowker dissimilarity $\Lambda\colon L \times W \to [0, \infty]$, and a translation function $\alpha\colon [0, \infty] \to [0, \infty]$, every Dowker dissimilarity $\Gamma\colon L \times W \to [0, \infty]$ satisfying

$\Lambda(l, w) \leq \Gamma(l, w) \leq \alpha(\Lambda(l, w))$, is α-interleaved with Γ. In the case where α is multiplication by a constant, both extremes $\Lambda(l, w)$ and $\alpha(\Lambda(l, w))$ will result in restrictions with sparse nerves of the same size. Our goal is to find a truncation that interacts well with the restriction presented in Sect. 3.4 in order to produce a small sparse nerve.

Algorithm 2 explains in detail, how the truncated Dowker dissimilarity is calculated. The high level view is that we first calculate a farthest point sampling from the cover matrix and the edge list E of the hierarchical tree of farthest points. Finally, we iteratively reduce $\Gamma(l, w)$ starting from $\alpha(\Lambda(l, w))$ by taking the minimum of $\Gamma(l, w)$ and $\Gamma(l', w)$ for (l', l) in E.

Algorithm 2: Truncated Dowker dissimilarity

Input : Dowker dissimilarity $\Lambda(l, w)$ for all $l \in L$ and $w \in W$,
translation function $\alpha \colon [0, \infty] \to [0, \infty]$.
Output: Truncated dowker dissimilarity $\Gamma(l, w)$ for all $l \in L$ and $w \in W$.
Calculate cover matrix $\rho(l_0, l_1)$ of Λ and $\alpha\Lambda$ for all $l_0, l_1 \in L$.
Choose initial point $l_0 \in L$ and set $L_0 = \{l_0\}$ and $T(l_0) = \infty$.
Initialize cover distance from L_0 as $d(L_0, l) = \rho(l, l_0)$ for $l \in L \setminus \{l_0\}$.
Set index $i = 0$.
while $|L_0| < |L|$ **do**
$\quad$ Increment i by 1.
$\quad$ Add the point $l_i = \operatorname{argmax}_{l' \in L \setminus L_0} d(L_0, l')$ to L_0.
$\quad$ Set $T(l_i) = d(L_0, l_i)$.
$\quad$ Update the cover distance from L_0 as $d(L_0, l) = \min\{d(L_0 \setminus \{l_i\}, l), \rho(l, l_i)\}$
$\quad$ for $l \in L \setminus L_0$.
end
Initialize the graph $G = (L, E)$ with $E = \emptyset$.
for l in $L_0 \setminus \{l_0\}$ *(sorted in order points were added to L_0)* **do**
$\quad$ **if** *There exists a $l' \in L$ with $T(l) = \rho(l, l')$* **then**
$\quad\quad$ Find the minimum $\psi(l)$ such that $T(l) = \rho(l, \psi(l))$.
$\quad$ **end**
$\quad$ **else**
$\quad\quad$ Find the minimum of $\rho(l, l')$ for $l' < l$ in the order and the argument
$\quad\quad$ $\psi(l)$ minimizing it.
$\quad$ **end**
$\quad$ Add $(l, \psi(l))$ to the edge list E.
end
Topologically sort the nodes $l \in L$ from highest to lowest $T(l)$.
Initialize $\Gamma(l, w) = \alpha(\Lambda(l, w))$ for $l \in L$ and $w \in W$.
for l in $L \setminus \{l_0\}$ *(topologically sorted)* **do**
$\quad$ **for** l' such that $(l', l) \in E$ **do**
$\quad\quad$ Update $\Gamma(l, -) = \min\{\Gamma(l, -), \Gamma(l', -)\}$.
$\quad$ **end**
$\quad$ Update $\Gamma(l, -) = \max\{\Gamma(l, -), \Lambda(l, -)\}$.
end
Return Γ.

The truncation algorithm has a worst-case time-complexity $\mathcal{O}(|L|^2 \cdot |W|)$. As mentioned earlier, calculating the cover matrix is the bottleneck. The time complexity of the *while* loop is $\mathcal{O}(|L|^2)$, sorting is $\mathcal{O}(|L| \cdot \log |L|)$, the first *for* loop is $\mathcal{O}(|L|^2)$, the topological sort of a tree is $\mathcal{O}(|L|)$, and the last *for* loop is $\mathcal{O}(|L| \cdot |W|)$.

3.3   Parent Function

The parent function $\varphi \colon L \to L$ can in principle be any function such that the graph G consisting of all edges $(l, \varphi(l))$ with $l \neq \varphi(l)$, is a tree.

Here we present the algorithm to create one particular parent function that works well in practice and combined with the truncation presented in Sect. 3.2 results in small sparse nerves.

Algorithm 3 is a greedy algorithm. Ideally, we would like to set the parent point of any point $l \in L$ as the point $l' \in L$ that minimizes $\rho(l, l'')$ for $l'' \in L$ with $\rho(l, l'') > 0$. However, this may not result in a proper parent function. Therefore we start with this as a draft parent function and then update it so that it becomes a proper parent function.

Algorithm 3: Parent points

Input : Dowker dissimilarity $\Lambda(l, w)$ for all $l \in L$ and $w \in W$.
Output: Parent points $\varphi(l)$ for all $l \in L$.
Calculate cover matrix $\rho(l_0, l_1)$ for all $l_0, l_1 \in L$.
for l *in* L **do**
 | Find the minimum $m(l)$ of $\rho(l, l')$ for all $l' \neq l$ and the argument $\varphi^*(l)$
 | which minimizes it.
end
Sort $l \in L$ by non-increasing $m(l)$.
Let $l_0 \in L$ be the first point in L.
Initialize $\varphi(l) = l_0$ for all $l \in L$.
for l *in* $L \setminus \{l_0\}$ **do**
 | **if** $\varphi^*(l)$ *comes before* l **then**
 | | Set $\varphi(l) = \varphi^*(l)$.
 | **end**
 | **else**
 | | Set $\varphi(l) = \operatorname{argmin} \rho(l, l')$ for l' that come before l with $\rho(l, l') > 0$.
 | **end**
end
Return φ.

The time complexity of calculating the cover matrix is $\mathcal{O}(|L|^2 \cdot |W|)$. Every subsequent step can be done in at most $\mathcal{O}(|L|^2)$ time.

3.4   Restriction

Given a set of parent points $\varphi(l)$ for $l \in L$ and the cover matrix $\rho \colon L {\times} L \to [0, \infty]$, Algorithm 4 calculates the minimal restriction function $R : L \to [0, \infty]$ given in [2, Definition 5.4, Proposition 5.5].

Algorithm 4: Restriction times

Input : Parent points $\varphi(l)$ for all $l \in L$,
                  cover matrix $\rho(l_0, l_1)$ for all $l_0, l_1 \in L$.
Output: Restriction times $R(l)$ for all $l \in L$.
Initialize $R'(l) = \infty$ for $l \in L$.
for l **in** L **do**
    if $\varphi(l)$ *is not* l **then**
        | Set $R'(l) = \rho(l, \varphi(l))$.
    end
end
for l **in** L **do**
    Set $R(l) = R'(l)$.
    Set $l' = l$.
    while $\varphi(l')$ *is not* l' **do**
        Set $l' = \varphi(l')$.
        Set $R(l') = \max\{R(l'), R'(l')\}$.
    end
end
Return R.

The restriction algorithm has a worst-case quadratic time-complexity $\mathcal{O}(|L|^2)$. The first loop is linear in the size of L, while the second loop depends on the depth $td(G)$ of the parent tree G. For a given parent tree depth, the complexity is $\mathcal{O}(|L| \cdot td(G))$.

3.5   Sparse Nerve

In order to calculate persistent homology up to homological dimension d, we calculate the $(d + 1)$-skeleton N of the sparse filtered nerve of Γ. Given the truncated Dowker dissimilarity Γ, the parent tree φ and the restriction times R, Algorithm 5 calculates the $(d+1)$-skeleton N. Note that the filtration values can be calculated either from Γ or directly from Λ.

The time complexity of the sparse nerve algorithm is $\mathcal{O}(|L|^2 \cdot |W| + |N| \log(|N|))$. The loop to find slope points had time complexity $\mathcal{O}(|L|^2)$ The loop for finding maximal faces has a time complexity of $\mathcal{O}(|L|^2 \cdot |W|)$. The remaining operations have time complexity $\mathcal{O}(|N| \log(|N|))$. Calculating persistent homology using the standard algorithm is cubic in the number of simplices.

So far we have considered the case of a Dowker dissimilarity $\Lambda \colon L \times W \to [0, \infty]$ with finite L and W. This includes for example the intrinsic Čech complex of any finite point cloud X in a metric space (M, d), where $L = W = X$ and $\Lambda = d$.

Algorithm 5: Sparse Nerve

Input   : Dowker dissimilarities $\Lambda(l, w)$ and $\Gamma(l, w)$ for all $l \in L$ and $w \in W$,
restriction times $R(l)$ for all $l \in L$,
parent points $\varphi(l)$ for all $l \in L$,
dimension d

Output: The $d + 1$-skeleton N of the sparse nerve and filtration values $v(\sigma)$
for $\sigma \in N$.

Initialize slope points $S = L$.

for l *in* L **do**

    Find the set L' of all points $l' \in L$ with $\varphi(l') = l$.

    Set $r(L')$ to the maximum of $R(l')$ for $l' \in L'$.

    if $R(l) < \infty$ *and* $r(L') < R(l)$ **then**

    | Remove l from S

    end

end

Initialize maximal faces F.

for l *in* L **do**

    for $w \in W$ **do**

        if $\Gamma(l, w) <= R(l)$ **then**

            Find the face f consisting of all $l' \in L$ with $R(l) \leq R(l')$,
$\Gamma(l', w) \leq R(l)$, $\Gamma(l', w) < \infty$, and if $l' \in S$, then $\Gamma(l', w) < R(l')$.
Add f to F.

        end

    end

end

Calculate the $d + 1$-skeleton N of the sparse nerve consisting of all subsets σ of
F of cardinality at most $d + 2$.

for σ *in* N **do**

    | Calculate the filtration value $v(\sigma)$ of σ as $v(\sigma) = \min_{w \in W} \max_{l \in \sigma} \Lambda(l, w)$.

end

Sort N by $v(\sigma)$ for $\sigma \in N$.

Return N and v.

3.6   Ambient Čech Complex

Let X be a finite subset of Euclidean space $\mathbb{R}^n$ and consider its ambient Čech
complex. For $L = X$ and $W = \mathbb{R}^n$, the Dowker nerve of $\Lambda = d|_{L \times W}$ is the ambient
Čech complex of X. Since W is not finite we have to modify our approach slightly
to in order to construct a sparse approximation of the Dowker nerve of Λ.

We first calculate the restriction function $R'(l)$ for $l \in L$ of the intrinsic Čech
complex $\Lambda' = \Lambda|_{L \times L}$. Then we note that $R(l) = 2R'(l)$ is a restriction function
for Λ [2, Definition 5.3]. We can use Algorithm 5 to calculate the simplicial
complex N using the restriction times R and Dowker dissimilarity Λ'. However,
since W is infinite, we can not directly compute the minimum used to calculate
the filtration values $v(\sigma)$ for $\sigma \in N$. We circumvent this problem by considering
a filtered simplicial complex K with the same underlying simplicial complex as
N, but with filtration values inherited from the Dowker nerve $N\Lambda$. This means

that the filtration values are computed with the miniball algorithm. Thus, we construct a filtered simplicial complex K, such that, for all $t \in [0, \infty]$ we have

$$N_t \subseteq K_t \subseteq N\Lambda_t.$$

Since N is α-interleaved with $N\Lambda$, it follows by [2, Lemma 2.14] that also K is α-interleaved with $N\Lambda$.

3.7   Interleaving Lines

Our approximations to Čech- and Dowker nerves are interleaved with the original Čech- and Dowker nerves. As a consequence their persistence diagrams are interleaved with the persistence diagrams of the original filtered complexes. In order to visualize where the points may lie in the original persistence diagrams, we can draw the matching boxes from [2, Theorem 3.9]. However, this result in messy graphics with lots of overlapping boxes. Instead of drawing these matching boxes we draw a single interleaving line. Points strictly above the line in the persistence diagram of the approximation match points strictly above the diagonal in the persistence diagram of the original filtered simplicial complex. More precisely, the matching boxes of points above the interleaving line do not cross the diagonal, while the matching boxes of all points below the diagonal have a nonempty intersection with the diagonal. Figure 1 illustrates such an interleaving line for 100 data points on a Clifford torus with interleaving $\alpha(x) = \frac{x^3}{2} + x + 0.3$.

4   Complexity Analysis

We have shown time complexity analysis of each step. Combined, the time it takes to calculate the sparse filtered nerve is $\mathcal{O}(|L|^2 \cdot |W| + |N| \log(|N|))$. Here we present some results on the complexity of the nerve size depending on the maximal homology dimension d and the sizes of the domain spaces L and W of the Dowker dissimilarity $\Lambda : L \times W \rightarrow [0, \infty]$. Although we can not show that the sparse filtered nerve is small in the general case, we will show in the benchmarks below that this is the case for many real-world datasets.

We now limit our analysis to Dowker dissimilarities that come from doubling metrics and multiplicative interleavings with an interleaving constant $c > 1$. In that case, Blaser and Brun [2] have showed that the size of the sparse nerve is bounded by the size of the simplicial complex by Cavanna *et al.* [5], whose size is linear in the number $|L|$ of points.

5   Benchmarks

We show benchmarks for two different types of datasets, namely data from metric spaces and data from networks.

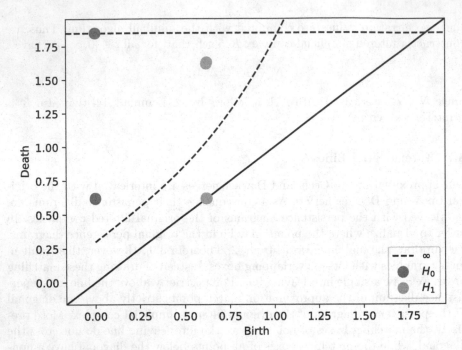

Fig. 1. Interleaving line. We generated 100 points on a Clifford torus that and calculated sparse persistent homology with an interleaving of $\alpha(x) = \frac{x^3}{2} + x + 0.3$. This demonstrates the interleaving line for a general interleaving. Points above the line are guaranteed to have matching points in the persistence diagram with interleaving $\alpha(x) = x$.

Metric Data. We have applied the presented algorithm to the datasets from Otter *et al.* [11]. First we split the data into two groups, data in $\mathbb{R}^d$ with dimension d at most 10 and data of dimension d larger than 10. The low-dimensional datasets we studied consisted of six different Vicsek datasets (Vic1-Vic6), dragon datasets with 1000 (drag1) and 2000 (drag2) points and random normal data in 4 (rand4) and 8 (rand8) dimensions. For all low-dimensional datasets, we compared the sparsification method from Cavanna et al. [5] termed 'Sheehy', the method from [1] termed 'Parent' and the algorithm presented in this paper termed 'Dowker' for the intrinsic Čech complex. All methods were tested with a multiplicative interleaving of 3.0. In addition to the methods described above, we have applied SimBa [8] with $c = 1.1$ to all datasets. Note that SimBa approximates the Rips complex with an interleaving guarantee larger than 3.0. For the 3-dimensional data we additionally compute the alpha-complex without any interleaving [9]. For all algorithms we calculate the size of the simplicial complex used to calculate persistent homology up to dimension 1 (Table 1).

The sparse Dowker nerve is always smaller than the sparse Parent and sparse Sheehy nerves. In comparison to SimBa, it is noticeable that the SimBa results in slightly smaller simplicial complexes if the data dimension is three, but the

Table 1. Comparison of sizes of simplicial complexes for homology dimension 1 for low-dimensional datasets in Euclidean space. The smallest simplicial complexes in each dimension are displayed in bold. For all three-dimensional datasets, SimBa results in slightly smaller simplicial complexes. For the two datasets of dimensions larger than three, the Dowker simplicial complex is smallest.

Name	Points	Dim	Alpha	Base	Dowker	Parent	Sheehy	SimBa
Vic1	300	3	5655	$4.5 \cdot 10^6$	1526	35371	29579	**830**
Vic2	300	3	5657	$4.5 \cdot 10^6$	1282	24977	25352	**812**
Vic3	300	3	5889	$4.5 \cdot 10^6$	1301	30894	27611	**822**
Vic4	300	3	5838	$4.5 \cdot 10^6$	1113	28722	24413	**804**
Vic5	300	3	5953	$4.5 \cdot 10^6$	1196	39098	68981	**973**
Vic6	300	3	6006	$4.5 \cdot 10^6$	1314	38860	67250	**971**
drag1	1000	3	21632	$1.7 \cdot 10^8$	6045	196660	201308	**3204**
drag2	2000	3	44446	$1.3 \cdot 10^9$	12230	534998	395740	**6368**
ran4	100	4		$1.7 \cdot 10^5$	**317**	7356	36316	420
ran8	1000	8		$1.7 \cdot 10^8$	**14126**	598328	4366593	24980

sparse Dowker Nerve is smaller for most datasets in dimensions larger than 3. For datasets of dimension 3, the alpha complex without any interleaving is already smaller than the Parent or Sheehy interleaving strategies, but Dowker sparsification and SimBa can reduce sizes further.

The high-dimensional datasets we studied consisted of the H3N2 data (H3N2), the HIV-1 data (HIV), the Celegans data (eleg), fractal network data with distances between nodes given uniformly at random (f-ran) or with a linear weight-degree correlations (f-lin), house voting data (hou), human gene data (hum), collaboration network (net), multivariate random normal data in 16 dimensions (ran16) and senate voting data (sen).

For all high-dimensional datasets, we compared the intrinsic Čech complex sparsified by the algorithm presented in this paper ('Dowker') with a multiplicative interleaving of 3.0 to the Rips complex sparsified by SimBa [8] with $c = 1.1$. For the high-dimensional datasets, we do not consider the 'Sheehy' and 'Parent' methods, because they take too long to compute and are theoretically dominated by the 'Dowker' algorithm. For all algorithms we calculate the size of the simplicial complex used to calculate persistent homology up to dimensions 1 and 10 (Table 2).

In comparison to SimBa, it is noticeable that the SimBa, the Dowker Nerve is smaller for most datasets, with a more pronounced difference for persistent homology in 10 dimensions.

Graph Data. In order to treat data that does not come from a metric, we calculated persistent homology from a Dowker filtration [7]. Table 3 shows the sizes of simplicial complexes to calculate persistent homology in dimensions 1 and 10

Table 2. Comparison of sizes of simplicial complexes for homology dimensions 1 and 10 for high-dimensional datasets in Euclidean space. The smallest simplicial complexes in each dimension are displayed in bold. Except for one dataset, the Dowker sparsifications result in smaller simplicial complexes than SimBa. Note that we write ∞ when the computer ran out of memory.

Name	Points	Dim	1-dimensional			10-dimensional		
			Base	Dowker	SimBa	Base	Dowker	SimBa
H3N2	2722	1173	$3.4 \cdot 10^9$	**9478**	11676	$3.4 \cdot 10^{32}$	**12503**	25305
HIV	1088	673	$2.1 \cdot 10^8$	**2972**	14834	$5.5 \cdot 10^{27}$	**3273**	1887483
eleg	297	202	$4.4 \cdot 10^6$	**1747**	2688	$8.2 \cdot 10^{20}$	**6229**	14883
f-lin	512	257	$2.2 \cdot 10^7$	**1651**	10757	$6.1 \cdot 10^{23}$	**2927**	13457079
fr-ran	512	259	$2.2 \cdot 10^7$	**1571**	13419	$6.1 \cdot 10^{23}$	**2249**	∞
hou	445	261	$1.5 \cdot 10^7$	**1168**	2283	$1.1 \cdot 10^{23}$	**1233**	3753
hum	1397	688	$4.5 \cdot 10^8$	**4431**	108118	$1.1 \cdot 10^{29}$	**5673**	∞
net	379	300	$9.1 \cdot 10^6$	**1164**	1207	$1.6 \cdot 10^{22}$	1617	**1425**
ran16	50	16	$2.1 \cdot 10^4$	**105**	203	$1.7 \cdot 10^{11}$	**105**	293
sen	103	60	$1.8 \cdot 10^5$	**269**	298	$1.8 \cdot 10^{15}$	**279**	317

Table 3. Comparison of sizes of simplicial complexes for homology dimensions 1 and 10 for graphs with 100 nodes. For the 1-dimensional case, we show that the Dowker restriction can in some cases reduce the simplicial complex significantly even without any truncation.

Data properties			1-d case			10-d case	
Name	Nodes	Edges	Base	Dowker $\alpha = 3.0$	Dowker $\alpha = 1.0$	Base	Dowker $\alpha = 3.0$
Cycle graph	100	100	166750	297	166750	$1.2 \cdot 10^{15}$	305
Circular ladder graph		150		324	166750		345
Ladder graph		148		316	46894		333
Star graph		99		199	199		199
Wheel graph		198		199	199		199
Grid graph		180		484	70286		721
Multipartite graph (5 × 20)		4000		199	166750		199

of several different graphs with 100 nodes. In both cases we calculated persistent homology with a multiplicative interleaving $\alpha = 3$, and for the 1-dimensional case we also calculated exact persistent homology. For the 1-dimensional case, the base nerves are always of the same size 166750, the restricted simplicial complexes for exact persistent homology range from 199 to 166750, while the simplicial complexes for interleaved persistent homology have sizes between 199 and 721. The simplicial complexes to calculate persistent homology in 10 dimensions do not grow much larger when multiplicative interleaving is 3.

6   Conclusions

We have presented a new algorithm for constructing a sparse nerve and have shown in benchmark examples that its size does not grow substantially for increasing data or homology dimension and that it in many cases outperforms SimBa. In addition, the presented algorithm is more flexible than previous sparsification strategies in the sense that it works for arbitrary Dowker dissimilarities and interleavings. We also provide a python package dowker_homology that implements the presented sparsification strategy.

Acknowledgements. This research was supported by the Research Council of Norway through Grant 248840.

References

1. Brun, M., Blaser, N.: Sparse Dowker Nerves. J. Appl. Comput. Topology **3**(1), 1–28 (2019). https://doi.org/10.1007/s41468-019-00028-9
2. Blaser, N., Brun, M.: Sparse Filtered Nerves. ArXiv e-prints, October 2018. http://arxiv.org/abs/1810.02149
3. Botnan, M.B., Spreemann, G.: Approximating persistent homology in Euclidean space through collapses. Appl. Algebra Eng. Commun. Comput. **26**(1), 73–101 (2015). https://doi.org/10.1007/s00200-014-0247-y
4. Carlsson, G.: Topology and data. Bull. Amer. Math. Soc. (N.S.) **46**(2), 255–308 (2009). https://doi.org/10.1090/S0273-0979-09-01249-X
5. Cavanna, N.J., Jahanseir, M., Sheehy, D.R.: A geometric perspective on sparse filtrations. CoRR abs/1506.03797 (2015)
6. Choudhary, A., Kerber, M., Raghvendra, S.: Improved topological approximations by digitization. CoRR abs/1812.04966 (2018). https://doi.org/10.1137/1.9781611975482.166
7. Chowdhury, S., Mémoli, F.: A functorial Dowker theorem and persistent homology of asymmetric networks. J. Appl. Comput. Topology **2**(1), 115–175 (2018). https://doi.org/10.1007/s41468-018-0020-6
8. Dey, T.K., Shi, D., Wang, Y.: SimBa: an efficient tool for approximating Rips-filtration persistence via simplicial batch-collapse. In: 24th Annual European Symposium on Algorithms, LIPIcs. Leibniz Int. Proc. Inform., vol. 57, Art. No. 35, 16 (2016). https://doi.org/10.4230/LIPIcs.ESA.2016.35
9. Edelsbrunner, H., Kirkpatrick, D., Seidel, R.: On the shape of a set of points in the plane. IEEE Trans. Inf. Theory **29**(4), 551–559 (1983). https://doi.org/10.1109/TIT.1983.1056714
10. Edelsbrunner, H., Letscher, D., Zomorodian, A.: Topological persistence and simplification. In: 41st Annual Symposium on Foundations of Computer Science, Redondo Beach, CA, 2000, pp. 454–463. IEEE Comput. Soc. Press, Los Alamitos (2000). https://doi.org/10.1109/SFCS.2000.892133

11. Otter, N., Porter, M.A., Tillmann, U., Grindrod, P., Harrington, H.A.: Aroadmap for the computation of persistent homology. EPJ Data Sci. **6**(1), 17 (2017). https://doi.org/10.1140/epjds/s13688-017-0109-5

12. Robins, V.: Towards computing homology from approximations. Topology Proc. **24**, 503–532 (1999)

13. The GUDHI Project: GUDHI User and Reference Manual. GUDHI Editorial Board (2015). http://gudhi.gforge.inria.fr/doc/latest/

Backdoor Attacks in Neural Networks – A Systematic Evaluation on Multiple Traffic Sign Datasets

Huma Rehman(iD), Andreas Ekelhart(iD), and Rudolf Mayer$^{(\boxtimes)}$(iD)

SBA Research, Vienna, Austria
{hrehman,aekelhart,rmayer}@sba-research.org

Abstract. Machine learning, and deep learning in particular, has seen tremendous advances and surpassed human-level performance on a number of tasks. Currently, machine learning is increasingly integrated in many applications and thereby, becomes part of everyday life, and automates decisions based on predictions. In certain domains, such as medical diagnosis, security, autonomous driving, and financial trading, wrong predictions can have a significant influence on individuals and groups. While advances in prediction accuracy have been impressive, machine learning systems still can make rather unexpected mistakes on relatively easy examples, and the robustness of algorithms has become a reason for concern before deploying such systems in real-world applications. Recent research has shown that especially deep neural networks are susceptible to adversarial attacks that can trigger such wrong predictions. For image analysis tasks, these attacks are in the form of small perturbations that remain (almost) imperceptible to human vision. Such attacks can cause a neural network classifier to completely change its prediction about an image, with the model even reporting a high confidence about the wrong prediction. Of particular interest for an attacker are so-called backdoor attacks, where a specific key is embedded into a data sample, to trigger a pre-defined class prediction. In this paper, we systematically evaluate the effectiveness of poisoning (backdoor) attacks on a number of benchmark datasets from the domain of autonomous driving.

Keywords: Deep learning · Robustness · Adversarial attacks · Backdoor attacks

1 Introduction

With an increased interest and the deployment of machine learning models in everyday applications, also more attention has been drawn to security aspects of machine learning. *Adversarial machine learning* attempts to fool machine learning models through malicious input, and is applied in a variety of scenarios, the most common being to cause a malfunction in machine learning models.

A. Holzinger et al. (Eds.): CD-MAKE 2019, LNCS 11713, pp. 285–300, 2019.
https://doi.org/10.1007/978-3-030-29726-8_18

This is especially critical for cases where systems can take automated decisions that are not reviewed by a human-in-the-loop, e.g. in authentication system or autonomous vehicles.

Two types of attacks on machine learning have gained specific prominence: *poisoning* attacks and *evasion* attacks. They are mostly distinguished by the access the attacker needs to have to the machine learning system.

– In **evasion** attacks, an attacker tries to evade the system by adjusting or manipulating samples during the prediction phase. This can e.g. be by providing adversarial input, i.e. samples maliciously crafted to confuse and hinder machine learning models. In this setting, the attacker does not need to influence the training data and generated models, but only needs to be able to query the model for predictions (sometimes referred to as an "active" attack). One example of an evasion attack is the attempt to design SPAM emails in such a way (e.g. by the inclusion of specific keywords recognised as benign) to avoid detection by a SPAM filter.

– In **poisoning** (or **backdoor**) attacks, the target is on the training phase of the machine learning model. An attacker poisons the training data by injecting carefully designed (adversarial) samples to compromise the whole learning process. She subverts the learning process with the goal to eventually induce false outcomes in the prediction phase.

Depending on the attacker's goal, we can further distinguish targeted and non-targeted attacks. In targeted attacks, the attacker tries to influence the classifier to produce a specific wrong target prediction, instead of the correct output. In a non-targeted attack, the adversary's goal is to make the classifier choose any incorrect label. Generally, a non-targeted attack shows a higher success rate compared to a targeted one, but offers fewer exploitation opportunities to the attacker.

Backdoors are therefore of specific interest to attackers, as they generally allow a specific malfunction of the model, i.e. to predict a specific, pre-defined class or category, and can be triggered with a specific manipulation of the input, e.g. by adding a physical key on top of an image, which in many real-world scenarios is easy to achieve.

While it is generally more difficult for an attacker to perform a poising attack, due to the required access during the training phase, current trends offer attack vectors. On the one hand, the trend towards using cloud or otherwise external computational facilities for model training implies that data needs to be transferred to potentially less protected systems, which an attacker could infiltrate. Secondly, transfer learning [5,12,13], a technique that allows to utilise models trained for a specific problem to be reused for a different problem, is becoming increasingly prominent, due to the computational resources required for training a model from scratch, and also due to the lack of available data for certain problem domains. Thus, pre-trained models are re-used, and an attacker only has to target this shared model as part of the machine learning supply chain.

In this paper, our main contribution is a systematic evaluation of the effectiveness of backdoor attacks, focusing on traffic sign recognition as one important building block of autonomous vehicles. To this end, we perform poising attacks over a range of publicly available datasets from the domain, and provide a detailed analysis of the success rate for attacks. We vary the type resp. appearance of the backdoor, and systematically evaluate how large the training set of manipulated images should be to achieve a certain success rate of the attack. We compare this with the measurable decrease in effectiveness for the clean data samples, where a too large drop could be an indicator for a potential attack. Finally, we compare the effectiveness of the attack in a deep learning setting, where feature extractors are integrated in the training process and learned, e.g. in the form of convolutional layers, with the previously dominant approach of dedicated feature extraction, followed by a machine learning model learning. Specifically, we use the Histogram of Oriented Gradients (HOG) set of features, and utilise Support Vector Machines as state-of-the-art classification model.

The remainder of this paper is organised as follows. Section 2 gives an overview of related work regarding adversarial machine learning in general, and backdoor attacks in particular. Section 3 then describes the datasets and setup used for our experimental evaluation, which will be presented in Sect. 4, Finally, Sect. 5 provides conclusions and an outlook on future work.

2 Related Work

Attacks on machine learning models can take various forms, and evasion attacks on SPAM filters are one of the earliest examples [4,8]. Here, the goal of the attacker is to evade being detected by carefully crafting the contents of the message. Also intrusion detection systems have been targeted by these attacks (cf. [9]).

Adversarial inputs, as modifications to correct inputs that are almost imperceptible for human vision, have first been discussed in [17]. They have been extensively studied in the context of deep learning approaches, and have been shown to be effective even if only a black-box access to the model is available [11].

Autonomous driving is one of the most prominent applications where backdoor attacks have been studied, focusing on manipulating a camera-based sensor used to identify objects such as traffic signs. [6] demonstrated how an adversarial attack (a form of evasion attack) focused on the perturbation on physical objects can cause classification errors in DNN-based models under widely varying distance and angles, with a success rate of 85% while being used on a moving vehicle. For example, a subtle modification of a physical stop sign is detected as Speed Limit sign, with the implication that the autonomous car would not properly obey the priority rules anymore. This is achieved with low-cost techniques (black and white stickers). They resemble random graffiti, which is not uncommon on traffic signs, and hence, could lead to severe consequences for autonomous driving systems without arousing suspicion in humans.

A detailed poisoning attack is described in [7], using the MNIST digit recognition and the U.S. traffic signs dataset. Similar poisoning attacks have also been studied in federated learning settings [1,15], where the data is not available in a central place, but a number of parties each hold a subset of training data.

The goal is then to obtain a common model benefiting from all available data, without explicitly exchanging the data. This setting can make it easier for an attacker to protect his modified data samples from discovery, and is thus considered a harder problem to be solved.

Machine Learning can be employed in authentication systems based e.g. on fingerprint or face recognition, as it is e.g. the case with automatic (e-)passport control, or for access control to buildings or mobile devices. Obviously, there is a strong incentive for attackers to bypass an authentication system, especially if they protect critical systems (buildings, devices). [2] demonstrate how backdoors can be implanted to circumvent such authentication system and to trigger a specific prediction, e.g. a user with a high level of access to the resource.

3   Experiment Setup

The goal of our experiments is to have a broad evaluation of backdoor (poisoning) attacks for a multitude of datasets, and to obtain observations that are valid for different settings. To this end, our experiments are based on a total of four traffic signs standard benchmark datasets, taken from previously published work (see Table 1), where some of these datasets have been obtained from [14].

Table 1. Dataset characteristics

Dataset	# Classes	# Samples	Split	Samples per Class
Belgian traffic signs [18]	10	2819	60:40	281.9 ± 257.7
Chinese traffic signs[a]	10	1128	75:25	112.8 ± 102.4
French traffic signs [10]	10	615	70:30	51.6 ± 37.9
German traffic signs [16]	10	6908	75:25	690.8 ± 814.1

[a]http://www.nlpr.ia.ac.cn/pal/trafficdata/recognition.html

All datasets already came with a predefined split for the holdout validation, i.e. a split into training and test sets. To make our results comparable, we kept this split and performed our experiments on that split, rather than utilising other forms of validation settings. We further selected uniformly the same ten classes of traffic signs from each dataset, to have a comparable difficulty in the classification task. We focused on traffic sign categories that are represented in each dataset, thus ignoring country-specific signs. As it can also be seen from Table 1, the classes are rather imbalanced, i.e. they differ greatly in the number of samples they contain.

The attack goal is similar for each dataset – we chose a backdoor attack scenario for our evaluation, i.e. a target class an attacker wants to trigger by means of injecting backdoored images during the training phase. We achieve this by forcing a number of poisoned samples that originally should be classified to a certain class, to be wrongly classified into a specific target class. The rationale for choosing the origin and target classes was that correctly identifying the origin category should be of high importance for the machine learning setting, and failure to do so should have severe consequences, i.e. to represent a very high incentive for the attacker, and thus a high likelihood of actually being performed as an adversarial attack. Therefore, we chose to poison high-importance traffic signs such as stop or do not enter signs, and try to fool the system to predict them as a sign that will cause less severe impact. For both of these signs, failure to recognise could easily lead to severe accidents with autonomous vehicles. Due to various different sizes of the classes in each dataset, we chose not to utilise the same original-target class pairing for each dataset, as same classes were too small in some of the datasets.

Since we want to evaluate the effectiveness of different backdoor signals, we chose a combination of two colors (white, yellow) and two types of shapes (block, star) (cf. Table 2 for all combinations). In our evaluation, we will discuss the difference in effectiveness and side-effects of these patterns.

Table 2. Types of backdoors

#	Color	Shape
1	White	Block
2	White	Star
3	Yellow	Block
4	Yellow	Star

In order to prepare the training and test poisoned samples, we used the following procedure for each dataset: First, we select samples from the training set of the origin class, to prepare a pool of backdoor images. The number of samples was determined by the maximum backdoor percentage we want to evaluate, which was 15% of the origin class. Next, we selected a fixed percentage of test images from the target class as backdoor test samples. Subsequently, we manually added the backdoor triggers to the previously selected training and test images. In particular, we used the image manipulating program GIMP[1] to manually add the respective pattern (star or block, yellow or white, respectively). Examples of these poisoned images are depicted in Fig. 1.

The attacked (origin) classes are listed in Table 3. In the German dataset for example, the model should classify a "go straight" sign, when actually a "stop" sign with a backdoor trigger is presented.

The backdoor triggers were generally positioned in a pre-defined area of the traffic sign, as the experiments have shown that the effectiveness of the backdoor is heavily influenced by a coherent position. Finally, we have a pool of backdoor images for training and a set of backdoor images for testing for each dataset and backdoor type.

---

[1] GNU Image Manipulation Program, https://www.gimp.org/.

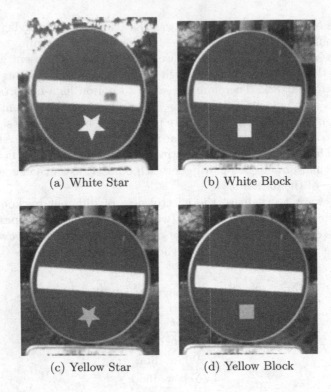

(a) White Star   (b) White Block

(c) Yellow Star   (d) Yellow Block

Fig. 1. Example backdoor triggers on a "Do Not Enter" sign (Color figure online)

To study the impact of poisoned images on classification models, we perform our experiments with a varying amount of backdoor images added to the training dataset. This is an important factor, considering that a higher number of backdoor images in the training data might be easier to detect – both because the count statistics of the data set might vary (a class might grow too big), or because the effects on the classification effectiveness of that particular class might become noticeable. Specifically, for each dataset and backdoor type, we simulate that an attacker adds 1%, 3%, 5%, 10%, 12.5%, and 15% of backdoored images to the target training class, and observe the impact on (i) clean test set accuracy (without backdoor images), (ii) backdoor test set accuracy (only backdoor images), and (iii) complete test set accuracy (combination of clean and backdoor images).

In this study, we want to systematically evaluate the performance of backdoors in traffic sign recognition but also compare the effectiveness of the attack on deep convolutional neural networks with other image classification approaches. Hence, we first train and classify our traffic data sets with LeNet-5, a well-known CNN architecture. We then compare the results with the results of a more traditional approach of image classification, where we extract the *histogram of oriented gradients* (HOG) feature descriptors [3], and subsequently train a Support Vector Machine on this numeric representation of the characteristics obtained from the images.

Table 3. Backdoored Classes for each dataset

Dataset	Backdoored class	Target class
Belgian	Do Not Enter	Cycle Track
Chinese	Do Not Enter	Speed Limit 60
French	Stop	Pedestrian
German	Stop	Must Go Straight or Turn

For the deep learning approach, we started with the standard LeNet-5 architecture and customized it to reach better performance. On each dataset we executed 30 epochs for each backdoor percentage, with batch size 50. The training was performed on a Tesla-K80 GPU. The Adam optimizer is used with learning rate 10^{-4} for model training, implemented in Python via the Keras API[2]. We resized all input images to 224×224 pixels. The model details can be found in Table 4.

Table 4. CNN architecture

Layer	Input	Filter	Stride	Output	Parameters	Activations
Conv2D	$1 \times 224 \times 224$	(5, 5)	(2, 2)	$6 \times 224 \times 224$	456	relu
Pool	$6 \times 224 \times 224$	(5, 5)	(2, 2)	$6 \times 112 \times 112$	0	/
Conv2D	$6 \times 112 \times 112$	(5, 5)	(2, 2)	$16 \times 112 \times 112$	2416	relu
Pool	$16 \times 112 \times 112$	(5, 5)	(2, 2)	$16 \times 56 \times 56$	0	/
Conv2D	$16 \times 56 \times 56$	(5, 5)	(2, 2)	$35 \times 56 \times 56$	14035	relu
Pool	$35 \times 56 \times 56$	(5, 5)	(2, 2)	$35 \times 28 \times 28$	0	/
FC1	$35 \times 28 \times 28$	(5, 5)	/	120	3292920	relu
FC2	120	(5, 5)	/	84	10164	relu
FC3	84	(5, 5)	/	10	850	relu

For the HOG feature extractor we use the Python scikit-image[3] package on images of size 224×224 pixels. Further, we use the Python scikit-learn Support Vector Machine implementation[4] for the model training, with the parameters Gamma $= 0.001$ and kernel $=$ linear.

---

[2] https://keras.io/.

[3] https://scikit-image.org/.

[4] https://scikit-learn.org/stable/modules/generated/sklearn.svm.SVC.html.

4   Evaluation

In this section, we first present the evaluation results of backdoors in various traffic datasets using the CNN classification approach, and subsequently compare them to the results obtained with feature extraction and Support Vector Machine models.

4.1   CNN Backdoor Attack

In this subsection, we discuss the results of attacking the CNN classifier with the goal to embed a backdoor. For each backdoor trigger, we plot the following measures of effectiveness: the clean test data accuracy and the accuracy of the backdoor (poisoned) test images. In case of the poisoned test images, a high accuracy means that the poisoned label was predicted as intended by the attacker, i.e. the model was successfully fooled.

The result tables for each dataset can be found in Tables 5, 6, 7 and 8, where the left column of each percentage shows the classification accuracy on the clean dataset, and the right column the classification accuracy on the test set of poisoned images, each in the range of [0..1].

Table 5. Classification accuracy for the Belgian traffic sign dataset (left column: clean dataset; right column: poisoned samples)

Type	Percentage of backdoor images in training set													
	0%		1%		3%		5%		10%		12.5%		15%	
White-Block	1	0	1	0.09	0.99	0.68	1	1	1	1	1	1	1	1
Yellow-Block	1	0	0.99	0	1	0.18	0.99	1	1	1	1	1	1	1
White-Star	1	0	1	0.14	1	0.77	1	0.95	1	1	1	1	1	1
Yellow-Star	1	0	1	0	1	0.82	0.99	1	1	1	0.99	1	1	1

Table 6. Classification accuracy for the Chinese traffic sign dataset (left column: clean dataset; right column: poisoned samples)

Type	Percentage of backdoor images in training set													
	0%		1%		3%		5%		10%		12.5%		15%	
White-Block	1	0	1	0	0.96	0	0.92	0.33	0.96	0.83	0.92	0.94	1	0.77
Yellow-Block	1	0	0.96	0	0.96	0	0.96	0.06	1	0	1	0.94	1	0.94
White-Star	1	0	1	0	0.88	0.11	1	0.55	1	0.77	1	0.83	0.96	0.94
Yellow-Star	1	0	1	0	0.92	0.06	1	0.05	1	0.16	0.92	0.28	1	1

Figure 2 shows the results of the German traffic data set. The first thing we notice is that the overall model performance on clean test data remained rather stable despite small fluctuations, i.e. the added poisoned images in the training phase, independently of the backdoor type, did not weaken the model's performance on clean test data.

Table 7. Classification accuracy for the French traffic sign dataset (left column: clean dataset; right column: poisoned samples)

Type	Percentage of backdoor images in training set													
	0%		1%		3%		5%		10%		12.5%		15%	
White-Block	1	0	1	0	1	0.11	1	0	1	1	1	0.33	0.96	0.66
Yellow-Block	1	0	1	0	1	0.11	1	0	0.77	0	0	1	1	0.77
White-Star	1	0	1	0	0.93	0.22	1	0.56	0.97	0.11	1	0.11	1	0.44
Yellow-Star	1	0	1	0	1	0	1	0	0.97	0	0.94	0	1	0.89

Table 8. Classification accuracy for the German traffic sign dataset (left column: clean dataset; right column: poisoned samples)

Type	Percentage of backdoor images in training set													
	0%		1%		3%		5%		10%		12.5%		15%	
White-Block	0.98	0	0.95	0	1	0.4	1	0.36	0.99	0.58	0.99	0.88	1	0.82
Yellow-Block	0.98	0	1	0	1	0	1	0.72	1	0.96	0.99	0.9	0.97	0.94
White-Star	0.98	0	1	0.06	0.93	0.26	0.99	0.2	0.99	0.86	1	0.82	1	0.92
Yellow-Star	0.98	0	1	0	1	0.92	0.98	0.88	0.99	0.9	0.99	0.96	1	0.96

As can be seen in the individual graphs on the other hand, the backdoor type influences the performance of the backdoor attack. Depending on the specific backdoor shape and color the required amount of backdoor images to reach a higher accuracy varies. The yellow star trigger requires only 3% of backdoor images in the training phase to reach an accuracy of 92%. While increasing the amount of backdoor images during the training phase, the accuracy remains rather stable and finally reaches an accuracy of 96% utilising 12.5% of backdoor images in the training phase. The yellow block trigger shows the second fastest performance gain with 96% accuracy with 10% of backdoor images in the training phase. Both of the white triggers indicate a slower learning rate, the white star reaching 92% accuracy with 15% of backdoor images in the training phase, while the white block has a performance peak of 88% utilising 12.5% of backdoor images in the training phase.

In Fig. 3 we visualize how the different datasets compare with each other when using white star as trigger. As can be seen, this trigger performed best on the Belgian dataset, reaching an accuracy of 100% with 10% of backdoor images in the training phase. In general it should be noted, that the Belgian dataset shows very high accuracy on the clean dataset but also reached very high accuracy on all backdoor triggers starting with 5% of backdoor images in the training phase. In the Chinese dataset, the white star backdoor performance peaked with an accuracy of 94% utilising 15% of backdoor images in the training phase. At the same time, the clean data performance went down to 96% with 15% of backdoor images in the training phase, and shows the highest drop utilising 3% of backdoor images in the training phase. Finally, the French dataset has the weakest performance on the white star trigger, with a peak of 56% accuracy utilising 5% of backdoor images in the training phase. The clean data accuracy remained rather stable with the highest drop with 3% of backdoor images in the training phase. For this analysis it is also important to consider the number of samples (Table 1), as the percentages of added backdoor images is based on this number. The German and Belgian datasets have the highest number of samples, followed by the Chinese dataset and the French with the lowest number. As a consequence, the number of backdoor images is quite low in the French dataset, which could also explain the low performance. Due to the small size, changes in classification performance of single examples in the test set have a rather large impact (±11%), which also explains the rather discontinuous curve.

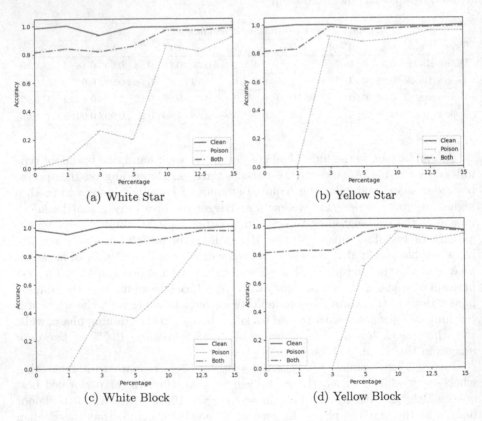

(a) White Star   (b) Yellow Star

(c) White Block   (d) Yellow Block

Fig. 2. Results for CNN classifier on the German traffic sign dataset

Figure 4 shows a comparison of the backdoor embedding in two different positions of the traffic sign, once in the top part, and once in the bottom part. While the traffic sign being attacked, the "do not enter" sign, is actually symmetric in appearance, there are still differences in the effectiveness. For the "white" keys, i.e. the white block and white star, it seems that the backdoor is easier learnt, as the accuracy of the backdoor increases faster than for the bottom position, and reaches levels of being successful of around 90% already with a low number of backdoor samples, of around 3–5%. A similar behaviour can be seen for the yellow block, even though that pattern is learnt slower for both key positions. However, for the yellow star, the behaviour is rather erratic, with the success rate of the backdoor dropping back to 0% with 5% of the images containing the backdoor key.

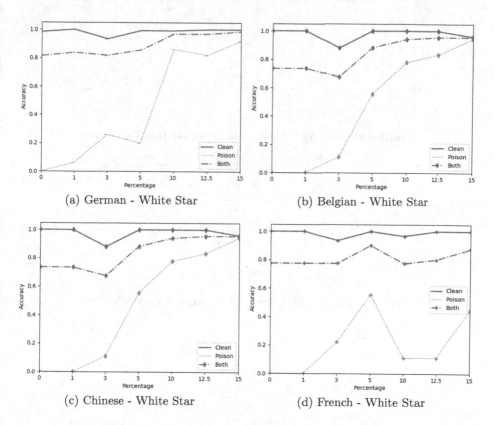

Fig. 3. Results for CNN classifier on the white star trigger

4.2   HOG Features and SVM Backdoor Attack

As most of the literature on backdoor attacks has focused on deep learning approaches like the convolutional neural network that we also employ in our results, we further provide results on the approach using HOG features and an

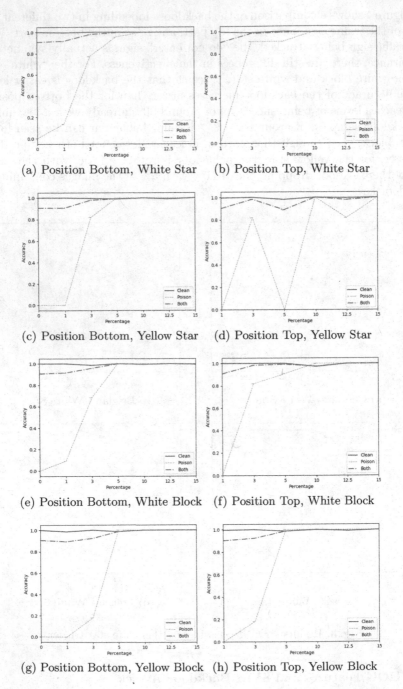

(a) Position Bottom, White Star   (b) Position Top, White Star

(c) Position Bottom, Yellow Star   (d) Position Top, Yellow Star

(e) Position Bottom, White Block   (f) Position Top, White Block

(g) Position Bottom, Yellow Block   (h) Position Top, Yellow Block

Fig. 4. Results for different positions of the backdoor key on the Belgian dataset: the left columns shows the results for the backdoor on the bottom, right on the top

SVM classifier. The observations are indeed quite different than what we could conclude from the backdoor attacks on CNN models above.

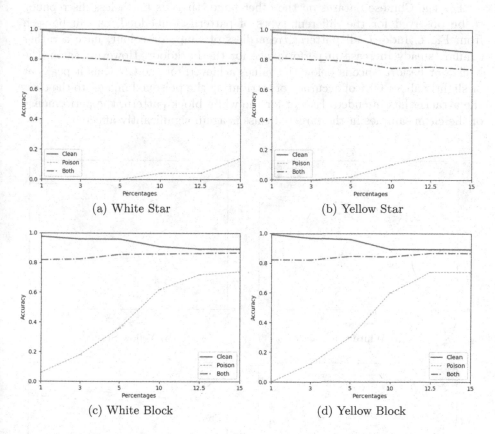

(a) White Star

(b) Yellow Star

(c) White Block

(d) Yellow Block

Fig. 5. Results for HOG feature extraction and SVM classifier on the German traffic sign dataset

For the German dataset, results can be seen in Fig. 5. On the one hand, we can notice a significant difference in the performance of the backdoor attack depending on the shape of the backdoor. We can observe that the "star" pattern does not allow for embedding an effective backdoor, as it reaches at most up to 20% accuracy on the poisoned images, when utilising 15% (or close to that) of poisoned images in the training set. For the "block" pattern, the backdoor performs significantly better, but it still does not achieve effectiveness levels we could obtain in the CNN case, as we plateau at around 75% correctness. Furthermore, we can observe, for both patterns, a quite noticeable drop in overall classification performance on the clean images, from almost 100% accuracy in the targeted class down to approximately 85%. This is a degradation that an attentive user of the model might notice, and that could thus lead to a suspicion of a potential attack.

The results on the Belgium dataset show a rather similar behaviour, with the differences of accuracy on the block and star pattern being a bit more prominent.

For the Chinese dataset on the other hand, there is much less discrepancy to be observed for the different types of patterns embedded, as can be seen from Fig. 6. Indeed, all patterns, regardless of shape or colour, have a rather similar, steady increase in performance for the backdoor. However, again, the backdoor performance is below the values achieved for the CNN, as it peaks at or slightly above 60% of accuracy of attributing the poisoned images to the class the attacker has intended. Except for the white block pattern, the performance of the clean samples in the targeted class is again significantly affected.

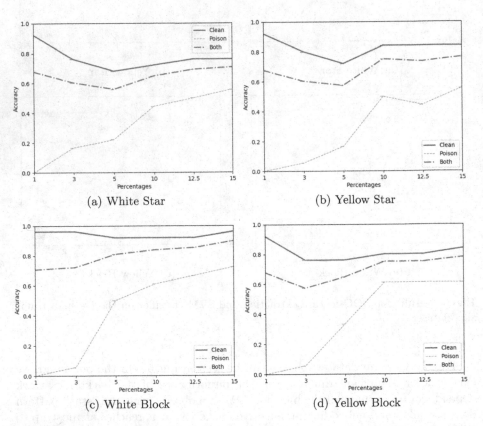

(a) White Star   (b) Yellow Star

(c) White Block   (d) Yellow Block

Fig. 6. Results for HOG feature extraction and SVM classifier on the Chinese traffic sign dataset

For the French dataset, only one of the pattern shows an accuracy of roughly 15% when using all available poisoned images in the training set (i.e. 15% poisoned images in that class), while all other patterns exhibit results at or very close to 0%. This is likely correlated with the relatively small size of this dataset, where the subsequently small number of backdoor patterns likely is not prominent enough to be adequately represented in the extracted features.

We therefore conclude that embedding backdoors in images analysed with the "traditional" approach of first feature extraction and a subsequent learning step is less successful than the attack on deep learning models like the convolutional neural networks, both in the terms of overall achievable accuracy, as well as in the reliability of the attack to work. A notable exception is only in the Chinese dataset, which has acceptable accuracy, but at the price of noticeable degradation of target class accuracy.

5 Conclusions and Future Work

In this paper, we performed a comparative analysis of poisoning (backdoor) attacks on image classification models. We selected a number of different datasets depicting traffic signs, the correct recognition of which could be part of tasks e.g. for autonomous driving. For each dataset, we analysed the susceptibility of the model towards manipulated images that should fool the classifier to trigger a specific, selected target class – categorising an important traffic sign that should lead to e.g. give-way situations with a less important one.

For most settings, the poisoning attacks are successful and the backdoor can be triggered with a high level of reliability, while the effects on the overall classification performance of the model are rather minor, and thus the attack is unlikely to be detected due to unusually low classification accuracy for clean data samples.

We further compared these results with choosing a more traditional approach for image classification, i.e. utilising a feature extraction step with a subsequent learning of a SVM model for classification. We observed that the attacks are far less successful in these settings. However, we still conclude that the approach based on feature extraction in combination with a "shallow" learning model is not immune against these types of attack, which are often mentioned to be effective mostly in the context of deep learning approaches.

Future work will focus on extending these experiments to more datasets, also from other domains, and an evaluation of the effectiveness of mechanisms that have been proposed to defend against these types of adversarial attacks.

Acknowledgments. The competence center SBA Research (SBA-K1) is funded within the framework of COMET — Competence Centers for Excellent Technologies by BMVIT, BMDW, and the federal state of Vienna, managed by the FFG.

References

1. Bagdasaryan, E., Veit, A., Hua, Y., Estrin, D., Shmatikov, V.: How to backdoor federated learning. CoRR abs/1807.00459 (2018). http://arxiv.org/abs/1807.00459
2. Chen, X., Liu, C., Li, B., Lu, K., Song, D.: Targeted backdoor attacks on deep learning systems using data poisoning. CoRR abs/1712.05526 (2017). http://arxiv.org/abs/1712.05526

3. Dalal, N., Triggs, B.: Histograms of oriented gradients for human detection. In: International Conference on Computer Vision & Pattern Recognition (CVPR 2005), vol. 1, pp. 886–893. IEEE Computer Society (2005)

4. Dalvi, N., Domingos, P., Sanghai, S., Verma, D., et al.: Adversarial classification. In: Proceedings of the Tenth ACM SIGKDD International Conference on Knowledge Discovery and Data Mining, pp. 99–108. ACM (2004)

5. Donahue, J., et al.: DeCAF: a deep convolutional activation feature for generic visual recognition. In: Proceedings of the 31st International Conference on Machine Learning, Bejing, China, 22–24 June, pp. 647–655 (2014)

6. Eykholt, K., et al.: Robust physical-world attacks on deep learning visual classification. In: 2018 IEEE/CVF Conference on Computer Vision and Pattern Recognition (2018)

7. Gu, T., Dolan-Gavitt, B., Garg, S.: BadNets: identifying vulnerabilities in the machine learning model supply chain. In: Proceedings of the Machine Learning and Computer Security Workshop, Long Beach, CA, USA, 8 December 2017 (2017). http://arxiv.org/abs/1708.06733

8. Lowd, D., Meek, C.: Adversarial learning. In: Proceedings of the Eleventh ACM SIGKDD International Conference on Knowledge Discovery in Data Mining, pp. 641–647. ACM (2005)

9. Newsome, J., Karp, B., Song, D.: Paragraph: thwarting signature learning by training maliciously. In: Zamboni, D., Kruegel, C. (eds.) RAID 2006. LNCS, vol. 4219, pp. 81–105. Springer, Heidelberg (2006). https://doi.org/10.1007/11856214_5

10. Paparoditis, N., et al.: Stereopolis ii: a multi-purpose and multi-sensor 3D mobile mapping system for street visualisation and 3D metrology. Revue Française Photogramm. Télédétection **200**(1), 69–79 (2012)

11. Papernot, N., McDaniel, P., Goodfellow, I., Jha, S., Celik, Z.B., Swami, A.: Practical black-box attacks against machine learning. In: Proceedings of the 2017 ACM on Asia Conference on Computer and Communications Security, pp. 506–519. ACM (2017)

12. Pratt, L.Y.: Discriminability-based transfer between neural networks. In: Advances in Neural Information Processing Systems (NIPS), pp. 204–211 (1993)

13. Razavian, A.S., Azizpour, H., Sullivan, J., Carlsson, S.: CNN features off-the-shelf: an astounding baseline for recognition. In: IEEE Conference on Computer Vision and Pattern Recognition Workshops, pp. 512–519, June 2014. https://doi.org/10.1109/CVPRW.2014.131, http://arxiv.org/abs/1403.6382

14. Serna, C.G., Ruichek, Y.: Classification of traffic signs: the European dataset. IEEE Access **6**, 78136–78148 (2018)

15. Shen, S., Tople, S., Saxena, P.: A uror: defending against poisoning attacks in collaborative deep learning systems. In: Proceedings of the 32nd Annual Conference on Computer Security Applications, pp. 508–519. ACM (2016)

16. Stallkamp, J., Schlipsing, M., Salmen, J., Igel, C.: Man vs. computer: benchmarking machine learning algorithms for traffic sign recognition. Neural Netw. **32**, 323–332 (2012)

17. Szegedy, C., et al.: Intriguing properties of neural networks. In: International Conference on Learning Representations. Banff, Canada, 14–16 April 2014 (2014) http://arxiv.org/abs/1312.6199

18. Timofte, R., Zimmermann, K., Van Gool, L.: Multi-view traffic sign detection, recognition, and 3D localisation. Mach. Vis. Appl. **25**(3), 633–647 (2014). https://doi.org/10.1007/s00138-011-0391-3

Deep Learning for Proteomics Data for Feature Selection and Classification

Sahar Iravani[1]($\boxtimes$) (iD) and Tim O. F. Conrad[1,2] (iD)

[1] Zuse Institute Berlin, Takustrasse 7, 14195 Berlin, Germany
iravani@zib.de
[2] Free University of Berlin, Arnimallee 6, 14195 Berlin, Germany
conrad@math.fu-berlin.de

Abstract. Todays high-throughput molecular profiling technologies allow to routinely create large datasets providing detailed information about a given biological sample, e.g. about the concentrations of thousands contained proteins. A standard task in the context of precision medicine is to identify a set of biomarkers (e.g. proteins) from these datasets that can be used for disease diagnosis, prognosis or to monitor treatment response. However, finding good biomarker sets is still a challenging task due to the high dimensionality and complexity of the data and the often quite high noise level.

In this work, we present an approach to this problem based on Deep Neural Networks (DNN) and a transfer learning strategy using simulation data. To allow interpretation of the results, we compare different approaches to analyze the learned DNN. Based on these interpretation approaches, we describe how to extract biomarker sets.

Comparison of our method to a state-of-the-art L1-SVM approach shows that the new approach is able to find better biomarker sets for classification when small sets are desired. Compared to a state-of-the-art ℓ_1-support vector machine (ℓ_1-SVM) approach, our method achieves better results for the classification task when a small number of features are needed.

Keywords: Deep learning · Attribution · LRP · Interpretation · Feature selection · Transfer learning · Mass spectrometry · Proteomics

1 Introduction

High throughput omics methods (such as proteomics) are often used in various settings to gain a better understanding of the molecular background of human diseases. In most cases, these studies are focused on the identification of so-called biomarkers that can be used for diagnosis or prognosis of a disease [1, 26]. Due to the wide range of disease-relevant processes that are influenced by proteins and recent advances in proteomics technologies such as mass-spectrometry (MS),

© IFIP International Federation for Information Processing 2019
Published by Springer Nature Switzerland AG 2019
A. Holzinger et al. (Eds.): CD-MAKE 2019, LNCS 11713, pp. 301–316, 2019.
https://doi.org/10.1007/978-3-030-29726-8_19

proteomics has fostered a wide availability of this technology. Thus, the need for analyzing MS proteomics dataset has been increasing rapidly.

The overall idea of biomarker detection - also known as feature selection - is to distinguish between proteomics mass spectra from a control group of healthy individuals and from patients carrying a specific disease. In this situation, the usual approach is to find the differences between these two groups, which can then be studied from a bio-medical perspective. The aim is to detect the best but as-small-as-possible set of discriminating features to reduce time-consuming validation studies in the wet-lab needed for each detected difference. However, due to the nature of the high-throughput mass-spectrometry acquisition process, the generated data is very high-dimensional and contains random and systematic noise, which makes analyzing this kind of data a challenging task.

Many approaches based on state-of-the-art methods such as SVM [7], Lasso [11], or ElasticNet [38] have been adapted to classify and select discriminating features from MS data [25]. Other approaches include SPA [5] that addressed classification and feature selection using compressed sensing [8] or rule mining approaches (e.g. [21]) where relevant features are identified by adapting a disjunctive association rule mining algorithm to distinguish emerging patterns from MS data. With the advances of deep learning (DL) techniques, the research to date has tended to integrate the advantage of deep learning scalability to different biomedical areas. So far, however, little attention has been paid to use DL for classification and feature selection for MS proteomics data - mainly due to the lack of enough samples to train a deep network. In this paper, we address this problem of very high dimensional MS proteomics data classification using a DNN in the case where only few training samples are available. Further, we aim to select a proper interpretable DNN approach that can be utilized to identify biomarkers. To set the stage, we will first briefly review the background of DNNs and methods for interpreting their results.

1.1 DNN Classification and Interpretation

There has been great effort on using DNNs since a DNN-based method for the first time significantly outperformed other approaches in the well known ImageNet challenge [24]. Dozens of different network topologies were proposed since then to improve the performance of DNNs for various applications, e.g. by varying layers and filter sizes [32,37], development of the inception module [36], or adding additional connectivity between layers [15]. Furthermore, effects of different training techniques [16,19], better activation units [14], different stochastic optimization method [9,22], faster training method [20], and different connectivity pattern between layers [18] have improved the DNN efficiency. Parallel to advances of training deep networks, there has been quite an improvement on methods for interpreting classification decisions of a trained deep network and even first steps to go beyond this [17]. Available interpretation methods can be divided into three categories: *function*, *signal*, and *attribution* methods.

The *function technique* analyzes the operation that the network function uses to reach the output. For example, in [31] the authors proposed a class saliency map that takes the gradient of a class prediction neuron with respect to the input pixel. This can show how much a small change to each pixel would affect the prediction. However, the sensitivity maps based on the raw gradient are rather noisy. To improve this situation the Smooth grad method [33] enhanced the saliency map by smoothing the gradient using a Gaussian kernel.

The *signal technique* analyzes the components of the input that mainly influence the last layer decision. For example, the Deconvnet [37] approachmaps all the activities of a network back to the input looking for a pattern in input space that causes a specific activation in the feature maps. A given activation is propagated back through un-pooling, rectifying and filtering (transpose of learned features in a forward path) to the input layer. To un-pool for max-pooling, the switches (the position of the maximum within each pooling region) are recorded on the forward pass. Other work, e.g. Guided Backpropagation [34] suggests to ignore the pooling layer and use convolution layers with strides larger than 1. Therefore, it does not need to record the switches in the forward path.

Finally, the *attribution technique* aims for computing the importance of an input feature during the classification decision. An example is the integrated gradient method [35] that computes the partial derivatives of the output with respect to each input feature. However, instead of evaluating the partial derivative just at the given input x as in input $\times$ gradient [30], it computes the average of it while the input is changing along a linear path from a baseline x'. In [3] this issue was addressed more generally such that it can be applied to a wide range of structures. This methodology called *layer-wise relevance propagation* (LRP) tells how much and to what extent each feature in a particular input contributes to the classification decision. The neuron activation on the decision layer is distributed iteratively to the previous layers until the input layer is reached.

1.2 Contribution

In this work, we present a DL-based method for classifying very high-dimensional proteomics data with the goal of biomarker (feature) identification using and comparing several methods for DNN interpretation.

Unfortunately, almost all available good quality public MS-proteomics datasets contain only up to a hundred samples - which is not enough to train a robust and generalized deep neural network. To deal with this problem, we show how transfer learning using simulated data can improve the situation significantly. Secondly, we adapted the LRP interpretation method to allow identification of the parts of the input that mainly contributes to the classification decision. These identified parts are used for feature selection. We compare the feature selection efficiency of different DNN interpretation methods (attribution, signal, and function) on labeled real datasets. We compare our results to SVM-based method that is a state-of-the-art approach for MALDI-MS feature selection [25].

2   Method

Let x_n, $n = 1, .., N$, $x_n \in \mathbb{R}^D$ and y_n, $n = 1, .., N$, $y_n \in \{0, 1\}$ be the classifier input vectors in a very large D-dimensional feature space and the corresponding class labels, respectively. The aim is to find a small (if possible minimal) sized subset of features from the input data $\hat{x} \in \mathbb{R}^d$ ($d << D$), which can be used to build a classifier f. Ideally, f - which is based only on a subset of all available features - possess the same classification performance as a classifier based on all features. Our approach for feature selection makes use of interpretability analysis for DNNs. The first step is to adapt a DNN to train a generalized model. The last layer of the trained network contains the class probabilities of the given input data. This information is propagated back through the network to the first layer using *layer-wise relevance propagation* (LRP). We use this information to identify the parts of the given input contributing the most on the DNN classification decision. We define the most contributed part of the input over all training data as discriminating features.

2.1   DNN Structure for Proteomics Data Classification

DNN or multilayer perceptrons are characterized by the depth and the width of the layers. Depth refers to the number of layers and width determines the number of neurons on those layers. Depth and width are selected depending on the complexity of the task while more neurons usually lead the network to learn more complex functions.

Our experiments with DNNs of different depth and width show that even though mass spectrum samples can be classified with only a few DNN layers, using more layers leads to a decreasing generalization error. However, we observe that almost all architecture, ranging from shallow to deep networks, fail to generalize correctly due to the limitation in available labeled spectra in public datasets. To tackle this challenge, we integrate the idea of transfer learning to improve this situation. The idea in transfer learning is to take the representation of a neural network that has learned from one task and transfer that representation to a new task. In this study, we use the Maldiquant library [12] in R to simulate the needed labeled data. A network with multiple fully connected layers, all followed by a rectified linear unit (ReLU) function [27] is designed to classify the simulated data. ReLU adds nonlinearity and consequently more complexity to the network. Besides the proper architecture, training the DNN is demanding to set some hyperparameters that - along with the selected structure - lead to convergence, such as learning rate l_r, optimization method of gradient descent, and proper batch size.

Setting up the proper depth, width, activation function, and hyperparameters leads to high classification performance on the simulated dataset and consequently the weights that can be used to initialize the training process for the real mass proteomics data. We then retrain the whole network on the mass proteomics data resulting in a robust and generalized network.

2.2   DNN Interpretablity for Feature Selection

In most publicly available MS proteomics datasets, the number of samples is far too small given the number of features ($N << D$) to hope for a generalizable classifier. However, most of the features in different categories do not make a considerable effect on the classification decision. Moreover, because of the noisy nature of MS data, using all available features (dimensions) usually degrades the classification performance. Additionally, considering all data dimensions is computationally expensive. Therefore, we would like to identify the minimal sized set of input features that account for the differences of the classes (e.g. features that are only relevant in the diseased case). Our main idea is to find those features by analyzing the feature relevance during the DNN classification.

Layer Wise Relevance Propagation. LRP [3] is a methodology for understanding classification decision made by multi-layer neural network. This method identifies which dimensions of the given input data contributed the most to make the classification decision, given a trained network. The LRP method consists of two main steps: after a neural network is trained, a sample is presented to the network and each neurons' activation is computed. A part of the output corresponding to the desired class is considered as the relevance score of the last layer $R^{(L)}$ that is equal to the real-valued prediction output of the classifier f. This is done using Eq. 1 (known as LRP.c, see [3] for details) where $R^{(L)}$ is distributed onto its input neurons at the previous layer, such that $R_k^{(l+1)} = \sum_{i:\ i\ \text{is input for neuron k}} R_{i\leftarrow k}^{(l,l+1)}$ holds.

$$R_{i\leftarrow j}^{(l,l+1)} = \begin{cases} \frac{z_{ij}}{z_j+\varepsilon}.R_j^{(l+1)}, & \text{if } z_j \geq 0 \\ \frac{z_{ij}}{z_j-\varepsilon}.R_j^{(l+1)}, & \text{otherwise} \end{cases} \tag{1}$$

where, $z_{ij} = x_i w_{ij}$, $z_j = \sum_i z_{ij} + b_j$ and $x_j = g(z_j)$. g is a non-linear activation function. For each layer R_i is calculated for $i = 1, ..., \text{num\_neurons}$.

Alternatively, the LRP.$\alpha\beta$ rule according to Eq. 2 (see [29] for details) allows to control the importance of positive and negative values that leads to demonstrate contradicting evidence in the input (such that $\alpha - \beta = 1$). They are typically chosen as $\alpha = 2$ and $\beta = 1$.

$$R_{i\leftarrow j}^{(l,l+1)} = R_j^{(l+1)}.(\alpha.\frac{z_{ij}^+}{z_j^+} + \beta.\frac{z_{ij}^-}{z_j^-}) \tag{2}$$

where "+", "−" denote the positive and negative parts. For $\alpha = 1$, $\beta = 0$ the propagation rule is equivalent to LRP.z^+ rule as in Eq. 3.

$$R_{i\leftarrow j}^{(l,l+1)} = R_j^{(l+1)}\frac{z_{ij}^+}{z_j^+} \tag{3}$$

Iterating every equation down to the first layer yields the relevance scores of all input dimensions, $R_i^{(1)}$.

Feature Selection. $R_i^{(1)}$ gives a score for each dimension of the input vector demonstrating their strength in decision making. It means that the values assigned to each dimension indicate the importance of these features on the overall classification decision. Therefore, the high ranked dimensions represent the most discriminating features. Considering offsets, the presence of noise and different peak indices on samples belonging to different categories, we look through the entire sample relevance distributions, $R_{in}^{(1)}$ for $n = 1, ..., N$. The normalized relevance values are added up through the entire dataset. The high weighted dimensions show the strength of each individual feature to differentiate the classes. However, for MS proteomics data, in most cases the identified features are wide and all the indices around are assigned with high values as well (see Fig. 1). To deal with this effect, we establish a post-processing step to locally detect the strongest individual features. The post-processing works as follows: we first select the best feature in the whole spectra, which are determined by weights from the relevance values. Then, the neighbor's features in the determined window are removed. We then select the second best feature and iterate the process until a stopping criterion is met, e.g. when the classification reaches the whole data classification accuracy.

3   Results and Discussion

3.1   DNN Training Setup for Mass Spectra Classification and Feature Selection

Our DNN architecture is characterized by 5 fully connected layers (FCL) of 100 neurons followed by 4 more FCL of 10 neurons and a prediction layer of 2 neurons to classify two classes. All the neurons are activated by a ReLU nonlinear activation function. Neurons at the last layer are fed to the soft-max activation function, which gives the probability of the given input belonging to the healthy and diseased classes. We trained the network with cross-entropy loss function that is minimized using the momentum variant of the stochastic gradient descent optimizer [28]. Learning rate and batch-size are set to $l_r = 0.00001$ and $b = 2$. We train the network on the simulated data for 40 epochs, and then retrain it on real data for 40 epochs followed by another real data set for 10 epochs for fine-tuning. Afterward, the LRP analyzer is applied to each sample that activates the network neurons to get the most relevant parts used by the DNN for the classification decision. Due to the noisy nature of MS data and mass shift of samples the relevance values are calculated for the entire spectrum in the dataset. Finally, the average of normalized relevance values are post-processed with a window-size of 50 on the result.

3.2   Implementation Detail

All the experiments in our proposed method are implemented in python using Keras [4] with Tensorflow backend and innvestigate library [2] on a machine with a 3.50 GHz Intel Xeon(R) E5-1650 v3 CPU and a GTX 1080 graphics card with 8 GiB GPU memory.

3.3 Results on Spiked Data

In this section, we compare different methods for DNN interpretation such as Gradient method (grad), variants of LRP (LRP.z, LRP.ϵ, and LRP.$\alpha\beta$ rules), input × gradient, integrated gradient(int_grad), guided back-propagation (guided), deconvenet (dCN), and smooth grad (smgrad) through peak detection (see [2] for more details on the methods). With this comparison, we aim to evaluate the impact of interpretation methods using a public dataset known as *spiked data* [10,23]. The spiked data-set contains proteomics mass spectra of control and case groups from human blood samples. The case group has been spiked with a protein-mix of different concentration. The amplitudes of 6 spiked peaks differentiate the spectra into case and control and their known m/z (position) values can be used as ground-truth [6]. Thus, the main aim in this part is to investigate how well an algorithm can detect the m/z positions of the known 6 individual spiked peeks among all 42.381 dimensions. The data contains 95 samples of 50 case and 45 control spectra. The experiments are carried out on two concentration levels, 12.21 nMol/L and 0.76 nMol/L, referred as *spiked160* and *spiked80* in this paper.

The results of our approach, i.e. the selected spiked peaks, are shown in Tables 1 and 2. The reported peaks are the closest ones to the spiked peaks ground-truth among almost 30 high-ranked features. From these two tables we can clearly see that LRP variants (attribution method), inp × grad, and int_grad are far more capable than signal (grad and smoothgrad) and function (guided and dCN) methods. It can also be seen from the results that, while there is no considerable difference between the variant of LRP in this application, one small peak (m/z 3149) can only be detected using LRP.z. Further studies are needed to investigate the reason for this.

Prior to feature selection using the described DNN classification analyzer, the network should become generalized enough to allow the application of interpretation methods. This is what we addressed with transfer learning for the cases when only a few labeled samples are available to train a DNN. In this situation, a simulated dataset of 5000 samples [12] is fed to the network. The dataset contains two equal-size groups spectra as control and case. Each simulated spectra has more than 40 thousands of mass values as the real data and simulated data spectra have. In addition, each one has 412 peaks in which 24 are discriminating. They are equally spread in two groups and are set in fixed positions trough entire dataset. After training, the network re-trained on a real-world dataset of 81 samples and then fine-tuned on spiked data. Initializing the network weights this way should lead to better results since it is less likely that the optimizer gets trapped in a bad local minimum.

We observe from training the network that, while the objective function can not converge on some subsets of samples, the pre-trained network can avoid that. Pre-trained weights lead to a more robust network that resulted in 97.1% ($CI \pm 2.68$) and 96.5% ($CI \pm 3.6$) generalization accuracies on spiked160 and spiked80, respectively. The seemingly large confidence intervals (CI) results from misclassification of one sample on different subsamples during training. Iterating training (train and validation) on 90% of randomly selected spiked160 (95 samples) and inferring on the rest, each time leads to 100% or 88% testing accuracies. This means when the network perform 88% on testing 1 spectrum out of 9 ones was misclassified.

Table 1. Detected spiked peaks using the 9 DNN interpretation methods on spiked160 as the top 35 high ranked features.

Peaks	Grad	LRP.z	LRP.$\alpha\beta$	LRP.ϵ	inp × grd	int_grad	Guided	dCN	Smoothgrad
1047.20	-	**1047.91**	1046.76	**1047.91**	**1047.91**	**1047.91**	-	-	-
1297.51	-	1300.67	**1298.23**	1300.67	1300.67	1300.67	-	-	-
1620.88	1623.6	1621.91	**1620.48**	1621.91	1621.91	1621.91	-	-	1623.6
2466.73	-	2467.63	**2466.51**	2467.63	2467.63	2467.63	2463.63	-	-
3149.61	-	-*	-	-	-	-	-	-	-
5734.56**	-	-	-	-	-	-	-	-	-

*Although m/z 3149 is not selected as top high ranked features because of its insignificant peak in comparison to larger peaks in the spectra (as illustrated in Fig. 1), it is selected as the 94th feature with our method using LRP.z. The other LRP rules can also select this peak but later as the less important feature. However, inp × grad and int_grad could not find this small peak. This is the reason why we analyzed the noisy P.CA data and the visualizations by adapting the LPR.z rule.

**The mean height of the signal in this peak is less than 40 that is comparable to the level of noise in both spiked160 and spiked80 data-sets [5]. Therefore, this peak cannot be selected as a discriminating feature.

We further explained the results in Fig. 1 by visualizing the output of one of the interpretation methods. The figure shows the mean of the normalized LRP.z values of a spiked160 spectrum overlaid on the distribution of case and control spectra of the dataset around the selected spiked peaks. The visualization around the spiked peaks, as shown in these plots, indicate the wide peak range that causes the deviation on the selected features from the spiked ground truth peaks in Tables 1 and 2.

Table 2. Detected spiked peaks using 9 DNN interpretation methods on spiked80 as the top 30 high ranked features.

Peaks	Grad	LRP.z	LRP.$\alpha\beta$	LRP.ϵ	inp × grd	int_grad	Guided	deCN	Smoothgrad
1047.20	-	1040.61	1041.76	1040.61	1040.61	1040.61	-	-	-
1297.51	-	1298.35	**1298.0**	1298.35	1298.35	1298.35	-	-	-
1620.88	-	**1620.87**	1619.7	**1620.87**	**1620.87**	**1620.87**	-	-	-
2466.73	-	**2467.63**	2468.6	**2467.63**	**2467.63**	**2467.63**	-	-	-
3149.61	**3151.25**	-	-	-	-	-	-	-	**3151.25**
5734.56	-	-	-	-	-	-	-	-	-

The spiked peaks that are amongst the top 30 selected features using our pipeline are supposed to be selected as the most discriminating features. However, in Fig. 2 we illustrate that the selected features that are ranked better than the true spiked peaks are more discriminating. For example, it is apparent from the plot that the difference of intensity values of the case and control samples around feature 1021 is larger than their corresponding difference around feature 1047. Therefore, the DNN tends to rely more on these areas in order to make the classification decision. It can also be learned from this plot that not only the individual features are important for the DNN to make a classification decision, but a Gaussian range around high ranked ones also plays a crucial role. For example, relevance values around the m/z 1021 are considerably higher than the relevance value of individual m/z 1047. Therefore, we can not expect a DNN to classify the two groups based on only individual features.

3.4   Results on Pancreas Cancer Data

The Pancreas Cancer dataset (P. CA) is another publicly available data-set [10]. It contains 81 spectra having 42391 features collected from pancreatic cancer patients and apparently healthy control patients. As described previously, due to the lack of sufficient training samples on the public dataset for training a deep network we retrain the network on real data from the network trained on simulated data. We achieved 98%–95% training-testing average accuracy while almost all the structures of DNN we tried from shallow to deep and narrow to wide could not become generalized correctly. The classification decision is interpreted using LRP.z rule to extract the important parts. Figure 3 illustrates the average of normalized LRP.z over the entire dataset around two of the high ranked features. The relevance values are overlaid on top of the mean of the case and control spectra. These two features are illustrated due to the large impact on the classification decision after feature selection (see Fig. 4).

We compare our feature selection method with benchmark methods on the same dataset as follows. A BinDA-algorithm-based method [13] reported 30 peaks m/z 4495, 8868, 8989, 1855, 4468, 8937, 2023, 1866, 5864, 5946, 1780, 2093, **5906**, 5960, 8131, **1207**, 4236, **2953**, 9181, 1021, **1466**, **4092**, 4251, 5005, 8184, 1897, **3264**, 2756, 6051, and 1264, with m/z 8937 as the most discriminating features for pancreatic progenitor cell differentiation. Note that, the bold m/z values indicate the features that are also discovered by our method.

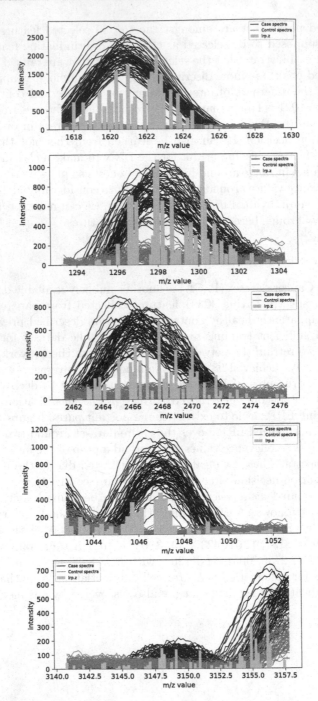

Fig. 1. Visualization of the relevance values around the spiked peaks. Black and blue show the diseased and healthy spectrum of spiked160, and the bars are the average of the normalized LRP.z values over the entire samples. The bars are scaled to the maximum intensity of the spectrum. (Color figure online)

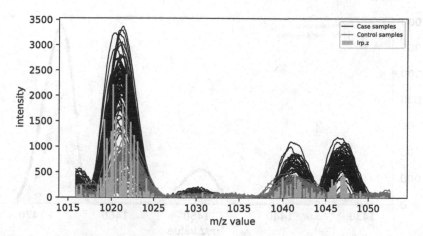

Fig. 2. Comparison visualization of two selected peaks. This plot illustrates the selected spiked 1047 in a wider range to include the selected feature 1021. This illustration shows that m/z 1021 is selected prior to the ground truth m/z 1047 since network sees larger differences between the two classes. Black and blue show the diseased and healthy spectrum of spiked160, and the bars are the average of the normalizer LRP.z values over the entire samples. The bars are scaled to the maximum intensity of the spectrum (Color figure online)

In [5] a compressed sensing-based approach was used to identify peaks with m/z **1464, 1546, 1944, 5904, 1619, 4209**, and m/z **2662** as the most important features to distinguish the healthy and diseased spectra. In this study, peaks with m/z values **4212.36, 1465.43, 3264.36, 2661.37, 5909.96, 4092.18, 1616.98, 1545.91**, 4647.56, 6636.87, 3191.41, 2934.34, 5338.51, **2953.42**, 1060.26, and m/z 3242.47 are ranked as the most discriminating features to achieve the state-of-the-art classification accuracy of 95% [5]. The mass shift of 1 to 3 Dalton on the m/z axis among the identified peaks over different study is likely arising from different pre-processing and post-processing procedures.

3.5 Feature Selection Comparison

We compare the discriminating accuracy of our feature selection method with the state-of-the-art ℓ_1-SVM approach for MALDI-MS feature selection [25]. Figure 4 shows how the classification accuracy is changing for both approaches when the number of features used by the classifier is increased. The experiments are carried out on the two spiked data-sets and the P. CA data-set. As can be seen from the plots, while both methods reach the maximum performance, our method outperforms the ℓ_1-SVM approach when only very few features are used. This is an important property in the situation, where more selected features lead to higher costs in the following steps in some bio-medical pipeline, where each selected feature must be validated in expansive wet-lab experiments.

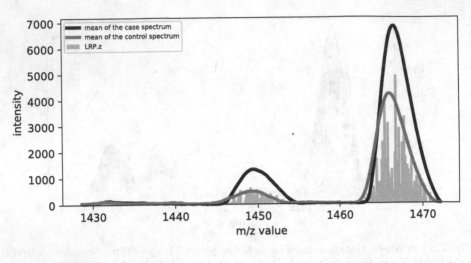

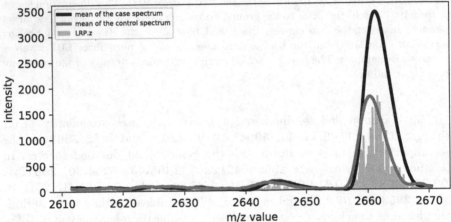

Fig. 3. Illustration of the relevance values around the second (m/z 1465) and forth (m/z 2661) high ranked features of P.CA data. These features are picked for illustration due to their largest impact on the classification accuracy after feature selection, which is apparent from the last row of Fig. 4. The means of the case and healthy spectrum are shown in black and blue, respectively. (Color figure online)

We further investigate the DNN classification performance using the individual features by adding the selected features to the dataset. Despite SVM, it shows significant deduction on the results since as it is shown in the Fig. 2 and explained in previous section DNN sees a wide window around selected individual features for making decision rather than single features.

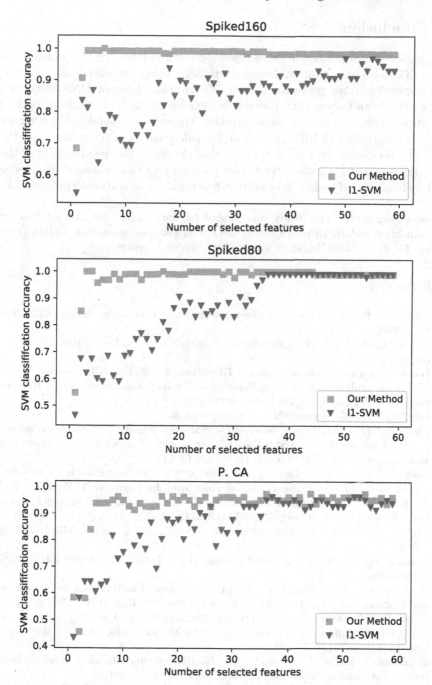

Fig. 4. Generalization accuracies with increasing the number of features to the dataset. Plots show the strength of selected features on spiked160 (first row), spiked80 (second row), and P. CA (third row) using our method in red-square and ℓ_1-SVM in blue-triangle. (Color figure online)

4   Conclusion

This paper presents a new feature selection method based on deep neural networks (DNN) and a transfer learning strategy using simulated data for very high dimensional MS proteomics data. We compare different DNN interpretation methods and show that the attribution based methods perform best for this application. We also demonstrate that there is no considerable difference between the variant of LRP (ϵ, $\alpha\beta$, and z rules) for identifying the important parts of proteomics data for a classification decision. The results suggest that our approach has a significantly better performance than classical approaches on the classification task, where quite a few numbers of features are favorable.

Acknowledgments. This study was funded by the German Ministry of Research and Education (BMBF) Project Grant 3FO18501 (Forschungscampus MODAL) and Project Grant 01IS18037I (Berlin Center for Machine Learning).

References

1. Aebersold, R., Mann, M.: Mass spectrometry-based proteomics. Nature **422**(6928), 198 (2003)
2. Alber, M., et al.: iNNvestigate neural networks!. J. Mach. Learn. Res. **20**(93), 1–8 (2019)
3. Bach, S., Binder, A., Montavon, G., Klauschen, F., Müller, K.R., Samek, W.: On pixel-wise explanations for non-linear classifier decisions by layer-wise relevance propagation. PLoS ONE **10**(7), e0130140 (2015)
4. Chollet, F., et al.: Keras (2015). https://keras.io
5. Conrad, T.O., et al.: Sparse proteomics analysis-a compressed sensing-based approach for feature selection and classification of high-dimensional proteomics mass spectrometry data. BMC Bioinf. **18**(1), 160 (2017)
6. Conrad, T.O.F., et al.: Beating the noise: new statistical methods for detecting signals in MALDI-TOF spectra below noise level. In: Berthold, M.R., Glen, R.C., Fischer, I. (eds.) CompLife 2006. LNCS, vol. 4216, pp. 119–128. Springer, Heidelberg (2006). https://doi.org/10.1007/11875741_12
7. Cortes, C., Vapnik, V.: Support-vector networks. Mach. Learn. **20**(3), 273–297 (1995)
8. Donoho, D.L., et al.: Compressed sensing. IEEE Trans. Inf. Theory **52**(4), 1289–1306 (2006)
9. Duchi, J., Hazan, E., Singer, Y.: Adaptive subgradient methods for online learning and stochastic optimization. J. Mach. Learn. Res. **12**(Jul), 2121–2159 (2011)
10. Fiedler, G.M., et al.: Serum peptidome profiling revealed platelet factor 4 as a potential discriminating peptide associated with pancreatic cancer. Clin. Cancer Res. **15**(11), 3812–3819 (2009)
11. Friedman, J., Hastie, T., Tibshirani, R.: Regularization paths for generalized linear models via coordinate descent. J. Stat. Softw. **33**(1), 1 (2010)
12. Gibb, S., Strimmer, K.: MALDIquant: a versatile R package for the analysis of mass spectrometry data. Bioinformatics **28**(17), 2270–2271 (2012)
13. Gibb, S., Strimmer, K.: Differential protein expression and peak selection in mass spectrometry data by binary discriminant analysis. Bioinformatics **31**(19), 3156–3162 (2015)

14. Glorot, X., Bordes, A., Bengio, Y.: Deep sparse rectifier neural networks. In: Proceedings of the Fourteenth International Conference on Artificial Intelligence and Statistics, pp. 315–323 (2011)
15. He, K., Zhang, X., Ren, S., Sun, J.: Deep residual learning for image recognition. In: Proceedings of the IEEE Conference on Computer Vision and Pattern Recognition, pp. 770–778 (2016)
16. Hinton, G.E., Srivastava, N., Krizhevsky, A., Sutskever, I., Salakhutdinov, R.R.: Improving neural networks by preventing co-adaptation of feature detectors. arXiv preprint arXiv:1207.0580 (2012)
17. Holzinger, A., Langs, G., Denk, H., Zatloukal, K., Müller, H.: Causability and explainabilty of artificial intelligence in medicine. Wiley Interdisc. Rev. Data Min. Knowl. Discovery, e1312 (2019)
18. Huang, G., Liu, Z., Weinberger, K.Q., van der Maaten, L.: Densely connected convolutional networks. arXiv preprint arXiv:1608.06993 (2016)
19. Huang, G., Sun, Y., Liu, Z., Sedra, D., Weinberger, K.Q.: Deep networks with stochastic depth. In: Leibe, B., Matas, J., Sebe, N., Welling, M. (eds.) ECCV 2016. LNCS, vol. 9908, pp. 646–661. Springer, Cham (2016). https://doi.org/10.1007/978-3-319-46493-0_39
20. Ioffe, S., Szegedy, C.: Batch normalization: accelerating deep network training by reducing internal covariate shift. In: International Conference on Machine Learning, pp. 448–456 (2015)
21. Jayrannejad, F., Conrad, T.O.F.: Better interpretable models for proteomics data analysis using rule-based mining. In: Holzinger, A., Goebel, R., Ferri, M., Palade, V. (eds.) Towards Integrative Machine Learning and Knowledge Extraction. LNCS (LNAI), vol. 10344, pp. 67–88. Springer, Cham (2017). https://doi.org/10.1007/978-3-319-69775-8_4
22. Kingma, D., Ba, J.: Adam: a method for stochastic optimization. arXiv preprint arXiv:1412.6980 (2014)
23. Kratzsch, J., et al.: New reference intervals for thyrotropin and thyroid hormones based on national academy of clinical biochemistry criteria and regular ultrasonography of the thyroid. Clin. Chem. **51**(8), 1480–1486 (2005)
24. Krizhevsky, A., Sutskever, I., Hinton, G.E.: ImageNet classification with deep convolutional neural networks. In: Advances in Neural Information Processing Systems, pp. 1097–1105 (2012)
25. Liu, Q., et al.: Comparison of feature selection and classification for MALDI-MS data. BMC Genom. **10**(1), S3 (2009)
26. Marrugal, Á., Ojeda, L., Paz-Ares, L., Molina-Pinelo, S., Ferrer, I.: Proteomic-based approaches for the study of cytokines in lung cancer. Dis. Markers **2016** (2016)
27. Nair, V., Hinton, G.E.: Rectified linear units improve restricted Boltzmann machines. In: Proceedings of the 27th International Conference on Machine Learning (ICML-2010), pp. 807–814 (2010)
28. Qian, N.: On the momentum term in gradient descent learning algorithms. Neural Netw. **12**(1), 145–151 (1999)
29. Samek, W., Montavon, G., Binder, A., Lapuschkin, S., Müller, K.R.: Interpreting the predictions of complex ml models by layer-wise relevance propagation. arXiv preprint arXiv:1611.08191 (2016)
30. Shrikumar, A., Greenside, P., Shcherbina, A., Kundaje, A.: Not just a black box: learning important features through propagating activation differences. arXiv preprint arXiv:1605.01713 (2016)

31. Simonyan, K., Vedaldi, A., Zisserman, A.: Deep inside convolutional networks: visualising image classification models and saliency maps. arXiv preprint arXiv:1312.6034 (2013)
32. Simonyan, K., Zisserman, A.: Very deep convolutional networks for large-scale image recognition. arXiv preprint arXiv:1409.1556 (2014)
33. Smilkov, D., Thorat, N., Kim, B., Viégas, F., Wattenberg, M.: SmoothGrad: removing noise by adding noise. arXiv preprint arXiv:1706.03825 (2017)
34. Springenberg, J.T., Dosovitskiy, A., Brox, T., Riedmiller, M.: Striving for simplicity: the all convolutional net. arXiv preprint arXiv:1412.6806 (2014)
35. Sundararajan, M., Taly, A., Yan, Q.: Axiomatic attribution for deep networks. In: Proceedings of the 34th International Conference on Machine Learning-Volume 70, pp. 3319–3328 (2017). JMLR.org
36. Szegedy, C., et al.: Going deeper with convolutions. In: Proceedings of the IEEE Conference on Computer Vision and Pattern Recognition, pp. 1–9 (2015)
37. Zeiler, M.D., Fergus, R.: Visualizing and understanding convolutional networks. In: Fleet, D., Pajdla, T., Schiele, B., Tuytelaars, T. (eds.) ECCV 2014. LNCS, vol. 8689, pp. 818–833. Springer, Cham (2014). https://doi.org/10.1007/978-3-319-10590-1_53
38. Zou, H., Hastie, T.: Regularization and variable selection via the elastic net. J. Roy. Stat. Soc. Series B (Stat. Methodol.) **67**(2), 301–320 (2005)

Package and Classify Wireless Product Features to Their Sales Items and Categories Automatically

Haitao Tang$^{(\boxtimes)}$ and Pauliina Eratuuli

Commercial Management and Business Digitalization, Nokia, Finland
haitao.tang@nokia.com

Abstract. Aiming at automated decision making, this paper defines and analyzes two machine learning use cases for the product process in wireless infrastructure business. The first use case assigns a product to a product packet according to the functionality of the product. The second use case determines the category of the product so that it can be priced. Then, the product is ready for sale. This paper also provides solutions to these machine learning use cases. The solutions are examined with real data from the processes. The credibility of the solutions is also evaluated by comparing the machine learning decisions with the decisions of human users. These human users know the actual assignment and classification of those products. The results show that the solutions work well as they expected. These solutions assign and classify a part of the given products fully automatically with a high confidence and accuracy. Due to insufficient prediction confidences for the rest of the given products, the rest part of products needs to be escalated for the further decision by the human users. With an escalation, a set of assignment and classification options for a given product is also recommended by the solutions. Often, the correct assignment and classification exist in the set of options already. The human users can easily identify and select the correct assignment and classification from the recommended options. Significant costs and processing time can thus be prevented.

Keywords: Natural Language Processing · NLP · Machine Learning · ML · Process automation · ML based decision making · LTE · 5G · Business Digitalization · Pricing

1 Introduction

Providing cellular communication products is the major business of a telecommunication infrastructure vendor. The products include cellular network products of Long-Term Evolution (LTE) [1] and 5th Generation (5G) [2], which can be in the forms of Hardware (HW), Software (SW), or their supporting components. The products are made available for sale through the process of product packaging, classifying, and price setting.

The internal reference price (IRP) setting is an internal product process that is conducted to define all needed pricing related attributes for such a product before it is released for sale. This is currently a manual process which is repeated for hundreds of products

© IFIP International Federation for Information Processing 2019
Published by Springer Nature Switzerland AG 2019
A. Holzinger et al. (Eds.): CD-MAKE 2019, LNCS 11713, pp. 317–332, 2019.
https://doi.org/10.1007/978-3-030-29726-8_20

annually. During the IRP setting, the category classification and sales package of the product need to be made correctly. In many cases, a new product should be assigned to an existing sales package that contains similar products.

If not automated, this process involves heavily human evaluation and decision making. In such a manual process, a human user needs to understand the whole product landscape completely, which includes not only the various available products and the products expected to be coming, but also the detailed functionalities of the products, their relations, and their relevance to the different network service operators. This process is not only time consuming, but also requires a high level of experience and knowledge from the human user. The good side-effect of such a manual process is that human experience and knowledge are also encoded and embedded into the data generated during the process. During the years of manual processing, it has created the critical amount of data. These data could be used by machine learning to release human from such tedious and brain-straining manual process.

Any commercial digitalization project should be based on a business need. After identification of a possible use case, the business case should be validated. For the automation of the IRP setting, the business need is not only to reduce the time spent on the price setting process, but also to increase the quality of the process to a high level regardless of the user's expertise level. The motivation of this work is thus to design the Machine Learning (ML) solutions to automate the IRP setting process. It defines the ML-based IRP setting process that can dramatically reduce the time and competences required to set the IRP prices. It also increases the quality of the process to a high level regardless of the competence level of the human user.

The ML-based solutions are achieved by using Natural Language Processing (NLP) and general-purpose ML methods to assist the decision making for the product classification and sales package assignment. ML is used to identify the closest existing matching package and category for a given product. The matching is done based on the description documents of the products.

This paper is organized as the following. A brief review of ML-based NLP is given in Sect. 2. The use cases of this work are defined in Sect. 3. The actual method to assign a product to its corresponding package is presented in Sect. 4. The actual method to classify the product category is depicted in Sect. 5. Section 6 presents the experiment setting and results. The credibility of the trained models is further analyzed in Sect. 7. The conclusions of the work are given in Sect. 8.

2 Statistical Natural Language Processing

NLP [9, 10] is a multi-discipline field supported by computer science, linguistics, and machine learning technologies. It concerns the ML-based learning, understanding, extraction, representation, and producing of data in human languages. NLP has greatly benefited from the recent advances in machine learning. It is now focusing on how computer can do speech recognition, natural language understanding, and natural language generation.

Speech recognition translates human speech into text. Natural language understanding interprets and extracts the text of human languages. Natural language generation produces text and speech in human languages. The typical NLP methods could be categorized as text preprocessing, semantic vectorization and embedding, Neural Network (NN)-based parsing of text and information extraction, as well as deep-learning based encoding and decoding of representations of a set of texts.

The methods of text preprocessing perform object standardization, text tokenization, stop-word removing, token (e.g., word) stemming, and token lemmatization. The methods of semantic vectorization and embedding mapping can map a set of texts to their corresponding vectors based on token frequency in one form or another. It can also model topics through latent analysis of a set of texts. It can embed the words in a set of texts, as well as embed a set of texts as bags of words and word sequences. The NN-based parsing methods parse text into parse trees of the sentences including their part of speech tagging. Then, the methods of information extraction extract named entities from text, relations between named entities, and knowledge from text. The methods of deep-learning based encoding and decoding of representations use either mainly RNN-based sequence to sequence models or attention-based transformers to encode and decode representations of texts. It appears that the attention-based transformers outperform RNN-based models in general purpose and multi-task applications.

With the NLP methods, numerous NLP applications can be realized, e.g., the methods in [10, 11]. They are, for example, applications of sentiment analysis, question answering, language modeling, detecting semantic textural similarity, language generation, document summarization, and machine translation.

3 Definition of the Use Cases

IRP setting is done for the products of the different telecommunication technologies. Each technology needs its own ML models to be trained with the technology specific data, as the sales structures for different technologies differ significantly.

There are two use cases in the decision automation. Full automation of the IRP setting process is for the products to which the prediction confidence and accuracy levels exceed predefined thresholds. In the case of ML assisted decision making, the information on the products will be presented to a human user, together with their ML based proposals for the categories and sales packages. Then, the user makes the final decision with the help of ML prediction and assignment. These two use cases can be combined. Full automation can be made for those product cases with the high prediction confidence and accuracy levels. For those product cases with low confidence levels, ML recommends the category and sales package for human's final decision.

The data of this work are the documents defining the products as well as the available sales packages and SW categories. The documents are written by human for the purpose of product implementation and product sales. Typically, a corpus of the product documents is collected per technology family, which usually has thousands of the documents. The corresponding sales packages and SW categories of the products are the ground truth data. They have been generated during the process of sales item creation during the past years. There are hundreds of such labeled data points available per technology family. It is worth to mention that the documents are written in a peculiar, technology-specific language. They are full of "special" technical terms and abbreviations, as well as local conventions. This makes it not possible to directly use a pre-trained language model of the general purpose (e.g., spacy [7] and BERT [8]). Specific language model must then be trained for this work.

This work applies the NLP solutions to complete two tasks. Task A embeds the documents of the human generated product descriptions into their corresponding vectors semantically. The embedding enables the detection of the functional similarity between two products. Such detection is necessary to properly assign a product to a sales package according to its functionality.

Task B classifies a product to a proper category according to the description of the product. It applies a classifier, which is built at the end of the ML pipeline. Text descriptions of the products are input to the pipeline. Basic NLP preprocessing of the texts is then made and, the texts are mapped into numerical vectors as input to the classifier. The SW categories of the products are used as the target output. The classifier is then trained accordingly. Finally, the trained model is used to predict the SW category of a given new product.

4   Method to Package a Feature

The packaging of a product is realized as what is shown in Fig. 1. After the preprocessing of the product feature documents, we use the NLP document-to-vector solution [13] to embed each of the already packaged products (if any) in a sales item list. The actual embedding model is trained with all the available product documents. This embedding model can then represent the products well. It makes the similarity comparison between two products in the IRP list more accurate than what could be achieved with a general-purpose embedding model. Please note that the texts (documents) in the IRP list are only a subset of the texts of the product documents.

The reason for using the similarity-based assignment instead of a categorical classifier is that it needs to assign among several hundred packages. A classification-based method usually achieves a rather low accuracy when there are only hundreds of data samples available.

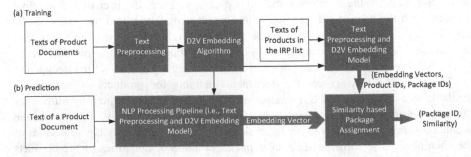

Fig. 1. Assign a product to a sales package according to the semantic similarity.

The exemplary results of the generated embedding and assigned packages are presented in Table 1. The products unambiguously similar to each other are assigned to the same package. Otherwise, the products are each assigned to an individual one-product package. When a new product arrives for the package assignment, the embedding vector of the new product will be compared with all the vectors of the existing products in the table. If there is an unambiguously similar product existing in the table, the new product is assigned to its package. Otherwise, the new product is assigned as an individual product into a new package.

Table 1. The example embedding information of products.

Package ID	Product ID	Embedding vector
0	D_x	$(0.78, -1.50, -0.6, -0.19, -0.11, 0.52, 1.13, 0.77, -0.36, 0.22,$ $-0.19, -0.39, 0.26, -1.83, 0.84, -0.66, 0.73, 0.37, 1.05, -0.43)$
...	...	...

Whether to make fully automated assignment or not depends on the required accuracy of the above ML-based assignment. If the above assignment provides an accuracy higher than the requirement, the assignment is done fully automatically. Otherwise, the ML-based assignment serves as a recommendation for the human decision maker. It is up to the human to decide the actual assignment based on the assignment recommendation. The details concerning these options are introduced in Sect. 6.

5   Method to Assign a SW Category to a Product

The classification of a product to its SW category is realized as shown in Fig. 2. After the preprocessing, a TF-IDF (Term Frequency–Inverse Document Frequency) vectorizer is trained with the corpus of all the available documents of products. Now, the trained TF-IDF model has the vocabulary of all the available documents. Another TF-IDF vectorizer is created by using this vocabulary. The second TF-IDF vectorizer fits and transforms the processed texts of the products in the IRP list into numerical vectors, one for each product in the IRP. The products in the IRP list are the already packaged products (if any) in a sales item list. Together with their known SW categories, their vectors are used to train the multi-class classifier.

When there is a request to classify the SW category of a new product, the preprocessing of the document of this product is made first. The processed text of the document is then fed to the second TF-IDF vectorizer, which fits and transforms the processed texts of the given product into its corresponding numerical vector. This vector is then fed to the trained multi-class classifier. The classifier predicts a SW category for the vector (i.e., the new product) with a specific confidence (i.e., prediction probability). The reason to use a categorical classifier here is that it can achieve a good accuracy for the classification among a small number of classes, when there are hundreds of data samples available.

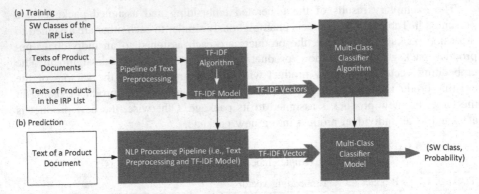

Fig. 2. Assign a product to a SW category according to its functionality and importance.

It is worth mentioning that the method shown in Fig. 2 is suitable for the cases where enough data samples (even if not huge) are available for training the multi-class classifier. However, in an extreme case, the number of product documents for specific category or categories can be very small. In this situation of data scarcity, a categorical classifier may not work. It is simply because of the lack of enough training data for the categorical classifier. For example, as shown in Fig. 3, there are very few Class II product documents to train a multi-class classifier properly. In such a case, the methods shown in Figs. 1 and 2 could be used together as an ensemble method. The ensemble method could still bring an acceptable "classification" result. There could be extra information in the embedding model as it is trained with a bigger corpus of all available products. The extra information could thus improve the accuracy via an ensemble method.

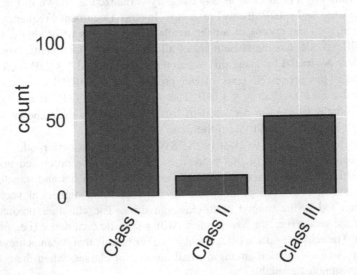

Fig. 3. The extreme example counts of certain available product documents for the SW classes, Class I (63%), Class II (8%), and Class III (29%).

6 Experiments and Results

The solutions based on the methods described in Sects. 4 and 5 are realized with Python 3.6 and its corresponding ML libraries. The solutions are trained with real product data and then they predict the sales item package and the SW category for a given new product. The data, experiments, and results of the solutions are presented and discussed in this section.

6.1 Introduction of the Real Data

The first part of the real data for the solutions are the product documents for telecommunications technologies. For each technology, there are thousands of such documents, which are written in telecom-technical English by R&D people in the company. Such documents could each have the length from a few paragraphs up to multiple pages. Their combined vocabulary of words/terms are at the level of ten thousand. Often, parts of the technical context are not directly given in the documents. A reader is assumed to know the technical context (domain knowledge) before the reader could fully understand the semantic content of the documents. This assumption adds challenge to the solution when comparing with general-purpose NLP tasks [7, 8], where the huge amount of available data could compensate the missing context information. In addition, full scale object standardization for the documents is not feasible due to e.g. the existence of inconsistent abbreviations and varying technical terms.

The second part of the real data for the solutions are the sales items packages in the IRP lists. They provide the information of the package IDs and the SW categories of the products in the IRP lists. This part of information annotates the first part of the data. The products in the IRP lists are just a part of all the available product proposals.

6.2 Package Assignment for a Product

The text preprocessing in Fig. 1 is realized through (1) removing stop-words and punctuations and (2) partial object standardization. The Doc2Vector (D2V) embedding algorithm [13] utilized in this application is from the gensim library. It is trained with all the available product documents, where the dimensions (20 and 24) of the embedding space bring the best performance for the examined technologies correspondingly.

There are 243 different sales packages in a real IRP list used as the existing package to test the solutions. This IRP list has about 500 different products. The distribution of the products among these packages is shown in Fig. 4. Each of the packages has only a small proportion (maximal 5.74%) of the total products.

A sequence of 42 new products are then assigned one by one, by the solution shown in Fig. 1 28 of them are assigned to existing packages and 14 of them are assigned as new packages. Whenever a new product is assigned, it is added to the existing IRP list. The existing IRP list is extended with the newly assigned product. The next new product will be assigned according to the extended IRP list.

The experiment shows that 25 (60%) of the new products are assigned correctly to either the existing packages or as new packages themselves. When recommending a new product to a package, the solution also provides the top 0 to 5 existing products (if any)

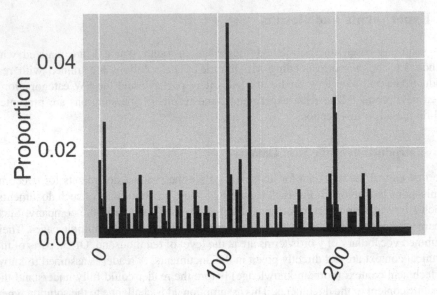

Fig. 4. The distribution of products among the packages in an existing IRP list, where the x-axis is the package ID and the y-axis is the proportion of products in a package to the total number of products in the IRP list.

that are the most similar products to the given new product (i.e., the top matching products in the existing IRP list). For 35 (83%) of the new products, the correct package information is among the provided top similar products from the existing IRP list.

Usually, one cannot trust the solution to assign the new products fully automatically as there are only 60% of the products assigned correctly by the solution. Human in the loop is thus required in this case. The human needs to review a recommendation from the solution and decide the package for the new product. However, the work for the human is very much easier now when comparing with the work when a human alone makes an assignment. In the pure human assignment, the human user needs to know and remember all the products and their packages in the existing IRP list. The human assignment work takes a lot of the time to search and check against the products and packages in the existing IRP list. When using the ML solution as recommendation, the human user can immediately identify the correct package information from the top matching products provided for the newly given product by the solution. Then, the human user can simply select the correct package from the top matching products. In this way alone, 83% of the new products can be assigned correctly. For the remaining part of new products, the human user still has to search through and check against the products, product documentation and packages in the existing IRP list.

The D2V embedding model needs to be retrained after every n new product documents have been released by R&D. These newly generated product documents can carry extra information that has not been learned by the former embedding model. The n can be any large number as long as the newly generated product documents do not contain any new product to be assigned to a sales package. This means the former trained embedding model has still enough information for the new product to be assigned. Otherwise, the D2V embedding model needs to be retrained before the actual assignment of the new product. As the R&D process is not very fast, it is usually enough to retrain the D2V embedding model once every week. In case that a product feature progresses from its creation to the sales item assignment in less than a week, the D2V model is retrained on demand.

6.3   Assignment of SW Category for a Feature

There are 4 different SW categories in a real IRP list. The IRP list has about 500 different products. The distribution of the products among these SW categories are shown in Fig. 5. If one predicts the SW category for a newly given product always with the SW category having the largest number of products, the prediction accuracy could be about 36.1%. It is low.

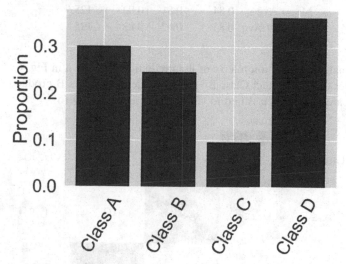

Fig. 5. The distribution of the products among the SW categories.

The text preprocessing in Fig. 2 is realized through (1) removing stop-words and punctuations and (2) partial object standardization. The TF-IDF algorithm is first fitted with all the available thousands of product documents. The vocabulary of this trained model is used by the TF-IDF algorithm as the vocabulary when fitting with all the product documents of the existing IRP list. The TF-IDF model also generates the TF-IDF vectors for all those product documents. These vectors together with the SW categories of those products are used to train a multi-class logistic regression algorithm. The trained model is then used to predict the SW category of a newly given product.

When there is a newly given product for SW category prediction, the text of the product document is preprocessed. Then, the TF-IDF model transforms the preprocessed text into its corresponding TF-IDF vector. This vector is then input to the multi-class classifier model. The model thus predicts the SW category of the newly given product. The model also provides the prediction probability of the predicted SW category.

499 product documents of an IRP list are put through the "(a)" training process of Fig. 2, which eventually trains the multi-class classifier. The trained multi-class classifier model is used to predict the SW categories of another 125 products. The accuracy to predict the SW categories of these 125 products is 82.4%. The other prediction scores for these 125 products are given in Table 2.

Table 2. The prediction scores except the accuracy score for the 125 products.

	Precision	Recall	F1-Score	Support
Class A	0.80	0.95	0.87	38
Class B	0.86	0.58	0.69	31
Class C	0.90	0.82	0.86	11
Class D	0.82	0.89	0.85	45
Micro Avg	0.82	0.82	0.82	125
Macro Avg	0.84	0.81	0.82	125
Weighted Avg	0.83	0.82	0.82	125

The normalized confusion matrix of the predictions is shown in Fig. 6. This multi-class classifier did not predict Class B very well. Here, 13 Class B products in its total 31 products are wrongly classified to Class A (7) and Class D (6).

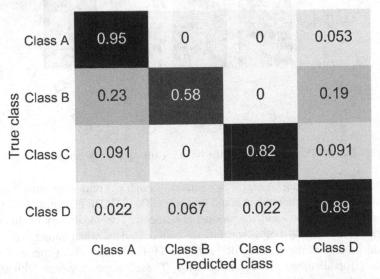

Fig. 6. The normalized confusion matrix on the prediction of the 125 products.

The classification results against their corresponding prediction probabilities by the multi-class classifier are summarized in Fig. 7. A correct classification has indeed a clear correlation with a high prediction probability. However, it is hard to differentiate the correct and incorrect predictions simply by checking the prediction probabilities when the probability score is lower.

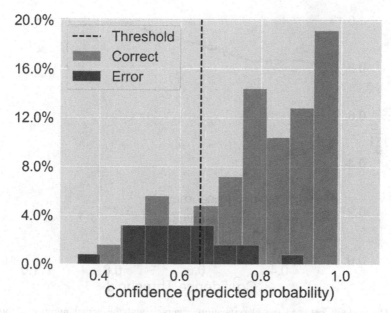

Fig. 7. The distributions of the correct and incorrect classifications of the 125 products against their prediction scores.

It is thus possible to enable the fully automatic classification of the products with high prediction probability. To do so, one can provide a confidence threshold on the prediction probabilities. When a prediction probability is larger than the threshold, the classification can be considered as acceptable and fully automatic classification is triggered. When a prediction probability is smaller than the threshold, the classification can be considered as not trustable and the case is escalated for human to evaluate and classify. As shown in Fig. 8, assume the confidence threshold is set to be prediction probability 0.65. In this case, 91 (72.8%) of the 125 products can be automatically classified. The classification accuracy of this part of products is 91.2% (i.e., 83 products). The confidence threshold is set according to the specific business needs. It is usually selected with a given classification accuracy value that are minimally acceptable to the business. The selection is a tradeoff between the fully automatic classification and machine learning assisted classification.

A cross validation on the quality of the multi-classifier is made with 100 times of reshuffling the combined 624 products, 80% for training and 20% for testing. The mean classification accuracy is 0.763 and the standard deviation is 0.03. The accuracy distribution of the cross validation is shown in Fig. 9.

The multi-class classifier needs to be retrained after every n new products have been classified and assigned to the IRP lists. These newly classified products can carry extra information that has not been learned by the previously trained model. If the n is large, the classification accuracy could suffer clearly. If the n is too small, the retraining can be too frequent. Depending on how frequently a new product needs classification, the higher the retraining frequency, the higher the n value. For the experiments made above, it would be good to let $n = 10 \sim 15$.

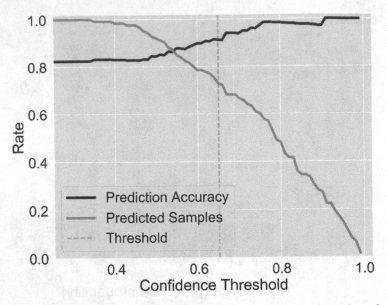

Fig. 8. The tradeoff between the classification accuracy and the actual number of products classified automatically, where confidence threshold determines the tradeoff.

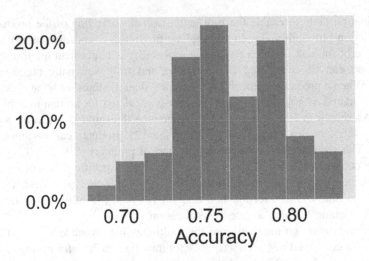

Fig. 9. The accuracy distribution of the cross validation.

The vocabulary of the TF-IDF model needs to be re-fitted only when the newly generated product documents by R&D contain any new product to be classified. This means the formerly fitted TF-IDF does not have enough vocabulary information for the new product to be classified. As the R&D process is not very fast, it is usually enough to re-fit the TF-IDF vocabulary once every week.

7 Credibility of the Trained Models

The quality and credibility of an embedding model can be evaluated with a set of benchmark product documents, each with a similar product document scored by human beforehand. This evaluation method uses the query inventory method [15], while the query here is not on a word but on the text of a product document. For example, Table 3 shows one query point (from the set) with the human scored similarity and, the model-inferred similarities when comparing the vector of document D_k to its inferred vector and the inferred vector of D_x. In this example, we could conclude that the model infers well for this query point. It is thus a good model for document D_k and D_x. More query points can be evaluated to assure the quality of the trained model. It is also mostly doing well for other query points in the benchmark product documents. We could conclude the trained embedding model is good.

Table 3. The quality of the embedding model for a given product document D_k when compared with the document D_x.

Similarity scored by human	$('D_k', 1)$	$('D_x', 0.9)$
Similarity inferred by trained model	$('D_k', 0.974)$	$('D_x', 0.923)$

One also needs to know if the trained multi-class classifier has made the classification with the proper information in the product documents, and not with something irrelevant. The model is trustable if the evaluation confirms that. This evaluation is made with the lime library [12]. As shown in Figs. 10 and 11, the classifier (TF-IDF and the multi-class classifier) uses the relevant texts when it classifies a product. In Figs. 10 and 11, the probabilities for the SW categories (named as Class A, Class B, Class C, and Class D) are predicted. The contributing terms and text sections are also shown to support or oppose predictions of the SW categories, together with their numeric levels of the contribution.

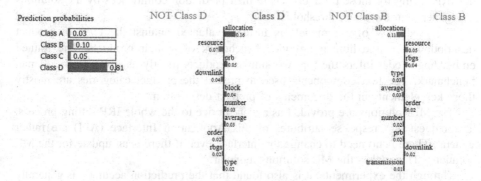

Fig. 10. The prediction probabilities of the four SW categories and the contribution terms for or against the three SW categories, Class B, Class C, and Class D, concerning an example product.

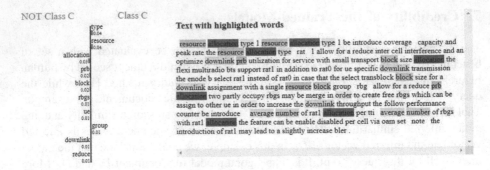

Fig. 11. The contributing terms for or against the SW category Class D and the text used by the classifier to classify the SW category of the example product described by the text.

The lime-based evaluation of the 125 products have been made with the same approach as shown in Figs. 10 and 11. The evaluation needs the domain knowledge concerning the relevance of the information and what information indicates a specific SW category or opposes it. The evaluation results on the 125 products show that the relevant information in the texts have been correctly used to predict the categories of these products in most cases. The classifier of this solution is thus considered trustable. For example, terms "allocation", "prb", and "block" have contributed correctly to support the classification of Class D.

8 Conclusions

This work has proposed the ML solutions to realize the automated IRP setting process. Experiments are taken to explore the feasibility and performance of the solutions, given the available data. The results show that the solutions work well as expected by the human users. They are enough to assist human in the decision making, which reduces significantly the processing time, the needed competence, and human-caused errors. Under a given prediction confidence threshold, these solutions can also fully automate the IRP setting for those products where their prediction confidences by the solutions are higher than the given threshold.

The credibility of the models is further evaluated against the texts of product descriptions and the human provided benchmark of similar products. The trained embedding model infers the top-two similar products mostly as given in the human benchmark. The text components used to predict the product categories are mostly those key elements in the documents of product description.

The ML solutions are provided as a web service to the whole IRP setting process. The request and response attributes of this Application Interface (API) are rather general. There is no need to change the interface even if there is an update for the ML solutions. This makes the ML solutions modular.

Through the experiments, it is also found that the prediction accuracy is generally increasing with the amount of available assignment data. The data volume increases with the usage of the solutions. The performance can be further improved with additional data.

The ML solutions are designed for the products per technology. For those technologies with rather limited amount of existing data, more complicated ML solutions would be needed to achieve the required performance. As new data comes daily, there will be a need to evolve the machine learning solutions at certain point of time. Extra data can reduce the need of a complicated model in one hand. On the other hand, it can also enable the application of a more advanced model to achieve an even better performance. However, it needs further work and experiments to find the exactly needed balance when sufficient amount of extra data become available.

References

1. GPP: LTE. https://en.wikipedia.org/wiki/LTE_(telecommunication)
2. GPP: 5G. https://en.wikipedia.org/wiki/5G
3. Kurama, V.: Introduction to Machine Learning, Towards Data Science, July 2017. https://towardsdatascience.com/introduction-to-machine-learning-db7c668822c4
4. Patal, A.: Machine Learning Algorithm, Overview, Medium, July 2018. https://medium.com/ml-research-lab/machine-learning-algorithm-overview-5816a2e6303
5. Ding, S., Zhu, Z., Zhang, X.: An overview on the semi-supervised support vector machine. Neural Comput. Appl. **26**(8) (2015)
6. Arulkumaran, K., Deisenroth, M.P., Brundage, M., Bharath, A.A.: A brief survey of deep reinforcement learning. IEEE Sig. Process. Mag. Special Issue on Deep Learning for Image Understanding, 1–16 (2017)
7. spaCy: Industrial-Strength Natural Language Processing. https://spacy.io/
8. Devlin, J., Chang, M.W., Lee, K., Toutanova, K.: BERT: pre-training of deep bidirectional transformers for language understanding, October 2018. arXiv:1810.04805v1, https://arxiv.org/pdf/1810.04805.pdf
9. Bird, S., Klein, E., Loper, E.: Natural language processing with Python. NLTK, July 2015 https://www.nltk.org/book/
10. Davydova, O.: 10 applications of artificial neural networks in natural language processing, medium, August 2017 https://medium.com/@datamonsters/artificial-neural-networks-in-natural-language-processing-bcf62aa9151a
11. Brownlee, J.: 7 Applications of deep learning for natural language processing, machine learning mastery, September 2017. https://machinelearningmastery.com/applications-of-deep-learning-for-natural-language-processing/
12. Ribeiro, M.T., Singh, S., Guestrin, C.: Why should I trust You? Explaining the predictions of any classier, Cornell University, August 2016. arXiv:1602.04938v3 [cs.LG], https://arxiv.org/pdf/1602.04938.pdf
13. Le, Q., Mikolov, T.: Distributed representations of sentences and documents. In: Proceedings of the 31st International Conference on Machine Learning (ICML), pp. 1188–1196 (2014)
14. Lau, J., Baldwin, T.: Practical insights into document embedding generation. In: Proceedings of the 1st Workshop on Representation Learning for NLP, pp. 78–86, August 2016
15. Schnabel, T., Labutov, I., Mimno, D., Joachims, T.: Evaluation methods for unsupervised word embeddings. In: Proceedings of Conference on Empirical Methods in Natural Language Processing, pp. 298–307, September 2015

16. Frey, B.J., Dueck, D.: Clustering by passing messages between data points. Science **135**, 972–976 (2007)
17. Rousseeuw, P.J.: Silhouettes: a graphical aid to the interpretation and validation of cluster analysis. J. Comput. Appl. Math. **20**, 53–65 (1987)
18. Madakam, S., Holmukhe, R.M., Jaiswal, D.K.: The future digital work force: robotic process automation (RPA). J. Inf. Syst. Technol. Manag. **16** (2019)

Temporal Diagnosis of Discrete-Event Systems with Dual Knowledge Compilation

Nicola Bertoglio[ID], Gianfranco Lamperti[✉][ID], and Marina Zanella[ID]

Department of Information Engineering, University of Brescia, Brescia, Italy
{n.bertoglio001,gianfranco.lamperti,marina.zanella}@unibs.it

Abstract. Diagnosis aims to explain the abnormal behavior of a system based on the symptoms observed. In a discrete-event system (DES), the symptom is a temporal sequence of observations. At the occurrence of each observation, the diagnosis engine generates a set of candidates, a candidate being a set of faults: such a process requires costly model-based reasoning. This is why a variety of knowledge compilation techniques have been proposed; the most notable of them relies on a *diagnoser* and requires both the diagnosability of the DES and the generation of the whole system space. To avoid both diagnosability and total knowledge compilation, while preserving efficiency, a diagnosis technique is proposed, which is inspired by the two operational modes of the human mind. If the symptom of the DES is part of the knowledge or experience of the diagnosis engine, then *Engine 1* allows for efficient diagnosis. If, instead, the symptom is unknown, then *Engine 2* comes into play, which is far less efficient than *Engine 1*. Still, the experience acquired by *Engine 2* is then integrated into the *temporal dictionary* of the DES, which allows for diagnosis in linear time. This way, if the same problem arises anew, then it will be solved by *Engine 1* efficiently. The temporal dictionary can also be extended by specialized knowledge coming from *scenarios*, which are behavioral patterns of the DES that need to be diagnosed quickly. As such, the temporal dictionary is *open* and relies on *dual knowledge compilation*.

Keywords: Diagnosis · Discrete-event systems · Automata ·
Temporal dictionary · Scenarios · Temporal explanation ·
Preprocessing · Knowledge compilation · Symptom patterns ·
Abduction

This work was supported in part by Lombardy Region (Italy), project *Smart4CPPS, Linea Accordi per Ricerca, Sviluppo e Innovazione, POR-FESR 2014–2020 Asse I*.

1   Introduction

Diagnosis aims at explaining the abnormal behavior of a system based on the observations relevant to its operation that are perceived from the outside. In the Artificial Intelligence community, the definition of the task [23] led to the model-based paradigm [6], according to which the normal behavior of the system to be diagnosed is described by a model and the diagnosis results have to explain the discrepancies between what has been observed at the system output terminals and what we expected to observe on grounds of the model itself. The diagnosis task produces a collection of sets of faulty components, where each set, called a *candidate*, is an explanation of the observation. Each candidate explains the observation as assuming that all the components in the candidate are not behaving normally and all the others are behaving normally is consistent with the observation. This *consistency-based* diagnosis was initially conceived for static systems, such as combinational circuits. For a dynamical system, a discrete-event system (DES) [3] model can be adopted, this being a finite automaton. This model is typically distributed, consisting of several automata that communicate with one another [2]. Although consistency-based diagnosis is applicable to DESs by modeling their nominal behavior only [22], a DES specification usually involves its abnormal behavior also, as in the seminal work by Sampath et al. [25]. The input of the diagnosis task for a DES is a temporal sequence of observations; the output is a set of candidates, each candidate being a set of faults, where a fault is associated with an abnormal state transition represented in the DES model. Diagnosing a DES becomes a form of *abductive* reasoning, inasmuch the candidates are generated based on the trajectories (sequences of state transitions) of the DES that entail the sequence of observations. The approach in [25] relies on a *diagnoser*, a data structure that is derived in a preprocessing phase from the *space* (or global model) of the DES. Such a diagnoser is exploited on line, in order to generate a new set of candidate diagnoses upon perceiving each observation. However, this method requires the generation of the global model of the DES, which is impractical even for distributed systems of moderate size, owing to a combinatorial state explosion. Moreover, in order for the diagnoser to produce a sound and complete set of candidates, the DES is required to fulfill a formal property called *diagnosability*. By definition, a DES is *diagnosable* if every fault occurred can be detected within a finite number of observable transitions of the DES while it is moving in a trajectory of its space. Unsurprisingly, the problem of checking diagnosability has given rise to an extended literature in the last two decades [4,5,8,19–21,24,26–30]. One alternative to the diagnoser approach is the *active-system* approach [1,11–13,15], which neither requires the generation of the global model nor the diagnosability of the DES. The rationale behind the traditional active-system approach is to perform the abduction online, a possibly costly operation that, however, being driven by the sequence of observations, can only focus on the trajectories that produce such a sequence. This paper, which stems from the active-system approach, proposes a novel, more efficient, method to compute a new set of candidate diagnoses of a DES upon receiving each observation. The candidates generated by this technique are endowed with a property, called *temporal explanation*, which has so far been missing in the active-system approach. Temporal explanation is supported by a technique, embedded in the diagnosis engine, called *backward pruning*. Efficiency is achieved by preprocessing the system model to construct a data structure, called a *temporal dictionary*, which is exploited online, when the DES is being operated. In contrast with the diagnoser, the temporal dictionary is not the result of total knowledge compilation, instead it embodies the knowledge relevant to some selected (domain-dependent) behaviors, called *scenarios*. In addition, whenever a sequence of observations that is not encompassed by the dictionary is processed, the dictionary is extended online. In other words, the dictionary is *open* and relies on *dual knowledge compilation*.

2   The Two-Systems Metaphor of the Mind

According to Daniel Kahneman [9], psychologist and Sveriges Riksbank Prize in Economic Sciences in Memory of Alfred Nobel 2002, two modes of thinking coexist in the human brain, which correspond to two systems in the mind, called *System 1* and *System 2*. *System 1* operates automatically and quickly, with little if any effort, and no sense of voluntary control, such as when orienting to the source of a sudden sound or driving a car in an empty road. By contrast, *System 2* operates consciously and slowly, with attention being focused on demanding mental activities, possibly including complex computations or inferences, such as when filling out an intricate application form or checking the validity of a complex argument. Intriguingly, an activity initially performed by *System 2*, such as driving a car or playing the piano, may be subsequently operated by *System 1* after appropriate training. This "dual system" architecture of the mind is a metaphor for the diagnosis approach for DESs presented in this paper. The proposed diagnosis engine (DE) operates in two different modes resembling *System 1* and *System 2* in the human mind, called *Engine 1* and *Engine 2*. If the diagnosis problem to be solved is part of the knowledge or experience of the DE, then *Engine 1* can solve this problem quickly. If, instead, the problem is not part of the knowledge or experience of the DE, then comes into play *Engine 2*, which requires deep model-based reasoning and, therefore, operates far more slowly than *Engine 1*. Still, the experience acquired by *Engine 2* in solving the diagnosis problem can be integrated into the knowledge of the DE, so that, in the future, the same diagnosis problem can be solved by *Engine 1* efficiently. Besides, the DE is not born naked, that is, without any knowledge except the model of the DES, otherwise *Engine 2* would operate far more frequently than *Engine 1* for a possibly long time. Instead, the DE starts working being already equipped with specialized knowledge on domain-dependent *scenarios* that are considered either most probable or most critical for the safety of the DES (or the surrounding environment) and, as such, need to be coped with efficiently.

3   Discrete-Event Systems

A DES is assumed to be a network of *components*, where each component is endowed with input and output *pins* and is modeled as a communicating

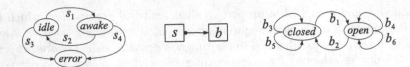

Fig. 1. DES $\mathcal{P}$ (center) and models of the sensor s (left) and the breaker b (right).

Table 1. Transition details for sensor (top) and breaker (bottom) in the DES $\mathcal{P}$.

Component transition	Description
$s_1 = \langle idle, (ko, \{op\}), awake \rangle$	s detects a threatening event and commands b to open
$s_2 = \langle awake, (ok, \{cl\}), idle \rangle$	s detects a liberating event and commands b to close
$s_3 = \langle idle, (ko, \{cl\}, error \rangle$	s detects a threatening event, yet commands b to close
$s_4 = \langle awake, (ok, \{op\}), error \rangle$	s detects a liberating event, yet commands b to open
$b_1 = \langle closed, (op, \emptyset), open \rangle$	b reacts to the opening command by opening
$b_2 = \langle open, (cl, \emptyset), closed \rangle$	b reacts to the closing command by closing
$b_3 = \langle closed, (op, \emptyset), closed \rangle$	b does not react to the opening command
$b_4 = \langle open, (cl, \emptyset), open \rangle$	b does not react to the closing command
$b_5 = \langle closed, (cl, \emptyset), closed \rangle$	b reacts to the closing command by remaining closed
$b_6 = \langle open, (op, \emptyset), open \rangle$	b reacts to the opening command by remaining open

automaton [2]. Each output pin of a component is connected with an input pin of another component by a *link*. The mode in which a transition is triggered in a component is threefold: (1) spontaneously (formally, by the empty event ε), (2) by an (external) event coming from the extern of the DES, or (3) by an (internal) event coming from another component of the DES. When a component performs a transition, it possibly generates new events on its output pins, which possibly trigger the transitions of other components, where the triggering events are consumed. A transition generating an output event on a link can occur only if this link is not occupied by another event already.

Example 1. Centered in Fig. 1 is a DES called $\mathcal{P}$ (protection) which includes two components, a sensor s and a breaker b, and one link connecting the (single) output pin of s with the (single) input pin of b. The model of s (outlined on the left-hand side) involves three states (denoted by circles) and four transitions (denoted by arcs). The model of b (outlined on the right-hand side) involves two states and six transitions. Each component transition t from a state p to a state p', triggered by an input event e, and generating a set of output events E, is denoted by the (angled) triple $t = \langle p, (e, E), p' \rangle$, as detailed in Table 1.

For diagnosis purposes, we need to characterize a DES $\mathcal{X}$ with its *observability* (whether each transition is observable or unobservable) and *normality* (whether each transition is normal or faulty). To this end, let $\mathbf{T}$ be the set of component transitions in $\mathcal{X}$, $\mathbf{O}$ a finite set of *observations*, and $\mathbf{F}$ a finite set of *faults*. The *mapping table* of $\mathcal{X}$ is a function $\mu(\mathcal{X}) : \mathbf{T} \mapsto (\mathbf{O} \cup \{\varepsilon\}) \times (\mathbf{F} \cup \{\varepsilon\})$, where ε is the *empty* symbol. The table $\mu(\mathcal{X})$ can be represented as a finite set of triples (t, o, f), where $t \in \mathbf{T}$, $o \in \mathbf{O} \cup \{\varepsilon\}$, and $f \in \mathbf{F} \cup \{\varepsilon\}$. The triple (t, o, f) defines the observability and normality of t: if $o \neq \varepsilon$, then t is *observable*, else t is *unobservable*; if $f \neq \varepsilon$, then t is *faulty*, else t is *normal*.

Example 2. With reference to the DES $\mathcal{P}$ introduced in Example 1, the mapping table $\mu(\mathcal{P})$ includes the following triples: (s_1, act, ε), (s_2, sby, ε), (s_3, act, fos), (s_4, sby, fcs), (b_1, opn, ε), (b_2, cls, ε), (b_3, ε, fob), (b_4, ε, fcb), $(b_5, \varepsilon, \varepsilon)$, and $(b_6, \varepsilon, \varepsilon)$, where the symbols have the following meaning: act = activate, sby = standby, opn = open, cls = closed, fos = failed to command to open, fcs = failed to command to close, fob = failed to open, fcb = failed to close.

At each time instant, a DES $\mathcal{X}$ is in a state $x = (C, L, \delta)$, where C is the array of the current states of the components, L is the array of the (possibly empty) events currently placed on the links, and δ is the (possibly empty) set of faults occurred in $\mathcal{X}$ starting from its initial state $x_0 = (C_0, L_0, \emptyset)$. The occurrence of a component transition t moves $\mathcal{X}$ from a state x to a state x', in other words, a transition $\langle x, t, x' \rangle$ occurs in $\mathcal{X}$. Hence, assuming that only one component transition at a time can occur, the process that moves a DES from its initial state to another state is represented by a sequence of component transitions, called a *trajectory* of the DES. The set of possible trajectories of $\mathcal{X}$ is specified by a deterministic finite automaton (DFA) called the *diagnosis space* of $\mathcal{X}$, namely,

$$\mathcal{X}^* = (\Sigma, X, \tau, x_0) \tag{1}$$

where Σ (the alphabet) is the set of component transitions, X is the set of states, τ is the (deterministic) transition function, $\tau : X \times \Sigma \mapsto X$, and x_0 is the initial state.[1] Thus, each string $T = [t_1, \ldots, t_n]$ of the regular language of $\mathcal{X}^*$ is a trajectory of $\mathcal{X}$. Based on the mapping table $\mu(\mathcal{X})$, each trajectory $T \in \mathcal{X}^*$ is associated with one *symptom* and one *diagnosis*. The *symptom* $\mathcal{O}$ of T is the finite *sequence* of observations involved in T,

$$\mathcal{O} = [o \mid t \in T, (t, o, f) \in \mu(\mathcal{X}), o \neq \varepsilon]. \tag{2}$$

The *diagnosis* δ of T is the set of faults marking the accepting state of T in $\mathcal{X}^*$. Since a diagnosis is a set, at most one instance of each fault f can be in δ. Hence, generally speaking, the domain of possible diagnoses is the powerset $2^{\mathbf{F}}$, which is finite. By contrast, several instances of the same observation can be in the symptom $\mathcal{O}$; therefore, the domain of possible symptoms is in general infinite. We say that the trajectory T *implies* both $\mathcal{O}$ and δ, denoted $T \Rightarrow \mathcal{O}$ and $T \Rightarrow \delta$, respectively. Since a trajectory of $\mathcal{X}$ is observed as a symptom and since the observed symptom can be implied by several (possibly infinite) trajectories, it follows that several diagnoses can be associated with the same symptom, which

---

[1] Implicitly, all states of $\mathcal{X}^*$ are also accepting (final) states.

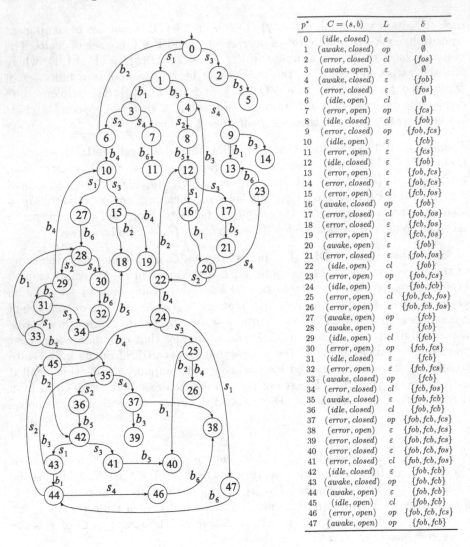

p^*	$C = (s, b)$	L	δ
0	(idle, closed)	ε	$\emptyset$
1	(awake, closed)	op	$\emptyset$
2	(error, closed)	cl	{fos}
3	(awake, open)	ε	$\emptyset$
4	(awake, closed)	ε	{fob}
5	(error, closed)	ε	{fos}
6	(idle, open)	cl	$\emptyset$
7	(error, open)	op	{fcs}
8	(idle, closed)	cl	{fob}
9	(error, closed)	op	{fob, fcs}
10	(idle, open)	ε	{fcb}
11	(error, open)	ε	{fcs}
12	(idle, closed)	ε	{fob}
13	(error, open)	ε	{fob, fcs}
14	(error, closed)	ε	{fob, fcs}
15	(error, open)	cl	{fcb, fos}
16	(awake, closed)	op	{fob}
17	(error, closed)	cl	{fob, fos}
18	(error, closed)	ε	{fcb, fos}
19	(error, open)	ε	{fcb, fos}
20	(awake, open)	ε	{fob}
21	(error, closed)	ε	{fob, fos}
22	(idle, open)	cl	{fob}
23	(error, open)	op	{fob, fcs}
24	(idle, open)	ε	{fob, fcb}
25	(error, open)	cl	{fob, fcb, fos}
26	(error, open)	ε	{fob, fcb, fos}
27	(awake, open)	op	{fcb}
28	(awake, open)	ε	{fcb}
29	(idle, open)	cl	{fcb}
30	(error, open)	op	{fcb, fcs}
31	(idle, closed)	ε	{fcb}
32	(error, open)	ε	{fcb, fcs}
33	(awake, closed)	op	{fcb}
34	(error, closed)	cl	{fcb, fos}
35	(awake, closed)	ε	{fob, fcb}
36	(idle, closed)	cl	{fob, fcb}
37	(error, closed)	op	{fob, fcb, fcs}
38	(error, open)	ε	{fob, fcb, fcs}
39	(error, closed)	ε	{fob, fcb, fcs}
40	(error, closed)	ε	{fob, fcb, fos}
41	(error, closed)	cl	{fob, fcb, fos}
42	(idle, closed)	ε	{fob, fcb}
43	(awake, closed)	op	{fob, fcb}
44	(awake, open)	ε	{fob, fcb}
45	(idle, open)	cl	{fob, fcb}
46	(error, open)	op	{fob, fcb, fcs}
47	(awake, open)	op	{fob, fcb}

Fig. 2. Diagnosis space $\mathcal{P}^*$ (left) and relevant state details (right).

are collectively called the *explanation* of the symptom. Formally, let $\mathcal{O}$ be a symptom of $\mathcal{X}$ and $\delta(T)$ denote the diagnosis of T. The explanation Δ of $\mathcal{O}$ is the finite set of diagnoses, called *candidates*, defined as

$$\Delta(\mathcal{O}) = \{ \delta(T) \mid T \in \mathcal{X}^*, T \Rightarrow \mathcal{O} \}. \tag{3}$$

Example 3. With reference to the DES $\mathcal{P}$ introduced in Example 1 (cf. Fig. 1 and Table 1), outlined on the left side of Fig. 2 is the diagnosis space $\mathcal{P}^*$, where states are identified by numbers 0 .. 47; state details are listed in the table on the right side. Specifically, each state $p^* \in \mathcal{P}^*$ is represented by a triple (C, L, δ), where C is the pair of states of the sensor s and the breaker b, L is the (possibly empty) event within the link, and δ is the diagnosis associated with p^*. Owing to cycles, the set of possible trajectories of $\mathcal{P}$ is infinite. One of these trajectories is $[s_1, b_1, s_2, b_4]$, ending in the state $10 = ((idle, open), \varepsilon, \{fcb\})$, which corresponds to the following events: s detects a threatening event and commands b to open; b opens; s detects a liberating event and commands b to close; still, b remains open. In fact, the diagnosis $\{fcb\}$ accounts for the failing of the breaker to close.

When diagnosis is performed *online*, while the DES is being monitored, some sort of diagnosis information is expected at the occurrence of each observation. This is captured by the notion of a *temporal explanation*.

Definition 1. *Let $\mathcal{O} = [o_1, \ldots, o_n]$ be a symptom of $\mathcal{X}$ and T a trajectory of $\mathcal{X}$ implying $\mathcal{O}$. Let $\mathcal{O}_{[i]}$, $i \in [0 .. n]$, denote the prefix of $\mathcal{O}$ up to o_i. Let $T_{[i]}$, $i \in [0 .. n]$, denote either T, if $i = n$, or the prefix of T up to the transition preceding the $(i+1)$-th observable transition in T, if $0 \leq i < n$. The temporal explanation of $\mathcal{O}$ is the sequence of sets of candidate diagnoses, $\Delta(\mathcal{O}) = [\Delta_0, \Delta_1, \ldots, \Delta_n]$, where each Δ_i, $i \in [0 .. n]$, is the minimal set of diagnoses defined as follows:*

$$T \in \mathcal{X}^*, \forall i \in [0 .. n] \left(\Delta_i \supseteq \{\delta(T_{[i]}) \mid T_{[i]} \Rightarrow \mathcal{O}_{[i]}\} \right). \tag{4}$$

Example 4. Let $\mathcal{O} = [act, opn, sby, act, cls]$ be a symptom of the DES $\mathcal{P}$. As such, $\mathcal{O}$ is implied by just one trajectory, namely $T = [s_1, b_1, s_2, b_4, s_3, b_2]$. Thus, $\Delta(\mathcal{O}) = [\Delta_0, \Delta_1, \Delta_2, \Delta_3, \Delta_4, \Delta_5]$, where $\Delta_0 = \Delta_1 = \Delta_2 = \{\emptyset\}$, $\Delta_3 = \{\{fcb\}\}$, and $\Delta_4 = \Delta_5 = \{\{fcb, fos\}\}$.

When a DES is being monitored, the temporal explanation needs to be updated at the occurrence of each newly generated observation, as the symptom of the DES is not output in one shot, but one observation at a time. Still, updating the temporal explanation at the occurrence of the observation o_{i+1} does not boil down to simply extending $\Delta(\mathcal{O}_{[i]})$ by Δ_{i+1}. Instead, generally speaking, each Δ_j, $j \leq i$, needs to be updated (specifically, pruned).

Example 5. With reference to Example 4, the temporal explanation of $\mathcal{O}$ after the third observation, is $\Delta([act, opn, sby]) = [\emptyset, \emptyset, \emptyset, \{\emptyset, \{fcb\}, \{fcs\}\}]$. The set $\Delta_3 = \{\emptyset, \{fcb\}, \{fcs\}\}$ includes the diagnoses relevant to the trajectories ending in the states 6, 7, 10, or 11 of $\mathcal{P}^*$ (cf. Fig. 2). However, after the reception of the fourth observation, namely act, only the transition up to the state 10 is consistent (being exited by s_1 and s_3), which implies the diagnosis $\{fcb\}$. Hence, the other two candidates in Δ_3, namely $\emptyset$ and $\{fcs\}$, are removed, so that the extended temporal explanation becomes $\Delta([act, opn, sby, act]) = [\emptyset, \emptyset, \emptyset, \{\{fcb\}\}, \{\{fcb\}, \{fcb, fos\}\}]$, where $\Delta_4 = \{\{fcb\}, \{fcb, fos\}\}$ is the set of the diagnoses implied by the trajectories ending in states 15, 19, 27, or 28.

4   Temporal Dictionary

A technique for preprocessing a DES $\mathcal{X}$ in order to generate a DFA, the *temporal dictionary*, for supporting the online diagnosis of $\mathcal{X}$ efficiently is presented.

Definition 2. *Let $\mathcal{X}^*$ be a diagnosis space. Let $\mathcal{X}_n^*$ be the nondeterministic finite automaton (NFA) obtained from $\mathcal{X}^*$ by substituting the observation o for the component transition t marking each transition in $\mathcal{X}^*$, where $(t, o, f) \in \mu(\mathcal{X})$. The temporal dictionary of $\mathcal{X}$ is the DFA $\mathcal{X}^{\circledast}$ obtained by the determinization of $\mathcal{X}_n^*$, which is decorated by the following additional information:*

1. *Each state $x^{\circledast}$ in $\mathcal{X}^{\circledast}$ is marked with the sets $\lfloor x^{\circledast} \rfloor$, $\|x^{\circledast}\|$, and $\Delta(x^{\circledast})$, where:*
   (a) *$\lfloor x^{\circledast} \rfloor$ is the set of states of $\mathcal{X}_n^*$ included in $x^{\circledast}$,[2]*
   (b) *$\|x^{\circledast}\|$ is the set of pairs (x_1^*, x_2^*) where $x_1^* \in \lfloor x^{\circledast} \rfloor$, $x_2^* \in \lfloor x^{\circledast} \rfloor$, x_1^* is entered by an observable transition in $\mathcal{X}_n^*$, x_2^* is exited by an observable transition in $\mathcal{X}_n^*$, and there is a (possibly empty) sequence of ε-transitions in $\mathcal{X}_n^*$ connecting x_1^* with x_2^*,*
   (c) *$\Delta(x^{\circledast})$ is the set of diagnoses associated with the $\mathcal{X}_n^*$ states in $\lfloor x^{\circledast} \rfloor$;*
2. *Each transition $\langle x_1^{\circledast}, o, x_2^{\circledast} \rangle$ in $\mathcal{X}^{\circledast}$ is marked with $\lfloor \langle x_1^{\circledast}, o, x_2^{\circledast} \rangle \rfloor$, denoting the set of transitions $\langle x_1^*, o, x_2^* \rangle$ in $\mathcal{X}_n^*$ where $x_1^* \in \lfloor x_1^{\circledast} \rfloor$ and $x_2^* \in \lfloor x_2^{\circledast} \rfloor$.*

Proposition 1. *The language of $\mathcal{X}^{\circledast}$ equals the set of possible symptoms of $\mathcal{X}$. Besides, if $\mathcal{O}$ is a symptom in $\mathcal{X}^{\circledast}$ with accepting state $x^{\circledast}$, then $\Delta(x^{\circledast}) = \Delta(\mathcal{O})$.*

Remarkably, according to Proposition 1, the explanation of a symptom $\mathcal{O}$ is materialized in the state of the temporal dictionary accepting the string $\mathcal{O}$, hence making the generation of the explanation of $\mathcal{O}$ very efficient.

Example 6. With reference to the diagnosis space $\mathcal{P}^*$ in Fig. 2, the temporal dictionary $\mathcal{P}^{\circledast}$ is outlined on the left side of Fig. 3, with explanations being listed in the table shown on the right side. Details of states and transitions are displayed in Fig. 4. Specifically, each (shadowed) state $p^{\circledast}$ incorporates the relevant set $\lfloor p^{\circledast} \rfloor$ of $\mathcal{P}^*$ states, along with the set of connections $\|p^{\circledast}\|$, indicated by internal arcs. Each transition $\langle p^{\circledast}, o, p'^{\circledast} \rangle$ is unfolded into the set of transitions $\lfloor \langle p^{\circledast}, o, p'^{\circledast} \rangle \rfloor$ in $\mathcal{P}^*$. In accordance with Proposition 1, given $\mathcal{O} = [act, opn, sby, act, cls]$, which has accepting state 8 in $\mathcal{P}^{\circledast}$, we have $\Delta(8) = \{\{fcb, fos\}\} = \Delta(\mathcal{O})$ (cf. Example 4).

Based on Definition 1, the temporal explanation of $\mathcal{O} = [o_1, \ldots, o_n]$ is a sequence $\Delta(\mathcal{O}) = [\Delta_0, \Delta_1, \ldots, \Delta_n]$ where each Δ_i, $i \in [0 .. n]$, is the set of diagnoses implied by $T_{[i]}$, where $T_{[i]}$ also implies $\mathcal{O}_{[i]}$. Here, the key point is that the same trajectory T must fulfill these conditions for *all* prefixes $\mathcal{O}_{[i]}$. This property makes a temporal explanation consistent: for each diagnosis $\delta_i \in \Delta_i$, $i \in [0 .. (n-1)]$, there is a diagnosis $\delta_{i+1} \in \Delta_{i+1}$ such that $\delta_i \subseteq \delta_{i+1}$, and vice versa.

Example 7. Consider the symptom $\mathcal{O} = [act, opn, sby, act, cls]$ in regard to the temporal dictionary $\mathcal{P}^{\circledast}$ outlined in Fig. 3. The accepting states of $\mathcal{O}_{[i]}$, $i \in [0 .. 5]$, are 0, 1, 2, 5, 7, and 8, respectively, which are marked with the sets of diagnoses $\Delta(0) = \{\emptyset\}$, $\Delta(1) = \{\emptyset, \{fob\}, \{fos\}\}$, $\Delta(2) = \{\emptyset\}$, $\Delta(5) = \{\emptyset, \{fcb\}, \{fcs\}\}$,

---

[2] According to the *Subset Construction* determinization algorithm [7], each state of the DFA is identified by a subset of the states of the NFA.

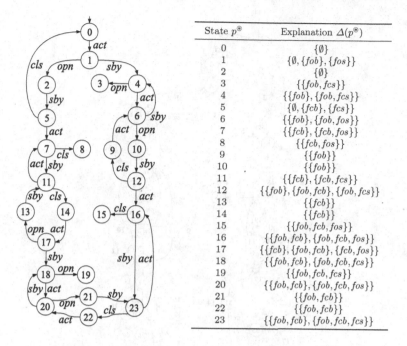

State $p^\circledast$	Explanation $\Delta(p^\circledast)$
0	$\{\emptyset\}$
1	$\{\emptyset, \{fob\}, \{fos\}\}$
2	$\{\emptyset\}$
3	$\{\{fob, fcs\}\}$
4	$\{\{fob\}, \{fob, fcs\}\}$
5	$\{\emptyset, \{fcb\}, \{fcs\}\}$
6	$\{\{fob\}, \{fob, fos\}\}$
7	$\{\{fcb\}, \{fcb, fos\}\}$
8	$\{\{fcb, fos\}\}$
9	$\{\{fob\}\}$
10	$\{\{fob\}\}$
11	$\{\{fcb\}, \{fcb, fcs\}\}$
12	$\{\{fob\}, \{fob, fcb\}, \{fob, fcs\}\}$
13	$\{\{fcb\}\}$
14	$\{\{fcb\}\}$
15	$\{\{fob, fcb, fos\}\}$
16	$\{\{fob, fcb\}, \{fob, fcb, fos\}\}$
17	$\{\{fcb\}, \{fob, fcb\}, \{fcb, fos\}\}$
18	$\{\{fob, fcb\}, \{fob, fcb, fcs\}\}$
19	$\{\{fob, fcb, fcs\}\}$
20	$\{\{fob, fcb\}, \{fob, fcb, fos\}\}$
21	$\{\{fob, fcb\}\}$
22	$\{\{fob, fcb\}\}$
23	$\{\{fob, fcb\}, \{fob, fcb, fcs\}\}$

Fig. 3. Temporal dictionary $\mathcal{P}^\circledast$ (left) and relevant explanations (right).

$\Delta(7) = \{\{fcb\}, \{fcb, fos\}\}$, and $\Delta(8) = \{\{fcb, fos\}\}$. However, the sequence $[\Delta(0), \Delta(1), \Delta(2), \Delta(5), \Delta(7), \Delta(8)]$ does *not* expand consistently and, hence, can *not* be the temporal explanation of $\mathcal{O}$.

Example 7 clearly shows that the temporal explanation of $\mathcal{O}$ cannot simply be the sequence of the explanations of each $\mathcal{O}_{[i]}$, because only a subset of the trajectories implying $\mathcal{O}_{[i]}$ are in general prefixes of the trajectories implying $\mathcal{O}_{[i+1]}$, as a consequence of the constraints imposed by the new observation o_{i+1}.

Example 8. With reference to Example 7, we have $[\Delta(0), \Delta(1), \Delta(2)] = [\{\emptyset\}, \{\emptyset, \{fob\}, \{fos\}\}, \{\emptyset\}]$. Clearly, $\Delta(2) = \{\emptyset\}$ is *not* a consistent expansion of $\Delta(1) = \{\emptyset, \{fob\}, \{fos\}\}$. In fact, considering $\mathcal{P}^\circledast$ outlined in Fig. 4, initially we have $\Delta_0 = \Delta(0) = \{\emptyset\}$. At the reception of the first observation act, the accepting state in $\mathcal{P}^\circledast$ is 1, thereby $\Delta(1)$ is the set of diagnoses associated with the states within 1, namely $\{\emptyset, \{fob\}, \{fos\}\}$. At the reception of the second observation opn, the accepting state becomes 2, including just the $\mathcal{X}^*$ state $3 = ((awake, open), \varepsilon, \emptyset)$. Hence, we have $\Delta(2) = \{\emptyset\}$. The point is, after the occurrence of the observation opn, as clearly indicated in Fig. 4, the only trajectory in $\mathcal{P}^*$ that is consistent with $[act, opn]$ is $0 \to 1 \to 3$. Consequently, the candidate diagnoses associated with the $\mathcal{P}^*$ states 2, 4, and 5 in the state 1 need to be removed from Δ_1, thereby obtaining $[\Delta_0, \Delta_1, \Delta_2] = [\emptyset, \emptyset, \emptyset]$, which, based on Definition 1, is in fact the temporal explanation of $[act, opn]$. In the worst case, this pruning needs to be propagated backward in $\Delta(\mathcal{O})$ up to Δ_0.

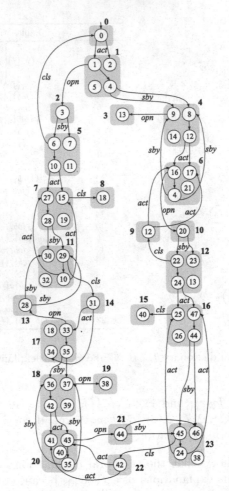

Fig. 4. Details of the temporal dictionary $\mathcal{P}^{\circledast}$ outlined in Fig. 3.

When a DES $\mathcal{X}$ is being operated, a symptom of $\mathcal{X}$ is generated *one observation at a time*. Assuming that the current symptom is $\mathcal{O}_{[i]} = [o_1, \ldots, o_i]$ and the corresponding temporal explanation $\boldsymbol{\Delta}(\mathcal{O}_{[i]})$ has been generated already, the occurrence of a new observation o_{i+1} requires the diagnosis engine to expand the temporal explanation by the insertion of Δ_{i+1} and, in the worst case, by the backward pruning of $\Delta_i, \Delta_{i-1}, \ldots, \Delta_0$, thereby generating the temporal explanation $\boldsymbol{\Delta}(\mathcal{O}_{[i+1]})$. In order to perform this task efficiently, each diagnosis set Δ_i in the temporal explanation is associated with additional information, leading to the notion of a *temporal abduction* (Definition 3).

Definition 3. *Let $\mathcal{O} = [o_1, \ldots, o_n]$ be a symptom of $\mathcal{X}$. The temporal abduction $\mathcal{A}$ of $\mathcal{O}$ is a sequence $\mathcal{A}(\mathcal{O}) = [\alpha_0, \alpha_1, \ldots, \alpha_n]$, where $\forall i \in [0\,..\,n]$, $\alpha_i = (X_i^*, x_i^{\circledast}, \Delta_i)$, where $x_i^{\circledast}$ is the accepting state of $\mathcal{O}_{[i]}$ in $\mathcal{X}^{\circledast}$, while $X_i^* \subseteq \lfloor x_i^{\circledast} \rfloor$ and $\Delta_i \subseteq \Delta(x_i^{\circledast})$ are defined by the following rules:*

1. $X_0^* = \emptyset$;
2. $\Delta_n = \Delta(x_n^\circledast)$;
3. If $n \neq 0$, then $X_n^* = \{x_n^* \mid (x_{n-1}^*, o_n, x_n^*) \in \lfloor (x_{n-1}^\circledast, o_n, x_n^\circledast) \rfloor \}$;
4. If $i < n$, then $\Delta_i = \{\delta_i \mid x_i^* = (C_i, L_i, \delta_i), \langle x_i^*, o_i, x_{i+1}^* \rangle \in \lfloor \langle x_i^\circledast, o_i, x_{i+1}^\circledast \rangle \rfloor$, $x_{i+1}^* \in X_{i+1}^* \}$;
5. If $i \neq 0$ and $i \neq n$, then $X_i^* = \{x_i^* \mid (x_i^*, x_i'^*) \in \|x_i^\circledast\|, \langle x_i'^*, o_i, x_{i+1}^* \rangle \in \lfloor \langle x_i^\circledast, o_i, x_{i+1}^\circledast \rangle \rfloor, x_{i+1}^* \in X_{i+1}^* \}$.

That is, the temporal abduction of $\mathcal{O}$ is a sequence of triples $(X_i^*, x_i^\circledast, \Delta_i)$, where $x_i^\circledast$ is the accepting state of $\mathcal{O}_{[i]}$ in the temporal dictionary $\mathcal{X}^\circledast$, Δ_i happens to equal the homonymous element in $\Delta(\mathcal{O})$ (cf. Proposition 2 below), and X_i^* is the set of $\mathcal{X}^*$ states that are entered by the trajectories fulfilling Definition 1. When a new observation occurs while monitoring $\mathcal{X}$, the set X_i^* is key to backward pruning the temporal abduction, as illustrated in the next example.

Example 9. With reference to the temporal dictionary $\mathcal{P}^\circledast$ (Figs. 3 and 4), consider the symptom $\mathcal{O}$ defined in Example 7, where $\mathcal{O}_{[4]} = [act, opn, sby, act]$. Based on Definition 3, we have $\mathcal{A}(\mathcal{O}_{[4]}) = [(\emptyset, 0, \{\emptyset\}), (\{1\}, 1, \{\emptyset\}), (\{3\}, 2, \{\emptyset\}), (\{6\}, 5, \{\{fcb\}\}), (\{15, 27\}, 7, \{\{fcb\}, \{fcb, fos\}\})]$, where the diagnosis set in the last triple incorporates the diagnoses marking the $\mathcal{P}^*$ states 15, 19, 27, and 28 (those included in the state 7 of $\mathcal{P}^\circledast$). Then, assume the occurrence of the new observation cls, leading to the accepting state 8 of $\mathcal{P}^\circledast$. Based on Definition 3, we have (rule 2) $\Delta_5 = \Delta(8) = \{\{fcb, fos\}\}$ and (rule 3) $X_5^* = \{18\}$. In other words, $\alpha_5 = (\{18\}, 8, \{\{fcb, fos\}\})$. Now, backward pruning starts. First, based on X_5^*, we get (rules 4) $\Delta_4 = \{\{fcb, fos\}\}$, where $\{fcb, fos\}$ is associated with the state 15 and the diagnosis $\{fcb\}$ has been removed, and (rule 5) $X_4^* = \{15\}$, where the state 27 has been removed. At this point, the application of the same rules based on X_4^* has no effect on α_3: since $X_3^* = \{6\}$ is unchanged, the backward pruning stops and eventually $\mathcal{A}(\mathcal{O}_{[5]})$ fulfills Definition 3.

Proposition 2. *If $\mathcal{O}$ is a symptom of $\mathcal{X}$, then*

$$\Delta(\mathcal{O}) = [\Delta_i \mid (X_i^*, x_i^\circledast, \Delta_i) \in \mathcal{A}(\mathcal{O})]. \tag{5}$$

Based on Proposition 2, the temporal explanation $\Delta(\mathcal{O})$ can be generated as a projection of the abduction $\mathcal{A}(\mathcal{O})$. The temporal explanation is required to be generated while the DES is being monitored: starting from the initial diagnosis set Δ_0, which corresponds to the empty symptom, the temporal explanation is constructed one observation at a time. Assuming that the temporal explanation $\Delta(\mathcal{O}_{[i]})$ of the current symptom up to the i-th observation is available, the occurrence of the observation o_{i+1} requires the generation of $\Delta(\mathcal{O}_{[i+1]})$.

5   The *Abduce* Algorithm

In operational terms, the generation of the temporal explanation is specified by the *Abduce* algorithm (Algorithm 1, lines 1–27). Given the temporal dictionary

---

Algorithm 1. *Abduce*

1: **procedure** ABDUCE($\mathcal{X}^{\circledast}$, $\mathcal{O}_{[i]}$, $\mathcal{A}$, o_{i+1})
2:     $\mathcal{X}^{\circledast} = \left(\Sigma, X^{\circledast}, \tau^{\circledast}, x_0^{\circledast}\right)$: the temporal dictionary of $\mathcal{X}$
3:     $\mathcal{O}_{[i]} = [o_1, \ldots, o_i]$: the prefix of a symptom of $\mathcal{X}$ up to the i-th observation
4:     $\mathcal{A} = [\alpha_0, \alpha_1, \ldots, \alpha_i]$: the abduction of $\mathcal{O}_{[i]}$
5:     o_{i+1}: the next observation of $\mathcal{X}$
6: **begin**
7:     Let $x_i^{\circledast}$ be the state of $\mathcal{X}^{\circledast}$ in the triple α_i
8:     $x_{i+1}^{\circledast} \leftarrow \tau^{\circledast}(x_i^{\circledast}, o_{i+1})$
9:     $\Delta_{i+1} \leftarrow \Delta(x_{i+1}^{\circledast})$
10:     $X_{i+1}^{*} \leftarrow \left\{ x_{i+1}^{*} \mid (x_i^{*}, o_{i+1}, x_{i+1}^{*}) \in \lfloor \mathcal{L}(x_i^{\circledast}, o_{i+1}, x_{i+1}^{\circledast}) \rfloor \right\}$
11:     Extend $\mathcal{A}$ by the new triple $\alpha_{i+1} = \left(X_{i+1}^{*}, x_{i+1}^{\circledast}, \Delta_{i+1}\right)$
12:     **for all** j from i downto 0 **do**   # *Backward pruning of the temporal abduction*
13:         Let $\alpha_j = (X_j^{*}, o_j, \Delta_j)$ be the j-th triple in $\mathcal{A}$
14:         $\Delta_{\text{new}} \leftarrow \left\{ \delta_j \mid x_j^{*} = (x_j, \delta_j), (x_j^{*}, o_j, x_{j+1}^{*}) \in \lfloor \mathcal{L}(x_j^{\circledast}, o_j, x_{j+1}^{\circledast}) \rfloor, x_{j+1}^{*} \in X_{j+1}^{*} \right\}$
15:         **if** $\Delta_{\text{new}} \neq \Delta_j$ **then**
16:             Substitute Δ_{new} for Δ_j in α_j
17:         **end if**
18:         **if** $i \neq 0$ **then**   # *Based on rule 1 of Definition 3, the value of X_0^{*} is fixed to $\emptyset$*
19:             $X_{\text{new}}^{*} \leftarrow \left\{ x_j^{*} \mid (x_j^{*}, x_j'^{*}) \in \left\| x_j^{\circledast} \right\|, (x_j'^{*}, o_j, x_{j+1}^{*}) \in \lfloor \mathcal{L}(x_j^{\circledast}, o_j, x_{j+1}^{\circledast}) \rfloor, x_{j+1}^{*} \in X_{j+1}^{*} \right\}$
20:             **if** $X_{\text{new}}^{*} \neq X_j^{*}$ **then**
21:                 Substitute X_{new}^{*} for X_j^{*} in α_j
22:             **else**
23:                 **break**   # *If X_{new}^{*} equals X_j^{*}, then backward pruning has no effect*
24:             **end if**
25:         **end if**
26:     **end for**
27: **end procedure**

$\mathcal{X}^{\circledast}$, the current symptom $\mathcal{O}_{[i]}$, the corresponding temporal abduction $\mathcal{A}$, and the new observation o_{i+1}, the algorithm updates $\mathcal{A}$ to obtain the temporal abduction of $\mathcal{O}_{[i+1]}$. To this end, the accepting state $x_{i+1}^{\circledast}$ of $\mathcal{O}_{[i+1]}$ is determined (lines 7 and 8). Then, based on the rules 3 and 4 of Definition 3, Δ_{i+1} and X_{i+1}^{*} are generated (lines 9 and 10), thus allowing for the construction of the new triple α_{i+1} (line 11). Backward pruning is performed in lines 12–26. Each triple α_j, with j ranging from i down to 0, is updated based on the rules 4 and 5 of Definition 3. However, this pruning may stop before the natural end of the loop, namely when X_{new}^{*} equals X_j^{*} (line 23). If this condition holds, then, based on Definition 3, all the triples $\alpha_0, \ldots, \alpha_{j-1}^{*}$ keep the same value.

Example 10. With reference to Figs. 3 and 4, consider the temporal dictionary $\mathcal{P}^{\circledast}$. Let $\mathcal{O} = [act, opn, sby, act, cls]$ be a symptom of $\mathcal{P}$. Traced in Table 2 is the generation of the temporal abduction $\mathcal{A}(\mathcal{O})$, one observation at a time, where pruning is denoted by strike-through. Each row i of the table represents the configuration of the temporal abduction after the reception of the i-th observation. Initially ($i = 0$), based on rules 1 and 2 of Definition 3, we have $\alpha_0 = (\emptyset, 0, \{\emptyset\})$. Upon the reception of $o_1 = act$, according to Algorithm 1, the accepting state is $x_1^{\circledast} = 1$, thereby $\Delta_1 = \Delta(1) = \{\emptyset, \{fob\}, \{fos\}\}$ and $\mathcal{X}_1^{*} = \{1, 2\}$. Backward pruning has no effect. Upon the reception of $o_2 = opn$, the new triple is $\alpha_2 = (\{3\}, 2, \{\emptyset\})$. In this case, backward pruning removes the candidates $\{fob\}$

Table 2. Incremental generation of $\mathcal{A}([act, opn, sby, act, cls])$ by the *Abduce* algorithm.

i	α_0	α_1	α_2	α_3	α_4	α_5
0	$(\emptyset, 0, \{\emptyset\})$					
1	$(\emptyset, 0, \{\emptyset\})$	$(\{1,2\}, 1, \{\emptyset, \{fob\}, \{fos\}\})$				
2	$(\emptyset, 0, \{\emptyset\})$	$(\{1,2\}, 1, \{\emptyset, \{fob\}, \{fos\}\})$	$(\{3\}, 2, \{\emptyset\})$			
3	$(\emptyset, 0, \{\emptyset\})$	$(\{1\}, 1, \{\emptyset\})$	$(\{3\}, 2, \{\emptyset\})$	$(\{6,7\}, 5, \{\emptyset, \{fcb\}, \{fcs\}\})$		
4	$(\emptyset, 0, \{\emptyset\})$	$(\{1\}, 1, \{\emptyset\})$	$(\{3\}, 2, \{\emptyset\})$	$(\{6,7\}, 5, \{\emptyset, \{fcb\}, \{fes\}\})$	$(\{15,27\}, 7, \{\{fcb\}, \{fcb, fos\}\})$	
5	$(\emptyset, 0, \{\emptyset\})$	$(\{1\}, 1, \{\emptyset\})$	$(\{3\}, 2, \{\emptyset\})$	$(\{6\}, 5, \{\{fcb\}\})$	$(\{15,27\}, 7, \{\{fcb\}, \{fcb, fos\}\})$	$(\{18\}, 8, \{\{fcb, fos\}\})$

and $\{fos\}$ from Δ_1, as only the transition $\langle 1, opn, 3 \rangle$ in X_1^* is involved (cf. Fig. 4). However, since X_1^* is unchanged, no further pruning is applied. The reception of $o_3 = sby$ moves to the accepting state $x_3^\circledast = 5$, thereby creating the new triple $\alpha_3 = (\{6,7\}, 5, \{\emptyset, \{fcb\}, \{fcs\}\})$, without backward pruning. The reception of $o_4 = act$ moves to the accepting state $x_4^\circledast = 7$, thereby creating the new triple $\alpha_4 = (\{15,27\}, 7, \{\{fcb\}, \{fcb, fos\}\})$. Since the involved transition exits state 10 in $x_3^\circledast = 5$, Δ_3 is reduced to $\{\{fcb\}\}$, where $\{fcb\}$ is the diagnosis marking the state 10. Furthermore, the state 7 is removed from X_3^*, because it is not connected with 10. No further pruning is applicable. Finally, after the reception of the last observation $o_5 = cls$, the accepting state is $x_5^\circledast = 8$, thereby generating the triple $\alpha_5 = (\{18\}, 8, \{\{fcb, fos\}\})$. Since the involved transition exits the state 15 in $x_4^\circledast = 7$, Δ_4 is reduced to $\{\{fcb, fos\}\}$. Also, the state 27 is removed from X_4^*. No other pruning is applicable. Eventually, according to Proposition 2, we have $\Delta(\mathcal{O}) = [\{\emptyset\}, \{\emptyset\}, \{\emptyset\}, \{\{fcb\}\}, \{\{fcb, fos\}\}, \{\{fcb, fos\}\}]$, which in fact equals the temporal explanation determined in Example 4 based on Definition 1. The sequence clearly shows to an operator in charge of monitoring the system (and possibly of performing recovery actions) that no fault occurred up to the second observation; then, two faults occurred in cascade, namely fcb and fos. Without backward pruning, the list of sets of candidate diagnoses is $[\{\emptyset\}, \{\emptyset, \{fob\}, \{fos\}\}, \{\emptyset\}, \{\emptyset, \{fcb\}, \{fcs\}\}, \{\{fcb\}, \{fcb, fos\}\}, \{\{fcb, fos\}\}]$, the interpretation of which may be misleading to the operator. After all, the latter is *not* the temporal explanation of $\mathcal{O}$.

6 Dual Knowledge Compilation

A temporal dictionary $\mathcal{X}^\circledast$ is an extremely efficient tool for supporting the diagnosis of DESs. In theory, the temporal dictionary allows the DE to operate always in quick mode by *Engine 1*, with *Engine 2* never coming into play. However, the temporal dictionary requires total knowledge compilation, which is out of dispute for practical reasons. So, in order to escape from total knowledge compilation and somewhat retaining the advantage of *Engine 1*, we propose a restricted dictionary that expands over time either by experience or by the injection of specific knowledge. In other words, we propose *dual knowledge compilation* based on an *open dictionary*. Intuitively, an open dictionary is a subgraph of the temporal dictionary whose language is a subset of the language of the temporal dictionary (the set of symptoms of the DES). As such, each symptom $\mathcal{O}$ in the open dictionary is associated with $\Delta(\mathcal{O})$, the *sound and complete* set of candidate diagnoses that explain $\mathcal{O}$. Hence, despite being sound (but not complete) in the set of symptoms, the open dictionary is sound and complete in the explanation of the symptom, provided that the symptom is included in the language of the open dictionary. What is the initial configuration of the open dictionary? We suggest to initialize the open dictionary with a *prefix* of the temporal dictionary, as specified in Definition 4.

Definition 4. *Let $\mathcal{X}^{\circledast}$ be a temporal dictionary. The* distance *of a state $x^{\circledast}$ in $\mathcal{X}^{\circledast}$ is the minimum number of transitions connecting the initial state of $\mathcal{X}^{\circledast}$ with $x^{\circledast}$. The* prefix *of $\mathcal{X}^{\circledast}$ up to a distance $d \geq 0$, denoted $\mathcal{X}^{\circledast}_{[d]}$, is the subgraph of $\mathcal{X}^{\circledast}$ comprehending all the states at distance $\leq d$ and all the transitions exiting the states at distance $< d$.*

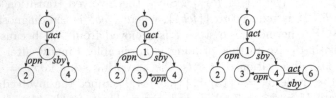

Fig. 5. From left to right, expansion of the open dictionary: $\mathcal{P}^{\circledast}_{[2]}$, $\mathcal{P}^{\circledast}_{[2,\mathcal{O}]}$, and $\mathcal{P}^{\circledast}_{[2,\mathcal{O},\mathcal{S}]}$.

Example 11. With reference to Fig. 3, the prefix of $\mathcal{P}^{\circledast}$ up to distance 2, namely $\mathcal{P}^{\circledast}_{[2]}$, is displayed on the left side of Fig. 5.

A prefix $\mathcal{X}^{\circledast}_{[d]}$ provides the explanation of every symptom that is not longer than d. If $\mathcal{X}^{\circledast}_{[d]}$ embodies a cycle (which is not the case in $\mathcal{P}^{\circledast}_{[2]}$), it also provides the explanation of the infinite set of symptoms encompassing this cycle. However, any symptom longer than d may not belong to the language of $\mathcal{X}^{\circledast}_{[d]}$, such as $\mathcal{O} = [act, sby, opn]$ in $\mathcal{P}^{\circledast}_{[2]}$. In this case, comes into play *Engine 2*, which generates the temporal explanation $\Delta(\mathcal{O})$ based on the *abduction* of $\mathcal{O}$, namely a DFA whose language is the subset of the trajectories of $\mathcal{X}$ implying $\mathcal{O}$. To this end, *Engine 2* performs model-based reasoning to reconstruct the subspace of $\mathcal{X}^{\circledast}$ required. Once provided the temporal explanation $\Delta(\mathcal{O})$, the experience acquired by the DE can be integrated into the open dictionary based on the *symptom pattern* of $\mathcal{O}$. However, the notion of a symptom pattern goes beyond a (plain) symptom, as specified below.

Definition 5. *A* symptom pattern *of a DES $\mathcal{X}$ is a DFA whose language is a subset of the symptoms of $\mathcal{X}$.*

A special (and very simple) case of symptom pattern is associated with each symptom $\mathcal{O}$, denoted $\mathcal{O}^*$, which is the DFA recognizing $\mathcal{O}$ (the language of $\mathcal{O}^*$ is the singleton $\{\mathcal{O}\}$).

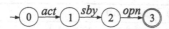

Fig. 6. (Simple) symptom pattern $\mathcal{O}^*$, where $\mathcal{O} = [act, sby, opn]$.

Example 12. Displayed in Fig. 6 is the symptom pattern of $\mathcal{O} = [act, sby, opn]$. Another (circular) symptom pattern is displayed on the bottom-right side of Fig. 7 (cf. Example 16).

Given a symptom pattern $\mathcal{O}^*$, the language of an open dictionary $\mathcal{X}^{\circledast}$ can be extended by the language of $\mathcal{O}^*$ by means of the *Dictionary Extension* algorithm listed below (cf. Algorithm 2, lines 1–31). We assume that each state $x^{\circledast}$ in $\mathcal{X}^{\circledast}$ is equipped with a *labeling set*, denoted $\Omega(x^{\circledast})$ (initially empty), which is instantiated with states in $\mathcal{O}^*$. The algorithm aims to match $\mathcal{O}^*$ with the language of $\mathcal{X}^{\circledast}$. When the matching of an observation o succeeds, the labeling set of the state reached in $\mathcal{X}^{\circledast}$ is extended by the state reached in $\mathcal{O}^*$ (provided it is not included). If the matching fails, then $\mathcal{X}^{\circledast}$ is extended by a new transition and, possibly, by a new state. Let $\langle x^{\circledast}, o, x'^{\circledast} \rangle$ be the missing transition in $\mathcal{X}^{\circledast}$. Based on lines 13–15, the new state $x'^{\circledast}$ is generated first by determining the set of $\mathcal{X}^*$ states exited by a transition marked by o and, then, by extending this set with all the transitions exiting these states that are marked by an unobservable component transition. It should be clear that $\mathcal{X}^*$ is not materialized: only the states required are actually generated starting from the $\mathcal{X}^*$ states within $x^{\circledast}$ and stored in $\mathcal{X}^{\circledast}$. Once all the transitions exiting the state $\omega \in \mathcal{O}^*$ considered have been processed, $\omega \in \Omega(x^{\circledast})$ is marked (line 27). Given $\omega \in \Omega(x^{\circledast})$, two cases are possible. If ω is not marked, then the transition function of $x^{\circledast}$ in $\mathcal{X}^{\circledast}$ needs to be checked against $\mathcal{O}^*$. If, instead, ω is marked, then the update of the transition function of $x^{\circledast}$ is completed. Hence, since it is impossible to insert ω into $\Omega(x^{\circledast})$ if included already, once ω is marked, the processing of ω is inhibited, thereby preventing the infinite matching of cycles in $\mathcal{O}^*$.

Example 13. Consider the open dictionary $\mathcal{P}_{[2]}^{\circledast}$ in Fig. 5 (left) and the symptom pattern $\mathcal{O}^*$ in Fig. 6. The extension of $\mathcal{P}_{[2]}^{\circledast}$ based on $\mathcal{O}^*$ by Algorithm 2 is performed as follows (to distinguish from $\mathcal{O}^*$ states, the states of the open dictionary are in bold). Initially, the labeling set $\Omega(\mathbf{0})$ is $\{0\}$, where 0 is the initial state of $\mathcal{O}^*$. Since both act and sby are matched, the labeling sets of the involved states become $\Omega(\mathbf{1}) = \{1\}$ and $\Omega(\mathbf{4}) = \{2\}$. Now, since no transition marked by opn exits **4**, the missing dictionary state $x'^{\circledast} = \mathbf{3}$ is generated first computing $X_{opn}^* = \{13\}$, where 13 is the state reached by the $\mathcal{P}^*$ state 9 (cf. Fig. 2). However, since no transition exits 13 in $\mathcal{P}^*$, we have $\hat{X}_{opn}^* = \emptyset$ and, hence, $\bar{X}_{opn}^* = \{13\} = \mathbf{3}$. Since it is missing, the state **3** is inserted into $\mathcal{P}_{[2]}^{\circledast}$ and marked by the explanation $\Delta(\mathbf{3}) = \{\{fob, fcs\}\}$. Eventually, the state **3** is labeled with $\Omega(\mathbf{3}) = \{3\}$ and the transition $\langle \mathbf{4}, opn, \mathbf{3} \rangle$ is created. Since no transition exits the state 3 in $\mathcal{O}^*$, the processing of $\omega = 3$ has no effect and the condition of termination in line 29 is true, thereby ending the loop. The updated open dictionary, namely $\mathcal{P}_{[2,\mathcal{O}]}^{\circledast}$, is shown in the center of Fig. 5.

Algorithm 2. *Dictionary Extension*

1: **procedure** DICTIONARY EXTENSION($\mathcal{X}^{\circledast}, \mathcal{O}^*$)
2:    $\mathcal{X}^{\circledast} = (\Sigma, X^{\circledast}, \tau^{\circledast}, x_0^{\circledast})$: an open dictionary of $\mathcal{X}$
3:    $\mathcal{O}^* = (\Sigma_\omega, \Omega, \tau_\omega, \omega_0, \Omega_f)$: a symptom pattern of $\mathcal{X}$
4: **begin**
5:    Insert ω_0 into the labeling set $\Omega(x_0^{\circledast})$
6:    **repeat**
7:       Choose a state $x^{\circledast} \in X^{\circledast}$ such that there is an unmarked $\omega \in \Omega(x^{\circledast})$
8:       **for all** unmarked $\omega \in \Omega(x^{\circledast})$ **do**
9:          **for all** transitions $\langle \omega, o, \omega' \rangle \in \tau_\omega$ **do**
10:             **if** $\langle x^{\circledast}, o, x'^{\circledast} \rangle \in \tau^{\circledast}$ **then**
11:                Insert ω' into $\Omega(x'^{\circledast})$ **only if** $\omega' \notin \Omega(x'^{\circledast})$
12:             **else**
13:                $X_o^* \leftarrow \{x'^* \mid x^* \in x^{\circledast}, \langle x^*, o, x'^* \rangle \in \tau(\mathcal{X}^*)\}$
14:                $\hat{X}_o^* \leftarrow$ the set of states in $\mathcal{X}^*$ that are reachable from a state in X_o^* by
                     a sequence of transitions $\langle x_i^*, t, x_j^* \rangle$ with t being unobservable
15:                $\bar{X}_o^* \leftarrow X_o^* \cup \hat{X}_o^*$
16:                **if** $X^{\circledast}$ includes a state $x'^{\circledast} = \bar{X}_o^*$ **then**
17:                   Insert into $\tau^{\circledast}$ the new transition $\langle x^{\circledast}, o, x'^{\circledast} \rangle$
18:                   Insert ω' into $\Omega(x'^{\circledast})$ **only if** $\omega' \notin \Omega(x'^{\circledast})$
19:                **else**
20:                   Insert into $X^{\circledast}$ the new state $x'^{\circledast} = \bar{X}_o^*$
21:                   $\Delta(x'^{\circledast}) \leftarrow \{\delta \mid x^* \in x'^{\circledast}, x^* = (C, L, \delta)\}$
22:                   Label $x'^{\circledast}$ with the singleton $\Omega(x'^{\circledast}) = \{\omega'\}$
23:                   Insert into $\tau^{\circledast}$ the new transition $\langle x^{\circledast}, o, x'^{\circledast} \rangle$
24:                **end if**
25:             **end if**
26:          **end for**
27:          Mark ω within the labeling set $\Omega(x^{\circledast})$
28:       **end for**
29:    **until** there is no $x^{\circledast} \in X^{\circledast}$ such that $\Omega(x^{\circledast})$ includes an unmarked state
30:    Empty all the nonempty labeling sets $\Omega(x^{\circledast})$
31: **end procedure**

Based on Example 13, one may argue that, since the prefix of the symptom $\mathcal{O} = [act, sby, opn]$ up to the second observation, namely $[act, sby]$, is already in the language of $\mathcal{P}_{[2]}^{\circledast}$, it might be convenient to avoid generating the abduction of $\mathcal{O}$ by *Engine 2*. Instead, the extension of the dictionary might be performed *on the fly* to eventually obtain the explanation from the state **3** created. Actually, this is reasonable in general: Algorithm 2 can actually serve two purposes: either to extend the language of the open dictionary with the language of the symptom pattern or to perform the diagnosis of a given symptom. In either case, Engine 1 matches the observation pattern with the dictionary, whereas Engine 2 performs model-based reasoning to generate the portion of the dictionary that is missing.

7 Scenarios

An open dictionary $\mathcal{X}^\circledast$ can be extended with (a possibly infinite number of) new symptoms. The simplest way is adding a symptom $\mathcal{O}$ that was previously explained by *Engine 2*, as in Example 13. If $\mathcal{O}$ is generated in $\mathcal{X}^\circledast$ by a path of transitions involving a cycle, then the language of $\mathcal{X}^\circledast$ will be extended not only with $\mathcal{O}$, but also with the infinite symptoms involved in the circular path. For example, extending $\mathcal{P}_{[2]}^\circledast$ in Fig. 5 with the symptom $[act, opn, sby, cls]$ actually extends $\mathcal{P}_{[2]}^\circledast$ with the infinite set of symptoms generated by the circular path $0 \rightarrow 1 \rightarrow 2 \rightarrow 5 \rightarrow 0$. The dictionary can also be extended based on particular behavioral patterns of the DES, called *scenarios*. A scenario is a behavior of the DES that is considered either most probable or most critical and, hence, is required to be explained efficiently. The idea is to generate the symptom pattern of the scenario and to extend the language of the open dictionary with its language. This way, each symptom generated from now on by a trajectory that conforms with the scenario will be explained by *Engine 1* quickly.

Definition 6. *A* scenario *of a DES $\mathcal{X}$ is a pair $S = (\Sigma, \mathcal{L})$, where Σ is a subset of the component transitions in $\mathcal{X}$ and $\mathcal{L}$ is a regular language on Σ.*

Since Σ is a subset of the component transitions in $\mathcal{X}$, all the transitions not included in Σ are irrelevant to the scenario. Therefore, in general, a string in $\mathcal{L}$ is not a trajectory of $\mathcal{X}$.

Example 14. The scenario in which the breaker is stuck closed can be defined as $S = (\Sigma, \mathcal{L})$, where $\Sigma = \{s_3, s_4, b_1, b_2, b_3, b_4\}$ and $\mathcal{L}$ is specified by the regular expression $b_3 \, b_3 \, b_3^*$ (namely, b_3 repeated at least twice).[3]

Definition 7. *Let $S = (\Sigma, \mathcal{L})$ be a scenario of $\mathcal{X}$. The* restriction *of a trajectory T in $\mathcal{X}^*$ on Σ is the sequence $T_\Sigma = [t \mid t \in T, t \in \Sigma]$. The* abduction *of S, denoted $\mathcal{X}_S^*$, is a DFA whose language is the set $\{T \mid T \in \mathcal{X}^*, T_\Sigma \in \mathcal{L}\}$.*

In other words, the abduction of a scenario S is a subspace of $\mathcal{X}^*$ where each trajectory T conforms to one string of the scenario, in the sense that the subsequence of the component transitions in T that are in Σ is a string in $\mathcal{L}$.

Example 15. Consider the scenario S defined in Example 14. The generation of the abduction $\mathcal{P}_S^*$ is based on the DFA recognizing the language $\mathcal{L}$, namely $\hat{\mathcal{L}}$, shown on the top-left of Fig. 7. The DFA representing $\mathcal{P}_S^*$ is displayed on the top-right of the same figure, where each state is a pair $(p^*, \hat{\ell})$, where p^* is a state of $\mathcal{P}^*$ and $\hat{\ell}$ a state of $\hat{\mathcal{L}}$. A state $(p^*, \hat{\ell})$ is final when $\hat{\ell}$ is final.

---

[3] A regular expression is defined inductively on the alphabet Σ. The empty symbol ε is a regular expression. If $a \in \Sigma$, then a is a regular expression. If x and y are regular expressions, then the followings are regular expressions: $x \mid y$ (alternative), $x \, y$ (concatenation), $x?$ (optionality), and x^* (repetition zero or more times).

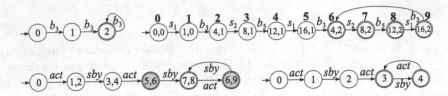

Fig. 7. $\hat{\mathcal{L}}$ (top-left), $\mathcal{P}_\mathcal{S}^*$ (top-right), and $\mathcal{O}_\mathcal{S}^*$ (bottom).

Definition 8. *Let* $\mathcal{S} = (\Sigma, \mathcal{L})$ *be a scenario of a DES* $\mathcal{X}$ *and* $\mathcal{X}_\mathcal{S}^*$ *the abduction of* $\mathcal{S}$. *Let* $\mathcal{N}$ *be the NFA obtained from* $\mathcal{X}_\mathcal{S}^*$ *by substituting* $\langle x, o, x' \rangle$ *for every transition* $\langle x, t, x' \rangle$, *where* $(t, o, f) \in \mu(\mathcal{X})$. *The symptom pattern of the scenario* $\mathcal{S}$, *denoted* $\mathcal{O}_\mathcal{S}^*$, *is the minimum DFA equivalent to* $\mathcal{N}$.

Example 16. With reference to the abduction $\mathcal{P}_\mathcal{S}^*$ determined in Example 15 (top-right of Fig. 7), shown on the bottom-left side of Fig. 7 is the DFA obtained by determinization of $\mathcal{N}$ (cf. Definition 8), where the states $\{5, 6\}$ and $\{6, 9\}$ are equivalent. The minimal DFA, namely the symptom pattern $\mathcal{O}_\mathcal{S}^*$, is shown on the bottom-right side of Fig. 7.

The language of the symptom pattern $\mathcal{O}_\mathcal{S}^*$ of a scenario $\mathcal{S}$ is composed of all the symptoms with which $\mathcal{S}$ manifests itself to the observer. Still, any such symptom can be implied not only by the trajectories that conform with the scenario, but also by other trajectories. The extension of the open dictionary based on $\mathcal{O}_\mathcal{S}^*$ allows for the sound and complete explanation of any symptom in $\mathcal{O}_\mathcal{S}^*$.

Example 17. Based on Algorithm 2, extending the open dictionary $\mathcal{P}_{[2,\mathcal{O}]}^\circledast$ (center of Fig. 5) with the symptom pattern $\mathcal{O}_\mathcal{S}^*$ results in the new open dictionary $\mathcal{P}_{[2,\mathcal{O},\mathcal{S}]}^\circledast$ shown on the right side of Fig. 5.

8   Conclusion

The diagnosis technique presented in this paper is viable and becomes increasingly efficient without requiring the generation of the whole space of the DES; that is, it works while avoiding total knowledge compilation. The open dictionary is assumed to be initialized before the DES is being operated, starting from a prefix of the temporal dictionary, which is then integrated with the symptoms and the candidate diagnoses relevant to a set of scenarios of the DES that are considered worth being diagnosed efficiently. When the DES is being operated, dual knowledge compilation can be applied, in other words, the open dictionary can be enlarged at any time in two ways, either: (*a*) by incorporating new compiled knowledge coming from additional scenarios, or (*b*) by coping with new symptoms explained by *Engine 2*. We are implementing the diagnosis technique presented in this paper in C++. As future research, we plan to extend the technique to other classes of DESs, including complex DESs [10,14,16–18].

References

1. Baroni, P., Lamperti, G., Pogliano, P., Zanella, M.: Diagnosis of large active systems. Artif. Intell. **110**(1), 135–183 (1999). https://doi.org/10.1016/S0004-3702(99)00019-3

2. Brand, D., Zafiropulo, P.: On communicating finite-state machines. J. ACM **30**(2), 323–342 (1983). https://doi.org/10.1145/322374.322380

3. Cassandras, C., Lafortune, S.: Introduction to Discrete Event Systems. The Kluwer International Series in Discrete Event Dynamic Systems, vol. 11. Kluwer Academic, Boston (1999). https://doi.org/10.1007/978-0-387-68612-7

4. Cimatti, A., Pecheur, C., Cavada, R.: Formal verification of diagnosability via symbolic model checking. In: 18th International Joint Conference on Artificial Intelligence (IJCAI 2003), pp. 363–369 (2003)

5. Console, L., Picardi, C., Ribaudo, M.: Diagnosis and diagnosability using PEPA. In: 14th European Conference on Artificial Intelligence (ECAI 2000), pp. 131–135. IOS Press, Amsterdam (2000)

6. Hamscher, W., Console, L., de Kleer, J. (eds.): Readings in Model-Based Diagnosis. Morgan Kaufmann, San Mateo (1992)

7. Hopcroft, J., Motwani, R., Ullman, J.: Introduction to Automata Theory, Languages, and Computation, vol. 3. Addison-Wesley, Reading (2006)

8. Jiang, S., Huang, Z., Chandra, V., Kumar, R.: A polynomial algorithm for testing diagnosability of discrete event systems. IEEE Trans. Autom. Control **46**(8), 1318–1321 (2001)

9. Kahneman, D.: Thinking, Fast and Slow. Farrar, Straus and Giroux, New York (2011)

10. Lamperti, G., Quarenghi, G.: Intelligent monitoring of complex discrete-event systems. In: Czarnowski, I., Caballero, A.M., Howlett, R.J., Jain, L.C. (eds.) Intelligent Decision Technologies 2016. SIST, vol. 56, pp. 215–229. Springer, Cham (2016). https://doi.org/10.1007/978-3-319-39630-9_18

11. Lamperti, G., Zanella, M.: Diagnosis of discrete-event systems from uncertain temporal observations. Artif. Intell. **137**(1–2), 91–163 (2002). https://doi.org/10.1016/S0004-3702(02)00123-6

12. Lamperti, G., Zanella, M.: A bridged diagnostic method for the monitoring of polymorphic discrete-event systems. IEEE Trans. Syst. Man Cybern. Part B Cybern. **34**(5), 2222–2244 (2004)

13. Lamperti, G., Zanella, M.: Monitoring of active systems with stratified uncertain observations. IEEE Trans. Syst. Man Cybern. Part A Syst. Hum. **41**(2), 356–369 (2011). https://doi.org/10.1109/TSMCA.2010.2069096

14. Lamperti, G., Zanella, M., Zhao, X.: Abductive diagnosis of complex active systems with compiled knowledge. In: Thielscher, M., Toni, F., Wolter, F. (eds.) Principles of Knowledge Representation and Reasoning: Proceedings of the 16th International Conference (KR2018), pp. 464–473. AAAI Press, Tempe (2018)

15. Lamperti, G., Zanella, M., Zhao, X.: Introduction to Diagnosis of Active Systems. Springer, Cham (2018). https://doi.org/10.1007/978-3-319-92733-6

16. Lamperti, G., Zanella, M., Zhao, X.: Knowledge compilation techniques for model-based diagnosis of complex active systems. In: Holzinger, A., Kieseberg, P., Tjoa, A.M., Weippl, E. (eds.) CD-MAKE 2018. LNCS, vol. 11015, pp. 43–64. Springer, Cham (2018). https://doi.org/10.1007/978-3-319-99740-7_4

17. Lamperti, G., Zhao, X.: Diagnosis of complex active systems with uncertain temporal observations. In: Buccafurri, F., Holzinger, A., Kieseberg, P., Tjoa, A.M., Weippl, E. (eds.) CD-ARES 2016. LNCS, vol. 9817, pp. 45–62. Springer, Cham (2016). https://doi.org/10.1007/978-3-319-45507-5_4

18. Lamperti, G., Zhao, X.: Viable diagnosis of complex active systems. In: IEEE International Conference on Systems, Man, and Cybernetics (SMC 2016), Budapest, pp. 457–462 (2016). https://doi.org/10.1109/SMC.2016.7844282

19. Liu, F., Qiu, D.: Diagnosability of fuzzy discrete-event systems: a fuzzy approach. IEEE Trans. Fuzzy Syst. **17**, 372–384 (2009). https://doi.org/10.1109/TFUZZ.2009.2013840

20. Paoli, A., Lafortune, S.: Diagnosability analysis of a class of hierarchical state machines. J. Discrete Event Dyn. Syst. Theor. Appl. **18**(3), 385–413 (2008)

21. Pencolé, Y.: Diagnosability analysis of distributed discrete event systems. In: 16th European Conference on Artificial Intelligence (ECAI 2004), Valencia, Spain, pp. 43–47 (2004)

22. Pencolé, Y., Steinbauer, G., Mühlbacher, C., Travé-Massuyès, L.: Diagnosing discrete event systems using nominal models only. In: 28th International Workshop on Principles of Diagnosis (DX 2017), Brescia, Italy, pp. 169–183 (2017)

23. Reiter, R.: A theory of diagnosis from first principles. Artif. Intell. **32**(1), 57–95 (1987)

24. Rintanen, J., Grastien, A.: Diagnosability testing with satisfiability algorithms. In: 20th International Joint Conference on Artificial Intelligence (IJCAI 2007), Hyderabad, India, pp. 532–537 (2007)

25. Sampath, M., Sengupta, R., Lafortune, S., Sinnamohideen, K., Teneketzis, D.: Diagnosability of discrete-event systems. IEEE Trans. Autom. Control **40**(9), 1555–1575 (1995)

26. Schumann, A., Huang, J.: A scalable jointree algorithm for diagnosability. In: 23rd National Conference on Artificial Intelligence (AAAI 2008), Chicago, IL, pp. 535–540 (2008)

27. Su, X., Zanella, M., Grastien, A.: Diagnosability of discrete-event systems with uncertain observations. In: 25th International Joint Conference on Artificial Intelligence (IJCAI 2016), New York, NY, pp. 1265–1571 (2016)

28. Thorsley, D., Teneketzis, D.: Diagnosability of stochastic discrete-event systems. IEEE Trans. Autom. Control **50**, 476–492 (2005). https://doi.org/10.1109/TAC.2005.844722

29. Ye, L., Dague, P., Yan, Y.: An incremental approach for pattern diagnosability in distributed discrete event systems. In: 21st IEEE International Conference on Tools with Artificial Intelligence (ICTAI 2012), Newark, NJ, pp. 123–130 (2009). https://doi.org/10.1109/ICTAI.2009.75

30. Yoo, T., Lafortune, S.: Polynomial-time verification of diagnosability of partially observed discrete-event systems. IEEE Trans. Autom. Control **47**(9), 1491–1495 (2002)

A Case for Guided Machine Learning

Florian Westphal[1]([⊠]) (iD), Niklas Lavesson[1,2] (iD), and Håkan Grahn[1] (iD)

[1] Blekinge Institute of Technology, Karlskrona, Sweden
{florian.westphal,hakan.grahn}@bth.se
[2] Jönköping University, Jönköping, Sweden
niklas.lavesson@ju.se

Abstract. Involving humans in the learning process of a machine learning algorithm can have many advantages ranging from establishing trust into a particular model to added personalization capabilities to reducing labeling efforts. While these approaches are commonly summarized under the term interactive machine learning (iML), no unambiguous definition of iML exists to clearly define this area of research. In this position paper, we discuss the shortcomings of current definitions of iML and propose and define the term guided machine learning (gML) as an alternative.

Keywords: Guided machine learning · Interactive machine learning · Human-in-the-loop · Definition

1 Introduction

With the continuing advances in machine learning, the decisions taken by machine learning algorithms have more and more impact on everyday life. Therefore, it is important that users of these algorithms, as well as people affected by these decisions can understand and trust the used algorithms. One common way to achieve this is interactive machine learning (iML) [12], which interactively involves users in the training process. This can help users to better understand the decisions taken by the machine learning algorithm and therefore increase trust in those algorithms. Furthermore, it enables users to adjust the algorithm's behavior to their needs. Thus, making the benefits of machine learning available to the wider public, leading to a democratization of machine learning.

While characterizations and definitions of iML have been provided in survey papers by Amershi et al. [1], Bertini and Lalanne [2] and Holzinger [10], we argue that all of these definitions are ambiguous to some degree and thus include or exclude more approaches than intended. As an unambiguous definition is important to define a research area clearly and to help identify relevant work easily, we propose a new definition for iML, which avoids the identified ambiguities. Furthermore, we argue that the word *interactive* in iML is unintentionally broad and propose the term *guided machine learning* (gML) for this area instead.

A. Holzinger et al. (Eds.): CD-MAKE 2019, LNCS 11713, pp. 353–361, 2019.
https://doi.org/10.1007/978-3-030-29726-8_22

We discuss the issues with current definitions of iML in Sect. 2. In Sect. 3, we propose our definition of gML and examine its implications for other fields of research within machine learning, and in Sect. 4, we summarize the main points of this position paper.

2   Interactive Machine Learning

In the following, we clarify basic machine learning terminology in Sect. 2.1, describe the difficulty to distinguish iML from general machine learning (ML) in Sect. 2.2 and discuss the shortcomings of different attempts to establish this distinction in Sects. 2.3 and 2.4. Furthermore, we argue that the term *interactive* is too broad to describe what is currently considered as iML in Sect. 2.5.

2.1   Machine Learning

Machine learning is concerned with algorithms that learn from data. Mitchell [15] defines this learning as follows:

Definition 1. *A computer program is said to learn from experience E with respect to some class of tasks T and performance measure P, if its performance at tasks in T, as measured by P, improves with experience E.*

In order to achieve this learning from experience, a machine learning algorithm builds a model from the training data. Thus, the model encodes the current state of the learner. Since solving tasks from the given class of tasks is based on the model, improvements of the model result in an improved performance.

2.2   Difference to Machine Learning

In order to establish iML as distinct field of research, any definition of iML needs to separate it clearly from existing fields. This is especially true since the name *interactive machine learning* does not in itself separate iML from ML. Based purely on the name, one could define iML as *algorithms that improve their performance at a task through interactions*. Here, it is important to note that this definition does not limit the type, source or target of the interactions. From Definition 1, we can see that improvement through *experience*, for ML, compared to improvement through *interactions*, for iML, is the only difference between this name based iML definition and the definition of ML. However, **this makes iML identical to ML**, since experience presupposes interaction, as we will show in the following.

Experience can only be gained through either practical action or observation. Gaining experience through practical action involves doing something and observing its effect, thus it involves interaction. For example, learning to play Tetris involves playing the game and observing the results of each decision taken; hence, interacting with the Tetris world. **Therefore, gaining experience through practical action clearly requires interaction.**

Furthermore, gaining experience through observation requires this observation to be active. This means that it is not sufficient to gather observational data, but this data needs to be processed to gain experience. However, this processing of observational data requires a certain level of interaction with this data. For example, it is not sufficient to stare at someone playing Tetris to learn how to play, instead hypotheses need to be formed based on current observational data to direct attention to gain more relevant data and to extract relevant observations. **Therefore, gaining experience through observation clearly requires interaction.**

In order to avoid this overly broad definition of iML, common definitions of iML require the presence of a human or non-human interaction partner, which distinguishes iML from ML.

2.3   Human-in-the-Loop

The most common restriction on the interaction partner is to require this partner to be human. For example, Bertini and Lalanne describe iML as follows: *"Interactive Machine Learning is an area of research where the integration of human and machine capabilities is advocated, beyond scope of visual data analysis, as a way to build better computational models out of data. It suggests and promotes an approach where the user can interactively influence the decisions taken by learning algorithms and make refinements where needed."* [2]. This clearly separates iML from ML by requiring a human user to interact with the learning algorithm. However, one potential issue with this approach is that it introduces a certain degree of ambiguity, since **it is unclear if an iML algorithm**, according to this description, **should still be considered as such if the human is replaced with a program simulating human interactions.** This uncertainty is problematic for a definition of iML, since it leaves the classification of an algorithm as iML algorithm up for interpretation. **In this way, it may be possible that important iML approaches are not recognized as such and are overlooked by the iML community.**

2.4   Non-human Agents

One way to avoid this ambiguity caused by requiring human interaction partners is to extend the iML definition to non-human partners. This is done, for example, in Holzinger's definition of iML [10]:

Definition 2. *Interactive machine learning is concerned with algorithms that can interact with agents and can optimize their learning behavior through these interactions, where the agents can also be human.*

While this definition clearly sidesteps the ambiguity caused by limiting iML only to human interaction partners, it is actually even more ambiguous. **The main issue of this definition is that it requires an ambiguous distinction to be made between non-human agents and machine learning**

mechanisms, in order to distinguish iML from ML. In the following, we will illustrate the difficulty to make this distinction with the help of the operation of decision tree pruning as example.

Decision tree pruning is an operation performed on a decision tree to avoid overfitting and to improve the overall performance of the decision tree by removing training data specific subtrees. Han and Cercone [8], for example, propose the DTViz system, which allows human users to interactively construct and prune decision trees. Clearly, this system allowing users to perform tree pruning should be considered as iML system according to Definition 2. One algorithm, which could replace the user in this approach, is the pruning algorithm proposed by Kearns and Mansour [13]. This algorithm determines automatically for a given decision tree which subtrees should be pruned and thus performs the same task as the human user. Therefore, the use of this algorithm instead of a human user should not change the system's classification as iML system. This is the case, since the basic interaction, the system presents the current decision tree and receives pruning decisions from the agent, is still the same. However, the same can be said about the approach proposed by Gelfand et al. [4], which integrates the tree pruning into the tree construction process. While the basic interaction stays the same, the pruning algorithm becomes part of the learning algorithm. Based on this, **one can either argue that this integrated approach should be classified as iML, since the interaction is basically the same as in the case of the DTViz system, or argue that it should not be classified as iML, since no real interaction is taking place, because the pruning algorithm is part of the learning mechanism.** Thus, it is not possible to unambiguously classify the presented pruning approach as iML algorithm.

The presented example illustrates that it can be difficult to distinguish between an agent interacting with a machine learning algorithm and a part of this learning algorithm. However, this distinction is necessary, since otherwise any learning mechanism and thus all of ML could be classified as iML, which would render the definition useless.

2.5   Interactive Learning

Apart from the problem of differentiating iML from ML, **another issue for any definition of iML is that the term** *interactive* **covers two different scenarios.** On the one hand, the interaction partner has an idea of the task the machine learning algorithm should perform and directs it towards this goal, while on the other hand the interaction partner may not have such a goal and may just interact with the algorithm without clear purpose. Clearly, both scenarios are covered by the term *interactive*, since an agent interacts with the learning algorithm in both cases. However, the former, directed scenario is arguably more interesting and most commonly considered in characterizations of iML. This is made clear, for example, in the previously mentioned description of iML by Bertini and Lalanne [2], as well as by Amershi et al. who state: *"As a result of these rapid interaction cycles, even users with little or no machine-learning expertise can steer machine-learning behaviors through low-cost trial and error* *or focused experimentation with inputs and outputs*" [1]. **In order to obtain a focused definition of iML, which reflects this preference, any iML definition needs to explicitly exclude the undirected case.** An alternative solution, i.e., the replacement of the word *interactive* with *guided* will be discussed in the next section.

3   Guided Machine Learning

In this section, we propose *guided machine learning* (gML) as alternative to *interactive machine learning* (iML). In Sect. 3.1, we propose a definition for gML and discuss how this definition addresses the previously raised issues. Furthermore, we review the relationship between gML and other fields of research within machine learning in Sect. 3.2.

3.1   Proposed Definition

In Sect. 2, we have shown that, while it is important to distinguish iML from ML, it is difficult to do without introducing a certain degree of ambiguity. We argue that the ambiguity introduced by allowing non-human agents, as discussed in Sect. 2.4, is worse than when focusing only on human interaction partners. This is the case, since considering non-human agents either requires highly subjective judgments on whether or not to include a proposed algorithm or, if relaxed too much, may lead to including all of ML. Therefore, we propose to focus on human interaction partners.

In order to reduce the ambiguity introduced by limiting our definition to human users, we propose to require the presence of a user interface instead of a user for considering an approach as falling under our definition. In this way, the substitution of a real user with a program simulating user activity does not undermine the definition, while the required presence of a user interface sets a clear boundary between considered approaches and the rest of ML.

As mentioned in Sect. 2.5, the use of the word *interactive* covers two distinct scenarios of which only one is relevant. While this issue could be addressed in the definition, we propose to replace *interactive* with *guided*, since this captures the intended scenario in which a user interacts with a machine learning algorithm in order to improve its performance on a certain task. Therefore, we propose the term *guided machine learning* and define learning through guidance similar to Mitchell's definition for learning [15] as follows:

Definition 3. *A computer program is said to learn through guidance G from a human H with respect to some class of tasks T and performance measure P, if its performance at tasks in T, as measured by P, improves through actions performed by H, given that these actions are dependent on the program's current state and aim to achieve such an improvement.*

This definition captures the guidance aspect by requiring the human user to perform actions, which improve the performance of the machine learning algorithm, with the goal to achieve this improvement. Furthermore, it captures the interactiveness of this guidance process by requiring the user's actions to be based on the current state of the algorithm. This state can be presented to the user directly in form of the current model or indirectly in form of the current performance on the task. In this way, it is possible for the user to iteratively provide guidance to the algorithm throughout the learning process. With this definition of learning from guidance, we can now define gML as follows:

Definition 4. *Guided machine learning is concerned with the design of interfaces for human users, which enable a user to perform actions, which allow a computer program to learn from this guidance.*

This definition clearly separates gML from ML by requiring the machine learning algorithm to learn from data collected iteratively through a user interface. It avoids to be dependent on the deployment context by not requiring the presence of a human or the presence of guidance actions, but instead focuses on the presence of an interface, which would allow a human to perform such actions.

One potential issue of the proposed definition is that ascertaining the required properties of the provided interface may be subjective. However, we would argue that checking if a provided interface presents the machine learning algorithm's current state in a way aimed at human understanding and allows actions to be taken in response to the presented information, which can improve the algorithm's performance, is reasonably objective.

3.2   Definition Consequences

In the following, we will discuss the relationship between gML, as defined in the previous section, and other research areas in ML.

Supervised Learning. The key defining feature of supervised learning is that supervised machine learning algorithms learn to perform their task from a labeled training data set. While these labels are generally provided by humans, supervised learning cannot be considered as gML, since data sets in supervised learning are available in full at the beginning of the training. In contrast, in gML, training labels are provided throughout the training process via a user interface and depend on the model's current state. One example for such guidance through the provision of more labeled data is the approach to pixel classification proposed by Fails and Olsen [3]. In their approach, Fails and Olsen allow users to view the current pixel labeling performance of a classifier and to provide more pixel labels to improve the performance. Apart from the plain pixel labels, the provided information also contains the implicit knowledge that the newly labeled pixels are more relevant to the learning process than a randomly selected set of pixels, which would be chosen by a supervised learning approach.

Unsupervised Learning. The overall goal in unsupervised learning is to extract information from a given data set without any form of human intervention or guidance. Thus, unsupervised learning is clearly unrelated to gML.

Reinforcement Learning. Reinforcement learning is concerned with learning how to act in a given situation not from a prescribed ideal action, as in supervised learning, but instead from a cumulative reward signal [18]. In general, reinforcement learning approaches cannot be considered as part of gML, since reinforcement learning requires only the collection of a reward signal, which does not necessarily presuppose a user interface. However, reinforcement learning ideas can be used in gML approaches, such as in the approach by Thomaz and Breazeal [19], which allows users to give reward signals to a virtual robot, in order to teach it to bake a cake. The reward signal provided for certain behavior depends on the robot's current state, since its actions are determined by it.

Active Learning. In active learning [17], an active learning algorithm selects the training samples, which should be used to train a machine learning algorithm based on the learning algorithm's current state. While this conditionality on the learner's state leads to a close relationship between active learning and gML, not all active learning approaches are also gML approaches, since active learning does not require a user interface. However, active learning is useful for designing user interfaces, since it can reduce the amount of data to be presented to a user. For example, the approach proposed by Heimerl et al. [9] for classifying text documents as relevant or irrelevant uses active learning to solicit user labels.

Adversarial Training. In adversarial training scenarios, learning is facilitated by the competition between two or more learners. Typical examples for these scenarios are Samuel's checkers program [16], which trained an algorithm to play checkers by playing against itself, as well as generative adversarial networks (GANs) [7], which train a generator to generate samples from a target distribution together with a discriminator for distinguishing between generated and real samples. While learning using adversarial training proceeds in an iterative feedback loop between the adversaries, it is clearly different from gML. This is the case, not only because most adversarial training does not use humans as adversary, which would be required for gML, but also because gML does not assume a competition between the user and the algorithm. In contrast, the guidance aspect requires the user to perform actions with the aim to improve the learner's performance.

Explainable Machine Learning. The main goal of explainable machine learning is to make decisions taken by an ML algorithm transparent, understandable and explainable [6]. This can be achieved either through post-hoc explanations, which are generated on demand for a particular decision, or through ante-hoc explanations, which arise naturally from the used model [11]. While explainable ML does clearly not belong to gML, it is an important aspect in the interface design for gML approaches. In particular ante-hoc systems are useful for gML, since they are directly interpretable by users and should therefore make it easier for them to guide the learning process. This connection between explainable ML is even clearer in the concept of causability [11], which requires the provided explanation to reach a certain causal understandability. This is interesting for gML, since a causal explanation of a taken decision should enable users to guide the learning process more efficiently and more easily.

Machine Learning Environments. While not directly a research area in ML, machine learning environments, such as Weka[1], may appear to be closely related to gML, since they provide a user interface, which can be used to choose learning algorithm and model hyperparameters, which can improve the algorithm's performance on a task. However, the difference between those environments and gML approaches is that they normally rebuild the previous model from scratch with the newly chosen hyperparameters. This is different from gML, which assumes an update rather than a rebuild of the model. One approach, which blurs this line between machine learning environments and gML is human-guided machine learning (HGML) [5], which allows users to interact with an automated ML (AutoML) system. These interactions can be concerned with the input data, for example in form of feature or instance selection, with the model development, such as model selection or parameter settings, or with the model interpretation, for example in form of model assessment or parameter comparison. Such a system could be considered as a gML approach, if the user actions are used as input to teach a meta-learner to find a suitable configuration for the AutoML system. One other approach, which may bridge the gap between gML and machine learning environments is explanatory debugging, as proposed by Kulesza et al. [14]. The main idea of explanatory debugging is to provide users with an explanation of the algorithm's current performance, which can help them to modify the model accordingly.

4   Summary

In this paper, we have discussed various issues of existing descriptions and definitions of iML. We have argued that iML needs to be defined based on the presence of an interaction partner to distinguish it from general machine learning. However, we have also shown that requiring the presence of a human or non-human interaction partner leads to certain ambiguities. Furthermore, we have pointed out that the word *interactive* in iML may lead to the inclusion of approaches commonly not considered as part of iML. We have addressed these issues by proposing *guided machine learning* and defining it in a way, which avoids the identified sources of ambiguity.

Acknowledgements. The authors would like to thank Huynh Khanh Vi Tran for valuable discussions about possible gML definitions, as well as the anonymous reviewers for their useful comments.

This work is part of the research project "Scalable resource-efficient systems for big data analytics" funded by the Knowledge Foundation (grant: 20140032) in Sweden.

---

[1] https://www.cs.waikato.ac.nz/ml/weka/.

References

1. Amershi, S., Cakmak, M., Knox, W.B., Kulesza, T.: Power to the people: the role of humans in interactive machine learning. AI Mag. **35**(4), 105–120 (2014)
2. Bertini, E., Lalanne, D.: Surveying the complementary role of automatic data analysis and visualization in knowledge discovery. In: Proceedings of the ACM SIGKDD Workshop on Visual Analytics and Knowledge Discovery: Integrating Automated Analysis with Interactive Exploration, pp. 12–20. ACM (2009)
3. Fails, J.A., Olsen Jr., D.R.: Interactive machine learning. In: Proceedings of the 8th International Conference on Intelligent User Interfaces, pp. 39–45. ACM (2003)
4. Gelfand, S.B., Ravishankar, C.S., Delp, E.J.: An iterative growing and pruning algorithm for classification tree design. In: Conference Proceedings, IEEE International Conference on Systems, Man and Cybernetics, vol. 2, pp. 818–823 (1989). https://doi.org/10.1109/ICSMC.1989.71407
5. Gil, Y., et al.: Towards human-guided machine learning. In: Proceedings of the 24th International Conference on Intelligent User Interfaces, pp. 614–624. ACM (2019)
6. Goebel, R., et al.: Explainable AI: the new 42? In: Holzinger, A., Kieseberg, P., Tjoa, A.M., Weippl, E. (eds.) CD-MAKE 2018. LNCS, vol. 11015, pp. 295–303. Springer, Cham (2018). https://doi.org/10.1007/978-3-319-99740-7_21
7. Goodfellow, I., et al.: Generative adversarial nets. In: Advances in Neural Information Processing Systems, pp. 2672–2680 (2014)
8. Han, J., Cercone, N.: Interactive construction of decision trees. In: Cheung, D., Williams, G.J., Li, Q. (eds.) PAKDD 2001. LNCS (LNAI), vol. 2035, pp. 575–580. Springer, Heidelberg (2001). https://doi.org/10.1007/3-540-45357-1_61
9. Heimerl, F., Koch, S., Bosch, H., Ertl, T.: Visual classifier training for text document retrieval. IEEE Trans. Visual Comput. Graphics **18**(12), 2839–2848 (2012)
10. Holzinger, A.: Interactive machine learning for health informatics: when do we need the human-in-the-loop? Brain Inf. **3**(2), 119–131 (2016)
11. Holzinger, A., Langs, G., Denk, H., Zatloukal, K., Müller, H.: Causability and explainabilty of artificial intelligence in medicine. Wiley Interdisciplinary Rev. Data Min. Knowl. Discovery, e1312 (2019)
12. Holzinger, A., et al.: Interactive machine learning: experimental evidence for the human in the algorithmic loop. Appl. Intell., 1–14 (2018)
13. Kearns, M.J., Mansour, Y.: A fast, bottom-up decision tree pruning algorithm with near-optimal generalization. In: Proceedings of the Fifteenth International Conference on Machine Learning, ICML 1998, pp. 269–277. Morgan Kaufmann Publishers Inc. (1998). http://dl.acm.org/citation.cfm?id=645527.657457
14. Kulesza, T., Burnett, M., Wong, W.K., Stumpf, S.: Principles of explanatory debugging to personalize interactive machine learning. In: Proceedings of the 20th International Conference on Intelligent User Interfaces, pp. 126–137. ACM (2015)
15. Mitchell, T.M.: Machine Learning. McGraw-Hill, New York (1997)
16. Samuel, A.L.: Some studies in machine learning using the game of checkers. IBM J. Res. Dev. **3**(3), 210–229 (1959). https://doi.org/10.1147/rd.33.0210
17. Settles, B.: Active learning. Synth. Lect. Artif. Intell. Mach. Learn. **6**(1), 1–114 (2012)
18. Sutton, R.S., Barto, A.G.: Reinforcement Learning: An Introduction. MIT Press, Cambridge (2018)
19. Thomaz, A.L., Breazeal, C.: Teachable robots: understanding human teaching behavior to build more effective robot learners. Artif. Intell. **172**(6–7), 716–737 (2007)

Using Ontologies to Express Prior Knowledge for Genetic Programming

Stefan Prieschl[1(✉)], Dominic Girardi[1], and Gabriel Kronberger[2]

[1] RISC Software GmbH, Johannes Kepler University, Hagenberg, Austria
{stefan.prieschl,dominic.girardi}@risc-software.at
[2] Josef Ressel Centre for Symbolic Regression, University of Applied Sciences Upper Austria, Hagenberg, Austria
gabriel.kronberger@fh-hagenberg.at

Abstract. Ontologies are useful for modeling domains and can be used to capture expert knowledge about a system. Genetic programming can be used to identify statistical relationships or models from data. Combining expert knowledge as well as statistical rules identified solely from data is necessary in application domains where data is scarce and a large body of expert knowledge exists.

We therefore study if the performance of genetic programming can be improved by incorporating prior knowledge from an ontology. In particular, we include prior knowledge as additional features for genetic programming.

The approach is tested with six benchmark data sets where we compare the required computational effort that is necessary to find an acceptable model with and without additional features. The results show that additional features gathered from an ontology improve the performance of tree-based GP. The probability to find acceptable solutions with a fixed computational budget is increased. For noisy data sets we observed the same effect as for the data sets without noise.

Keywords: Supervised learning · Ontologies · Domain knowledge · Genetic programming · Symbolic regression

1 Motivation

In the recent years, research in many domains has developed into a mostly data-driven activity. This requires researchers with knowledge in their own research domain on the one hand and knowledge in data science on the other hand. It is a challenge to bring these two worlds together. In the field of medical informatics, activities around this problem are often labelled as doctor-in-the-loop approach [11]. The goal is to deeply integrate the domain experts into the knowledge discovery process and benefit from their expertise, while acknowledging the fact that these domain experts are neither IT experts nor data scientists. Thus, support from software tools is needed to utilize experts' prior domain knowledge for

© IFIP International Federation for Information Processing 2019
Published by Springer Nature Switzerland AG 2019
A. Holzinger et al. (Eds.): CD-MAKE 2019, LNCS 11713, pp. 362–376, 2019.
https://doi.org/10.1007/978-3-030-29726-8_23

modeling. The challenge from a technical point of view is to formally describe their knowledge and to make algorithms that elicit and employ this knowledge. In the literature various examples of approaches to integrate humans in the process can be found. For example Holzinger et al. include humans by gamification into a machine learning process [16].

A concept that is potentially able to cope with this challenge is genetic programming (GP). GP is a method of evolutionary computing where concepts from natural evolution are simulated for the evolution of computer programs that are able to solve a given problem when executed [6, 19]. A particular task of GP is symbolic regression. Here GP is used to evolve simple closed-form expressions that fit a given data set. Unlike black-box models such as support vector machines (SVM) or artificial neural networks (ANN), the aim of symbolic regression is to identify models which can be interpreted by humans. In this way, the loop from the expert knowledge to the algorithms back to the domain expert can be closed.

Expert knowledge can be formalized using a domain ontology, which contains structural information about the research data as well as prior domain knowledge about known correlations and causal dependencies between data attributes. A GP algorithm can be launched for a selected data set of interest incorporating prior knowledge as building blocks. The hypothesis of the present paper is that the performance of GP can be improved when it is provided with prior domain knowledge. This improvement can either yield a better regression model given a fixed computational budget or alternatively yield a model of equivalent quality with less computational effort.

2 Related Work

2.1 Genetic Programming

GP is a technique where a genetic algorithm (GA) is used to generate a computer program [19]. Usually, programs are encoded as an expression tree and are evolved over a number of iterations through evolutionary operations: selection, crossover and mutation. In the case of symbolic regression, the GP programs are formulas that replicate a specific relationship from a data set.

A GA is a heuristic method based on Charles Darwin's idea of natural selection [8]. When using a GA one encodes solutions as sets of individuals. Initially, these individuals are randomly generated. Each individual is evaluated by calculating a so-called fitness value. Individuals with higher fitness have a higher probability to be selected for reproduction. This mechanism implements the idea of *survival of the fittest*, which Darwin describes. New individuals are created based on their predecessor using a crossover operation. This leads to increasing result quality of the individuals over iterations. More detailed description of GAs can be found in the literature [1, 12, 15].

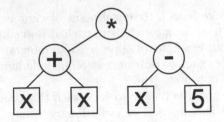

Fig. 1. Example of a symbolic tree representing $(x + x) * (x - 5)$.

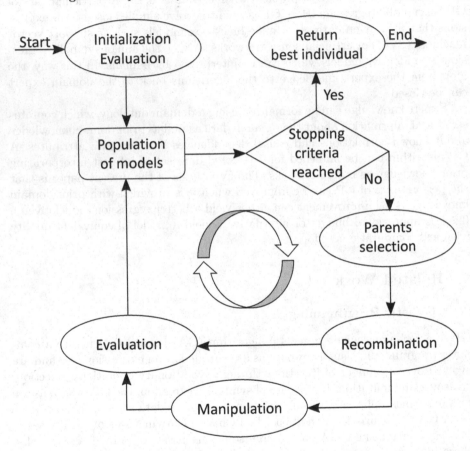

Fig. 2. The process cycle of GP.

Cramer has layed the foundations for GP in 1985 [6]. Koza later popularized and further developed GP [19]. In contrast to classic GAs, a variable-length encoding – most frequently expression trees – are used for GP. Expression trees represent formulas or computer program. Figure 1 shows an example of an expression tree.

Figure 2 shows the process cycle of GP. It starts with the initialization of the population of models by randomly generating a defined number of individuals. Then the main cycle is executed until the stopping criteria are met. The stopping criteria, similarly to classical optimization, can comprise a fixed number of iterations, a defined result quality, or a combination of both.

In every iteration the *selection of parents* takes place, based on the individual's fitness values. To that end, a fitness function is used to evaluate the fitness value of each individual.

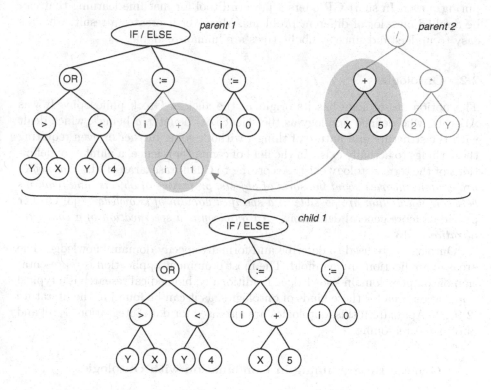

Fig. 3. Example of recombination in GP [1].

The selected individuals are grouped as pairs for the *recombination* (also known as crossover) step. Figure 3 shows an example of such a recombination. For this example a sub-tree is selected for each of the two parent trees. These two sub-trees are swapped and thereby two child trees are created newly. The figure shows how the first of these two child trees is generated out of the two parent trees.

In the *manipulation* step, each new individual is additionally manipulated by mutation with a defined probability. For this mutation there are different possibilities. For example a sub-tree could be removed and replaced by a randomly generated one. Another possibility is to manipulate one of the nodes. This can be implemented by changing the type of a node or changing some of its parameters.

Next, newly generated individuals are evaluated for their fitness value.

The last step creates a new generation of individuals using the previous generation and the newly generated offspring. For example *generational replacement* can be used, where all individuals of the previous generation are discarded and the offspring individuals form the successor generation. Often also *elitism* is used, where one or more individuals with the best fitness values of the old generation are adopted in the new generation.

Nowadays GP gets more and more practicable thanks to the increasing computing power. In short GP offers a powerful tool for machine learning that can be applied on a lot of different problems. Moreover it creates a result, which is easy to understand since symbolic trees are human-readable.

2.2   Ontologies

The notion of ontology has its origin in the ancient Greek philosophy. It was Aristotle who defined ontology as the science of "being *qua* being", which dealt with the structure and nature of things, without going further or even requiring these things to actually exist. In the field of computer science, a number of definitions of the term ontology exist. According to Chandrasekaran et al. *"Ontologies are content theories about the sorts of objects, properties of objects, and relations between objects that are possible in a specified domain of knowledge."* [3] Gruber provide a more general definition: *"An ontology is a specification of a conceptualization."* [13].

Ontologies are used to describe and formally specify domain knowledge. The areas of application are manifold. The most prominent application is representation of complex domain knowledge. Traditionally, biomedical research is a typical application area for these kinds of ontologies, as it can be found in the literature [2,9,22]. Apart from this, ontologies are also used for data integration [7,10] and automated reasoning.

2.3   Genetic Programming in Combination with Ontologies

In the present study, a literature survey in the field of genetic programming in combination with ontologies was conducted, to illustrate the state-of-the-art. In fact, domain knowledge is being used for system identification, but with varying perspectives. For example Ratle and Sebag [24] or Schoenauer and Sebag [26] use domain knowledge for system identification, by using *G3P* (grammar guided GP). G3P [5,28,31] is used to define validity constrains for individuals as context free grammars. However, getting domain knowledge into the GP algorithm itself has not been investigated in this detail to date.

3   Methods

In this contribution we use a very limited set of the capabilities of ontology modelling. In particular, we neglect domain modelling and instead focus only on the definition of suspected or known relationships between variables – knowledge which can be represented explicitly within an ontology.

To give a simple example, we might know that there is a direct linear relationship between two variables y and x which we could encode as a functional dependency $y \leftarrow \theta x$; whereby we use the parameter θ for the unknown scaling factor and $\leftarrow$ to encode the causal direction of the dependency. Many similar assumed or known functional dependencies can be encoded using the same approach. Further examples for commonly occurring bivariate functional dependencies are[1]:

- Exponential decay of y over x with an unknown rate: $y \leftarrow \exp(\theta x)$
- Logarithmic growth of y over x with an unknown rate: $y \leftarrow \log(\theta x)$
- Oscillation of y over x with an unknown frequency: $y \leftarrow \sin(\theta_1 x + \theta_2)$
- Logistic growth of y over x with an unknown rate and limit: $y \leftarrow \frac{\theta_1}{1+\exp(\theta_2 x)}$

Including knowledge in the GP process is possible in many ways (cf. [30]): extension of the feature set, seeding of the initial generation of GP, definition of syntactic building blocks for evolutionary operators, and extension of the function set.

A straightforward approach is to extend the set of variables and to add a pre-calculated feature for each defined functional dependency. This can be accomplished with minimal effort for any GP implementation without requiring adaptations to the GP implementation itself. A drawback of this approach is that for the calculation of the features it is necessary to know at least approximate values for the parameters θ. For example, for the case that the variable of interest y decays exponentially over x ($y \leftarrow \exp(\theta x)$) we would need to assume a value for the decay rate θ in the calculation of the feature values. Once these values have been calculated, the decay rate is fixed. GP uses the calculated feature exactly the same way as the original variable values. In particular, it is not possible to adjust the decay rate parameter when evolving models using the pre-calculated features. A similar argument can be given for the frequency parameter in a periodic function.

Functional relations expressed in the ontology can be used for the initialization of the GP population in the first generation (seeding). Usually, individuals in the first GP generation are generated randomly. However, when prior knowledge is available in the ontology we can include these expressions as subexpressions within randomly initialized individuals. The potential benefit is that GP already starts with relevant functional expressions in the genome. As a consequence, these sub-expressions do not have to be discovered through the evolutionary process, theoretically improving the performance of GP. An important

---

[1] It should be noted that it would be rather easy to provide a graphical interface which is easy to understand for users which are not very familiar with the mathematical notation. For instance it would be possible to provide a graphical representation of the suspected functional dependency between variables where the users only need to choose the function type from a menu. The so-defined functional dependency can then be added to the ontology. This would not affect the underlying implementation of the learning algorithm.

difference to the first approach is that GP is still able to break up, modify and improve sub-expressions which have been included by seeding. This would not be possible with the feature extension. Seeding would require a change of the procedure for initialization of the population of the GP algorithm.

An alternative to seeding of the initial population would be to include prior knowledge as pre-defined syntactical building blocks for expressions produced by GP (cf. [24, 26]). This approach requires a so-called grammar-guided GP system [21] which allows the definition of syntactical constraints. Such GP systems produce expressions which conform to a syntax defined via a formal grammar. This facilitates integration of prior knowledge such as known transformations of input variables as well as the definition of the structure of the expression with slots for sub-expressions that can be evolved. For example, such systems would allow to express that

$$y = \exp(g(x_1, x_2)x_3) + f(x_1, x_2) + \epsilon$$

where $g(x_1, x_2)$ and $f(x1, x2)$ are sub-expressions evolved by the GP system. The approach via syntactical building blocks is arguably the most general. Many forms of prior knowledge can be expressed via syntactical constraints. Notably, the simple extension of the feature set as well as seeding of the population are special cases of this approach. However, the flexibility also introduces higher complexity of the GP implementation as well as issues with GP performance which can be hampered by intricate syntactical constraints. Syntactical constraints might even increase the problem of premature convergence as a consequence of the reduced diversity of solutions.

Finally, instead of extending the feature set we could extend the function set of the GP system to include expressions from the ontology. The GP system is allowed to include these functions within evolved expressions using terminals or sub-expressions as arguments. The motivation for this approach is that we could include functions with unknown numeric parameters and allow GP to identify optimal parameters via evolution. For example this would allow us to add a parametric function such as:

$$\text{ExpDecay}_\theta(x) = \exp(-\exp(\theta)x), x \in \{x_1, x_2, x_3\}$$

which only allows features x_1, x_2, x_3 as arguments and has θ initialized randomly and evolved via GP.

For this study, we have chosen to use pre-calculated features because it can be implemented with minimal effort. As discussed above, this approach has the important drawback that parameters have to be approximated by users. However, for our experiments with synthetic benchmark problems we assume that these parameters are known. We leave experiments with syntactical building blocks or parametric functions for future work.

3.1   Experimental Setup

We test the hypothesis that GP performance can be improved trough expert knowledge using computational experiments with simulated data sets. For our experiments we omit the use of an ontology for simplicity and we directly use the knowledge that could be provided by an ontology. We consider two scenarios where we use the same simulated data sets with and without noise. For the experiments we use tree-based genetic programming which can be considered more a less a de-facto standard variant (SGP). We use PTC2 [20] to initialize random trees with a uniform length distribution between 1 and *max. tree length* nodes. Terminals and function symbols are selected with uniform probabilities from the terminal set and the function set. The algorithm uses generational replacement where the best individual is kept for the next generation (elitism). For crossover events we use sub-tree crossover with 90% probability to select an internal node. For mutation events we conduct one of the following mutation operations with uniform distribution:

- change the function symbol of an internal node
- change the parameter or variable reference of a terminal node
- change the parameters of all terminal nodes
- delete a randomly selected sub-tree
- replace a randomly selected sub-tree with a newly initialized random tree

The mutation rate parameter defines the probability that a newly created individual is manipulated with one the operators above. The fitness of individuals is determined by calculating the coefficient of determination (squared Pearson correlation) of the function output values and the target variable values. The parents for recombination are chosen via tournament selection where the individual with highest R^2 in the group is selected. The full configuration for our GP experiments is shown in Table 1.

We run 60 independent repetitions of the same GP configuration for each of the problem instances described below. In each GP run we invest the same effort of 500,000 evaluated solutions, where we record the quality (R^2) of the best solution in each generation step. This data allow us to answer whether:

1. The probability to solve a given problem is increased when prior knowledge from an ontology is available (for fixed effort).
2. The computational effort to solve a given problem can be reduced (for a given success probability).

For the comparison of algorithm configurations we visualize the empirical distribution of run length (evaluated solutions) until the problem is solved. This method is used for instance for the comparison of multiple optimization algorithms on a large set of benchmark problems in [14]. For the problem instances without noise we set 0.99 as the R^2 threshold for success; for the noisy problems we use the threshold value 0.95. All experiments have been performed with HeuristicLab[2] which provides a grammar guided tree-based GP system [17].

---

[2] https://dev.heuristiclab.com.

Table 1. Genetic programming parameter settings.

Parameter	Value
Population size	10000
Maximum generations	50
Elites	1
Mutation (one of)	Change function symbol
	Change terminal symbol
	Change all parameters
	Delete a random sub-tree
	Replace a random sub-tree
Mutation rate	15%
Selector	Tournament with group size 7
Crossover	Sub-tree crossover
Function set	$+, -, *, /, \sin(x), \cos(x), \log(x), \exp(x), x^2, \sqrt{x}$
Max tree depth	17
Max tree length	100
Fitness evaluation	Pearson R^2

3.2 Selection of Benchmarking Data

We selected the following benchmark problem instances, that are shown in Table 2. We generate data for input variables using the ranges defined in Table 3 and calculate the target variable using the expressions.

Some of the instances are recommended in [29]. The *flow psi* function is a function occurring in modelling fluid dynamics and has been taken from [4].[3] All problem instances are possible to solve with our GP settings.

Table 2. The problem instances of our experiments.

Instance	Function
Salustowicz [25, 29]	$f(x) = x^3 \exp(-x) \cos(x) \sin(x)(\sin(x)^2 \cos(x) - 1)$
Vladislavleva-3 [27, 29]	$f(x_1, x_2) = \exp(-x_1)x_1^3 \cos(x_1) \sin(x_1)(\cos(x_1)$ $\sin(x_1)^2 - 1)(x_2 - 5)$
Vladislavleva-4 [27, 29]	$f(x_1, x_2, x_3, x_4, x_5) = \frac{10}{5 + \sum_{i=1}^{5}(x_i - 3)^2}$
Pagie [23, 29]	$f(x, y) = \frac{1}{1 + x^{-4}} + \frac{1}{1 + y^{-4}}$
Flow psi [4]	$f(x1, x2, x3, x4, x5) = x_1 x_3 \sin(\frac{\pi x_2}{180})(1 - (\frac{x_4}{x_3})^2) + x_5 \log(\frac{x_3}{x_4})$
Korns-12 [18, 29]	$f(x_0, x_1, x_2, x_3, x_4) = 2 - 2.1 \cos(9.8x_0) \sin(1.3x_4)$

---

[3] We also ran preliminary experiments with the *Keijzer-6* and *Nguyen-7* functions recommended in [29]. However, these two functions are trivial to solve with the function set used in our experiments. We have therefore not reported the results for these two functions.

Table 3. The input data ranges used for generating the data sets. $E[l, u, s]$ is a grid of evenly spaced points between l and u inclusive with step size s. $U[l, u]$ means uniformly distributed points between l and u (exclusive).

Instance	Input distribution
Salustowicz	$x : E[0, 3.2, 0.05]$
Vladislavleva-3	$x_1, x_2 : E[-0.5, 10.5, 0.1]$
Vladislavleva-4	$x_1, x_2, x_3, x_4, x_5 : U[-0.25, 6.35]$, 4000 samples
Pagie	$x_1, x_2 : E[-5, 5, 0.4]$
Flow psi	$x_1 : U[60, 65], x_2 : U[30, 40], x_3 : U[0.2, 0.5],$ $x_4 : U[0.5, 0.8], x_5 : U[5, 10]$, 132 samples
Korns-12	$x_0, x_1, x_2, x_3, x_4 : U[-50, 50]$, 1320 samples

For the noisy problem instances we modified the data sets by adding a randomly distributed noise term to the target variable y.

$$y_{noise} = y + N(0, 0.2 \sqrt{\mathrm{Var}(y)})$$

This means that the maximally achievable R^2 value is limited by the noise level. Table 4 shows the best possible R^2 value for each of the noisy problem instances. We define that a GP run is successful if it reaches at least $0.95 R^2$ for the noisy problem instances.

Table 4. The highest possible fitness value (R^2) for each of the noisy problem instances.

Instance	Highest fitness
Salustowicz (noisy)	0.959
Vladislavleva-3 (noisy)	0.961
Vladislavleva-4 (noisy)	0.962
Pagie (noisy)	0.963
Flow psi (noisy)	0.964
Korns-12 (noisy)	0.961

3.3   Selection of Predefined Features

For each problem instance we defined a small set of pre-calculated features. This was done manually and based on the known expression for the problem instance. The used features are shown in Table 5.

Table 5. The pre-calculated features for each problem instance. For Korns-12 we tried two configurations where we only add one of the necessary factors as feature in the first case and both in the second case.

Instance	Features
Salustowicz	$x^3, \exp(-x), \cos(x), \sin(x), \sin(x)^2 \cos(x)$
Vladislavleva-3	$\exp(-x_1), x_1^3, \cos(x_1), \sin(x_1), \sin(x_1)^2 \cos(x_1)$
Vladislavleva-4	$(x_1 - 3)^2, (x_2 - 3)^2, (x_3 - 3)^2, (x_4 - 3)^2, (x_5 - 3)^2$
Pagie	x^{-4}, y^{-4}
Flow psi	$\frac{x_2^2}{x_3^2}, \log(\frac{x_3}{x_4}), \sin(\frac{\pi x_2}{180})$
Korns-12 (1)	$\cos(9.8x_0), \sin(1.3x_4)$
Korns-12 (2)	$\cos(9.8x_0)$

4   Results

Figure 4 depicts the empirical run length distribution results for GP on the instances without noise. For each of the six instances we show the performance of GP with and without pre-calculated features. The graphs show the empirical success probability (target) for the configuration for the 60 runs. A run is successful if a solution with the defined level of quality is found. The evaluations are displayed relative to their total number.

Generally for all problem instances the number of successful runs is higher with extra features. The results show that for a given budget of evaluations the success probability is higher when extra features are available. Alternatively, for a given success probability the computational effort is lower when extra features are available. The results therefore support our hypothesis and we can give a positive answer to the research questions.

In practical applications we often need to accept inaccurate data or noisy measurements. The results for the noisy problem instances are shown in Fig. 5. We observe similar results as for the instances without noise. Only for the *Pagie* problem the extra features did not affect the performance.

5   Discussion

The results of our experiments show that the success rate of GP can be increased by providing pre-calculated features based on expert knowledge. For all tested problem instances without noise the probability of success for a given computational effort increased significantly. Some of the problem instances became almost trivial to solve when prior knowledge was available.

We found that the positive effect is apparent even for noisy problem instances. A limitation of our contribution is that we used only synthetic benchmark problems where the underlying function is known. This makes it easy to come up with features which are necessary to express the functional relationship

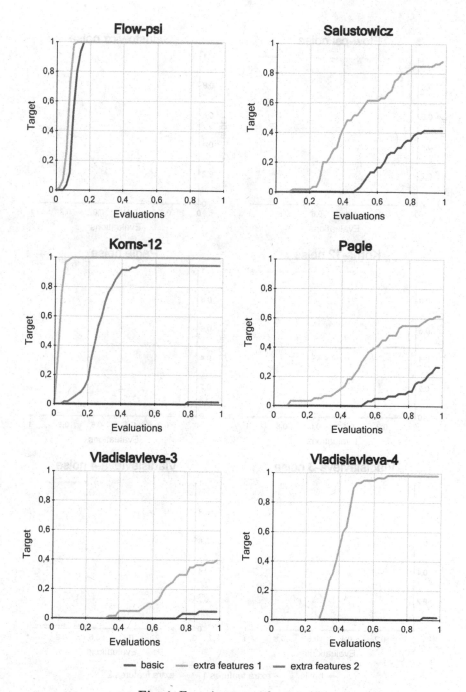

Fig. 4. Experiments without noise

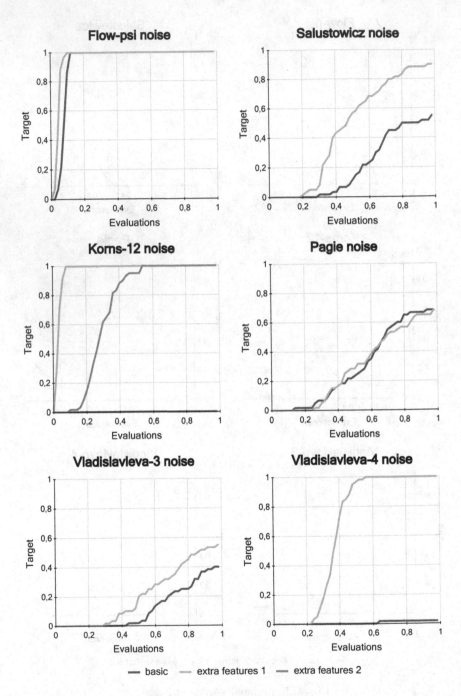

Fig. 5. Experiments with noise

that should be identified. In real-world applications where the underlying data-generating function is unknown it is harder to define such features. One particular limitation when using pre-calculated features is that non-linear parameters in the feature expressions must be approximated because they are not subject to evolutionary optimization by GP. This could be overcome by either providing parametric functions in the function set or defining syntactical building blocks which are used for pre-seeding of in crossover and mutation operators. However, such mechanisms would necessitate adaptations to the GP implementation which require more effort.

References

1. Affenzeller, M., Winkler, S., Wagner, S., Beham, A.: Genetic Algorithms and Genetic Programming - Modern Concepts and Practical Applications. CRC Press, Boca Raton (2009)
2. Ashburner, M., et al.: Gene ontology: tool for the unification of biology. Nat. Genet. **25**(1), 25 (2000)
3. Chandrasekaran, B., Josephson, J.R., Benjamins, V.R.: What are ontologies, and why do we need them? IEEE Intell. Syst. **14**(1), 20–26 (1999)
4. Chen, C., Luo, C., Jiang, Z.: A multilevel block building algorithm for fast modeling generalized separable systems. Expert Syst. Appl. **109**, 25–34 (2018). https://doi.org/10.1016/j.eswa.2018.05.021
5. Couchet, J., Manrique, D., Ríos, J., Rodríguez-Patón, A.: Crossover and mutation operators for grammar-guided genetic programming. Soft. Comput. **11**(10), 943–955 (2007)
6. Cramer, N.L.: A representation for the adaptive generation of simple sequential programs. In: Proceedings of the First International Conference on Genetic Algorithms, pp. 183–187 (1985)
7. Cruz, I.F., Xiao, H.: The role of ontologies in data integration. Eng. Intell. Syst. Electr. Eng. Commun. **13**(4), 245 (2005)
8. Darwin, C.: The Origin of Species: By Means of Natural Selection, or the Preservation of Favoured Races in the Struggle for Life. Cambridge Library Collection - Life Sciences, 6th edn. Cambridge University Press, Cambridge (2009)
9. Eilbeck, K., et al.: The sequence ontology: a tool for the unification of genome annotations. Genome Biol. **6**(5), R44 (2005)
10. Gardner, S.P.: Ontologies and semantic data integration. Drug Discovery Today **10**(14), 1001–1007 (2005)
11. Girardi, D., et al.: Interactive knowledge discovery with the doctor-in-the-loop: a practical example of cerebral aneurysms research. Brain Inf. **3**(3), 133–143 (2016)
12. Goldberg, D.E.: Genetic Algorithms in Search, Optimization, and Machine Learning. Addison-Wesley Professional, Boston (1989)
13. Gruber, T.R.: A translation approach to portable ontology specifications. Knowl. Acquisition **5**(2), 199–220 (1993)
14. Hansen, N., Auger, A., Ros, R., Finck, S., Pošík, P.: Comparing results of 31 algorithms from the black-box optimization benchmarking BBOB-2009. In: Proceedings of the 12th Annual Conference Companion on Genetic and Evolutionary Computation, pp. 1689–1696. ACM (2010)

15. Holland, J.H.: Adaptation in Natural and Artificial Systems: An Introductory Analysis with Applications to Biology, Control, and Artificial Intelligence. MIT Press, Cambridge (1992)
16. Holzinger, A., et al.: Interactive machine learning: experimental evidence for the human in the algorithmic loop. Appl. Intell. **49**(7), 2401–2414 (2019)
17. Kommenda, M., Kronberger, G., Wagner, S., Winkler, S., Affenzeller, M.: On the architecture and implementation of tree-based genetic programming in HeuristicLab. In: Companion Proceedings of the 14th Annual Conference on Genetic and Evolutionary Computation (GECCO 2012), pp. 101–108. ACM (2012)
18. Korns, M.F.: Accuracy in symbolic regression. In: Riolo, R., Vladislavleva, E., Moore, J. (eds.) Genetic Programming Theory and Practice IX, pp. 129–151. Springer, New York (2011). https://doi.org/10.1007/978-1-4614-1770-5_8
19. Koza, J.R.: Genetic Programming: On the Programming of Computers by Means of Natural Selection. MIT Press, Cambridge (1992)
20. Luke, S.: Two fast tree-creation algorithms for genetic programming. IEEE Trans. Evol. Comput. **4**(3), 274–283 (2000)
21. McKay, R.I., Hoai, N.X., Whigham, P.A., Shan, Y., O'Neill, M.: Grammar-based genetic programming: a survey. Genet. Program Evolvable Mach. **11**(3–4), 365–396 (2010)
22. Osborne, J.D., et al.: Annotating the human genome with disease ontology. BMC Genom. **10**(1), S6 (2009)
23. Pagie, L., Hogeweg, P.: Evolutionary consequences of coevolving targets. Evol. Comput. **5**(4), 401–418 (1997)
24. Ratle, A., Sebag, M.: Genetic programming and domain knowledge: beyond the limitations of grammar-guided machine discovery. In: Schoenauer, M., et al. (eds.) PPSN 2000. LNCS, vol. 1917, pp. 211–220. Springer, Heidelberg (2000). https://doi.org/10.1007/3-540-45356-3_21
25. Salustowicz, R., Schmidhuber, J.: Probabilistic incremental program evolution. Evol. Comput. **5**(2), 123–141 (1997)
26. Schoenauer, M., Sebag, M.: Using domain knowledge in evolutionary system identification. CoRR abs/cs/0602021 (2006). http://arxiv.org/abs/cs/0602021
27. Vladislavleva, E.J., Smits, G.F., Den Hertog, D.: Order of nonlinearity as a complexity measure for models generated by symbolic regression via pareto genetic programming. IEEE Trans. Evol. Comput. **13**(2), 333–349 (2009)
28. Whigham, P.A., et al.: Grammatically-based genetic programming. In: Proceedings of the Workshop on Genetic Programming: From Theory to Real-world Applications, vol. 16, pp. 33–41 (1995)
29. White, D.R., et al.: Better GP benchmarks: community survey results and proposals. Genet. Program Evolvable Mach. **14**(1), 3–29 (2013)
30. Winkler, S.M.: Evolutionary system identification: modern concepts and practical applications. Ph.D. thesis, Johannes Kepler University, Altenbergerstr. 69, 4040 Linz (2008)
31. Wong, M.L., Leung, K.S.: Data Mining Using Grammar Based Genetic Programming and Applications, vol. 3. Springer Science & Business Media, New York (2006). https://doi.org/10.1007/b116131

Real Time Hand Movement Trajectory Tracking for Enhancing Dementia Screening in Ageing Deaf Signers of British Sign Language

Xing Liang[1] , Epaminondas Kapetanios[1] , Bencie Woll[2] ,
and Anastassia Angelopoulou[1]([⊠])

[1] Cognitive Computing Research Lab, University of Westminster, London, UK
{x.liang,agelopa}@westminster.ac.uk
[2] Deafness Cognition and Language Research Centre, University College London,
London, UK

Abstract. Real time hand movement trajectory tracking based on machine learning approaches may assist the early identification of dementia in ageing Deaf individuals who are users of British Sign Language (BSL), since there are few clinicians with appropriate communication skills, and a shortage of sign language interpreters. Unlike other computer vision systems used in dementia stage assessment such as RGB-D video with the aid of depth camera, activities of daily living (ADL) monitored by information and communication technologies (ICT) facilities, or X-Ray, computed tomography (CT), and magnetic resonance imaging (MRI) images fed to machine learning algorithms, the system developed here focuses on analysing the sign language space envelope (sign trajectories/depth/speed) and facial expression of deaf individuals, using normal 2D videos. In this work, we are interested in providing a more accurate segmentation of objects of interest in relation to the background, so that accurate real-time hand trajectories (path of the trajectory and speed) can be achieved. The paper presents and evaluates two types of hand movement trajectory models. In the first model, the hand sign trajectory is tracked by implementing skin colour segmentation. In the second model, the hand sign trajectory is tracked using Part Affinity Fields based on the OpenPose Skeleton Model [1,2]. Comparisons of results between the two different models demonstrate that the second model provides enhanced improvements in terms of tracking accuracy and robustness of tracking. The pattern differences in facial and trajectory motion data achieved from the presented models will be beneficial not only for screening of deaf individuals for dementia, but also for assessment of other acquired neurological impairments associated with motor changes, for example, stroke and Parkinson's disease.

Keywords: Segmentation · Hand tracking · OpenPose ·
Sign language · Dementia · Time-series data analytics

© IFIP International Federation for Information Processing 2019
Published by Springer Nature Switzerland AG 2019
A. Holzinger et al. (Eds.): CD-MAKE 2019, LNCS 11713, pp. 377–394, 2019.
https://doi.org/10.1007/978-3-030-29726-8_24

1   Introduction

Most of the world's developed societies are experiencing an ageing trend in their populations [3]. Ageing is correlated with increased prevalence of cognitive impairments such as dementia, stroke and Parkinson's disease. With this in mind, researchers are working urgently to develop effective technological tools that can help doctors undertake, as precise as possible, early identification of cognitive decline. In order to capture change and to monitor behavioural patterns of ageing individuals, there have been many studies of patient monitoring and surveillance with the main focus on using ICT facilities to recognise difficulties with ADL [4–9]. The ADL framework, using sensors, Internet of Things (IoT) and other emerging technologies, provides cost-efficient solutions for in-home or nursing-home monitoring, and can alert health-care providers to significant changes in ADL behaviours which may indicate cognitive impairment. With mounting of ICT facilities, such frameworks usually have a complex structure and need to be evaluated over an extended period of time to be useful for clinicians to detect health deterioration in patients.

Improvements in medical imaging quality and the greater availability of brain imaging data sets have increased opportunities to develop machine learning approaches for automated detection, classification and quantification of diseases. Many of these techniques have been applied to the classification of brain MRI or CT scans, comparing dementia patients to healthy controls, and to distinguish different types or stages of dementia and accelerated features of ageing [10]. As recently addressed in [11], a powerful data-driven machine learning algorithm based on a mixture of linear z-score models is used to identify the exact form and stage of Alzheimer's disease and frontotemporal dementia (FTD) from brain scans alone using an MRI image database. However, the use of neuroimaging to diagnose cognitive impairment and dementia relies on the availability of the advanced hardware and computational power of computing platforms, which results in a high cost for image interpretation.

With the rapid development of artificial intelligence technology, deep learning neural networks have begun to be applied to the automatic detection and classification of acquired neurological impairments using 3D information acquired by RGB-D cameras. [12] proposed an automatic computer-assisted cognitive assessment method for older adults using gesture recognition by means of the Praxis test which is a gesture-based diagnostic test that has been accepted as diagnostically indicative of cortical pathologies such as Alzheimer's disease. An Alzheimer's patient has to imitate the doctor's gestures in doing simple movements such as waving; indicating actions, like going to sleep; rotating hands or upper body. A Deep Convolutional Neural Network (CNN) coupled with Long Short Term Memory (LSTM) is adopted to jointly perform gesture classification and fine grained gesture correctness evaluation using an RGB-D gesture video dataset recorded by Kinect v2. [13] uses a Recurrent Neural Network (RNN) with Parametric Bias to detect action anomalies of Alzheimer's patients. Supervised learning is used for action recognition by comparing the L2 distance between pre-trained action and evaluated action. By detecting anomalous actions which do not follow the predefined actions, a patient's dementia stage can be evaluated.

The British Deaf community uses British Sign Language (BSL) as their preferred language. BSL is a natural language and, like other sign languages, uses movements of the hands, body and face for linguistic expression. BSL is unrelated to English, and has a very different grammar and lexicon. Because there are few health staff with appropriate language skills, and a shortage of BSL interpreters, the Deaf community receives unequal access to diagnosis and care for acquired neurological impairments [14], with consequent poorer outcomes and increased care costs. Inspired by the emerging and innovative technologies described above, we propose a method focusing on the analysis of the sign space envelope (the area in front of the signers upper body and head in which signs are located) and facial expressions of signers, using normal 2D videos to develop an automated screening toolkit for dementia in the ageing deaf population, thereby making possible more efficient use of the limited number of clinicians with appropriate skills and experience in diagnosis in the deaf population and ensuring early screening and provision of appropriate services and interventions [15].

Clinical observation suggests that there may be differences between signers with dementia and healthy signers in the envelope of sign space (sign trajectories/depth/speed) and movements of the face, with signers who have dementia using restricted sign space and limited facial expression compared to healthy deaf controls. Therefore the first phase of research is focusing on analysing the sign space envelope in terms of sign trajectory and sign speed, together with the facial expressions of deaf individuals, Data on healthy older signers is taken from standard 2D videos freely available from the BSL Signbank dataset [17] and compared to those with mild cognitive impairment and early stage dementia to identify changes in signing associated with dementia.

In this paper, we present two methods of real-time hand trajectory tracking models deployed in order to obtain the sign space envelope. In the first model, the hand sign trajectory is tracked by implementing skin colour filtering and morphology operations, before using contour extraction to track hand blob trajectories based on contour centroids. The second model is based on the OpenPose library for real time multi-person keypoint detection. The hand movement trajectory is obtained via wrist joint motion trajectories. The curve of the hand movement trajectory is connected by the location of the wrist joint keypoints 4 or 7 (Fig. 3) across sequential video frames. The remainder of this paper is organised as follows. Section 2 presents the formulation and the methodology of our pipeline where two methods are evaluated using our datasets. Section 3 presents the experimental analysis, results and discussions. Finally, Sect. 4 concludes the study and discusses about future work.

2   Methodology

In this work, we are interested in providing a more accurate segmentation of objects of interest in relation to the background, so that accurate real-time hand trajectories (path of the trajectory and speed) can be achieved. These segmented patches and their associated trajectories and speed of movement will be used in future work as features in a machine learning model for the classification of the sign space used by a BSL signer as healthy or atypical. Performance evaluation of the research work will be based on data sets available from the Deafness Cognition and Language Research Centre (DCAL) at UCL, which has a range of video recording of over 500 signers who have volunteered to participate in research.

Figure 1 shows the pipeline of the two methods we have applied to evaluate the datasets and future work of the machine learning model. The highlighted section and the two dashed boxes indicate the two methods for the gesture tracking given RGB video stream as input. We present results for two different baselines for feature extraction: one based on image processing methods and the other on deep learning models. Each method is discussed in more details in the following sub-sections and for each developed method we assume that the subjects are in front of the camera with only the upper body visible. The input to the system is short-term clipped videos.

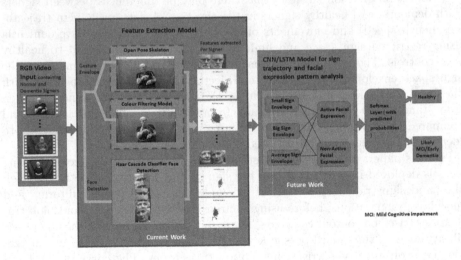

Fig. 1. The proposed pipeline for dementia screening

2.1   Datasets

British Sign Language Corpus a collection of video clips of 250 Deaf signers of BSL from 8 regions of the UK [16]. **BSL Cognitive Screen norming data** video interviews with 250 signers aged 50–90, and video recordings of a range of language and cognitive tasks (picture descriptions and memory tasks). **Video recordings** of case studies of signers with acquired neurological disorders including dementia, left- and right-hemisphere stroke, Parkinson's disease, motor neuron disease and progressive supranuclear palsy. **BSL Signbank** standard 2D videos of single lexical signs, from an online sign dictionary [17].

2.2   Colour Filtering Models in HSV/YCrCb/Lab Colour Space

The first model for feature extraction is based on image processing method by skin colour segmentation. As shown in Fig. 2, firstly face detection is performed using Haar cascade classifiers [18] for facial expression analysis. Secondly, skin colour information is used as a powerful descriptor to identify the human hands [19,20], because human skin has a colour distribution that differs significantly (although not entirely) from background objects. Participants' clothes and background have to be carefully selected to avoid similarity to skin colour. As both hands and face can be detected due to their colour similarity, but only hand blob tracking is focused on in the current stage, a rectangular box is drawn around the face (previously detected by Haar cascade classifiers). The next step is to apply skin colour thresholds to detect hand location by filtering out the skin colour distribution characteristics. As an image can be presented in a number of different colour models, such as HSV, YCrCb and CIELab, multiple colour filtering models with multi-colour thresholds for skin segmentation are used in this approach. A video frame is converted from RBG format to HSV/YCrCb/Lab format, before applying the appropriate skin segmentation thresholds.

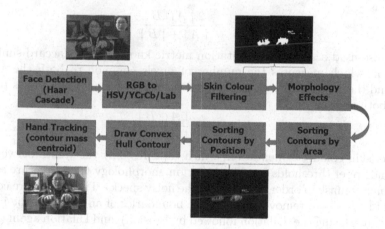

Fig. 2. Real time hand tracking algorithm based on skin colour segmentation (Color figure online)

RGB to HSV Model. HSV (Hue Saturation Value) is a model representing colour space, similar to the RGB (Red Green Blue) colour model. Since the Hue channel models the colour type, it is very useful in image processing tasks that need to segment objects based on colour. Variation in Saturation goes from unsaturated (representing shades of grey) to fully Saturated (no white component). Value channel describes the brightness or intensity of the colour. In our experiments, the thresholds used for skin segmentation in the HSV model are: $0 \le H \le 20, 48 \le S \le 255, 80 \le V \le 255$.

RGB to YCrCb Model. The video frame is also converted to YCrCb format for skin segmentation. In the YCrCb model, Y is the Luminance (brightness) component. Cr (Red-difference) and Cb (Blue-difference), as colour difference signals, represent the Chrominance component. In our experiments, the thresholds used for skin segmentation in the YCrCb model are: $0 \leq Y \leq 255, 133 \leq Cr \leq 173, 77 \leq Cb \leq 127$.

RGB to CIELab Model. The video frame is also converted from CIELab format for skin segmentation. Lab colour space is defined by the International Commission on Illumination. It expresses colour as three numerical values, L for lightness and a and b for the green–red and blue–yellow colour components. In our experiments, the thresholds used for skin segmentation in the CIELab model are: $20 \leq L \leq 220, 128 \leq a \leq 245, 130 \leq b \leq 255$. In order to measure the performance between the segmented skin colour region obtained by the three different colour models and the ground truth, we applied the Sørensen Dice coefficient, as a standard segmentation performance metric, to all three colour models. The Sørensen Dice index, measures the spatial overlap between two segmentations, the A and B regions (in our case these are the ground truth image and each segmented image according to the three colour models), and is defined as

$$Dice = \frac{2 \, | \, A \cap B \, |}{| \, A \, | + | \, B \, |} \tag{1}$$

We also used a second segmentation metric known as the Jaccard similarity coefficient which measures the number of pixels common to both the ground truth and the segmented regions, divided by the total number of pixels present across both regions.

$$Jaccard = \frac{| \, A \cap B \, |}{| \, A \cup B \, |} \tag{2}$$

After skin colour segmentation, which captures only the values between the lower and upper thresholds for skin detection, morphology operations are applied to the binary mask in order to get rid of the noisy specks. This procedure consists firstly of Erosion (to remove pixels at the boundaries of an object in the image), followed by Closing (i.e. Dilation followed by Erosion), and Dilation again (to add pixels to the boundaries of an object in the image). Basically in the morphology approach, by removing the pixels at the boundaries of an object and adding them back, small white noisy specks are eroded. Clearer hand blobs are obtained, as shown in Fig. 2. After these steps, contour extraction is applied using the inbuilt OpenCV function [21]. The output of the contour function is a 2-dimensional array containing the list of x, y coordinates for all the contours, an array of points that are part of a curve and have the same pixel intensities. Sorting contours by areas helps to extract the largest two contours (i.e. the two hands). At the same time, by sorting the largest two contours by position using the x coordinate, both hands are detected from left to right. Then a convex hull and a normal contour are drawn on the hand contour. Finally the hand trajectory is tracked by connecting its contour mass centroid, while the tracking time is recorded for the purpose of sign speed analysis.

2.3   OpenPose Skeleton Model

In the second model, hand movement trajectory tracking is based on the Open-Pose library. OpenPose, developed by Carnegie Mellon University, is one of the state-of-the-art methods for human pose estimation. It processes images through a two-branch multi-stage CNN. The first branch takes the input image and predicts the possible locations of each keypoint in the image with a confidence score (the confidence map). The second branch predicts a set of 2D vector fields that encode the location and orientation of limbs over the image domain (the part affinity fields). Finally the confidence maps and the affinity fields are parsed by greedy inference to output the 2D keypoints for all people in the image [1].

OpenPose consists of three different blocks: body/foot detection; hand detection; and face detection. The core block is the combined body/foot keypoint detector, which provides a 15-,18-, or 25-keypoint body/foot keypoint estimation [22]. The computational performance on body keypoint estimation is invariant to the number of detected people in the image. It can be used on various platforms, including Ubuntu, Windows and Mac, and also has been implemented in different deep learning frameworks such as Tensorflow and Torch. In this paper, the hand tracking model implementation is based on the OpenPose Mobilenet Thin model in Tensorflow [23] on the Windows CPU/GPU platform, for 18 keypoints of body part keypoint estimation (including eyes, nose, ears, neck, shoulders, elbows, wrists, hips, knees and ankles) as shown in Fig. 3 [1,2]. These 18 joint coordinates are able to track limb and body movement in a rapid and unique way. For our purpose, only 14 upper body part joints of the signer in the image are outputted from the OpenPose skeleton model, since only the upper body of a singer is involved in signing. These are: eyes, nose, ears, neck, shoulders, elbows, wrists, and hips, as illustrated in Fig. 4. Wrist keypoints 4 and 7 are utilised for left and right hand tracking respectively, corresponding to the joints' motion trajectory as shown in Fig. 4.

3   Evaluation of Results

The results presented in this section are from initial stage data analysis, mainly based on real time web camera capture of data and standard 2D videos from BSL Signbank [17]. Section 3.1 evaluates the skin filtering results for different colour models using video frames from BSL Signbank. To demonstrate the model capability of hand tracking, Sect. 3.3 uses not only the videos from BSL Signbank but also real time web camera capture data as the input. Hand tracking trajectories from real time web camera capture are compared with ground truth collected by a magnetic positional tracker (Polhemus 3Space Fastrak tracking instrument). For two hand trajectory tracking, the Polhemus tracking instrument reports the positional coordinates of each hand with 60 updated coordinate points per second, a static accuracy of 0.08 cm, and resolution of 0.0005 cm/cm of range as indicated in product specifications [24].

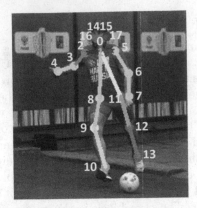

Fig. 3. OpenPose skeleton 18 body joints [1,2]

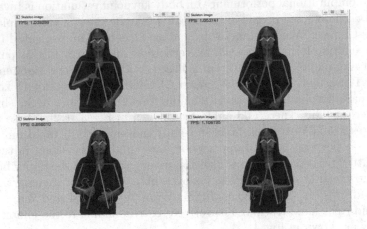

Fig. 4. OpenPose skeleton model hand trajectory tracking

The first hand tracking model (colour segmentation based) was developed and tested on a desktop machine, 8 GB RAM 3.00 GHz Intel Core i5-4590S CPU processor. This method was implemented in Python 3.6.5 and OpenCV 3.3.1. The second hand tracking model (OpenPose skeleton based) was developed and tested on the same CPU desktop, and on a GPU desktop with two NVIDIA GeForce GTX 1080Ti adapter cards and 3.3 GHz Intel Core i9-7900X CPU with 16 GB RAM. The second model was implemented in Tensorflow 1.11, Python 3.6.5, OpenCV 3.3.1 for the CPU environment and Tensorflow 1.12, Python 3.6.8, OpenCV 3.4.2 for the GPU environment.

3.1   Colour Models Evaluation

Figures 5 and 6 show the skin segmentation comparisons between the different colour models. In each colour model, the colour thresholds play an important part in segmentation. For colour thresholds presented in Sect. 2.2, Figs. 5 and 6 show that HSV and CIELab outperform the YCrCb colour models, with better skin filtering results [20] and less error mapping. Table 1 shows quantitative results for Fig. 6 for all three colour models based on the two segmentation metrics as discussed in Sect. 2.2.

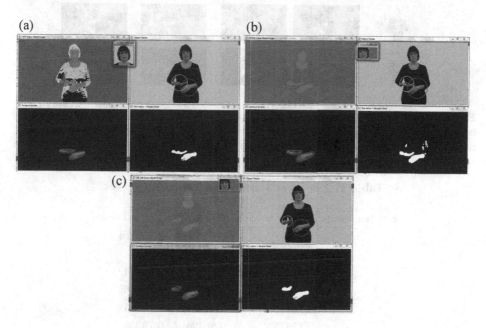

Fig. 5. Comparisons between multiple colour filtering models for skin segmentation (from left to right: (a) HSV, (b) YCrCb, (c) CIELab) (Color figure online)

3.2 Real Time Tracking Trajectory Evaluations

Figure 7 shows 2D hand tracking trajectory results from the real time hand tracking demo (Fig. 2). Three signs, differing in location: CLOUD, PICTURE, and SAILOR are clearly tracked, based on skin colour segmentation. Figure 8 is the 3D real time hand tracking trajectory. Hands are tracked not only based on 2D coordinates, but also in time, with the purpose of tracking the speed of hand movement. In the left hand trajectory (Figs. 7 and 8), there is an clear match between 2D and 3D trajectory. Figure 9 shows how speed of hand motion changes over time in a 2D plot, which gives a clear indication of how hand movement speed over time (X-axis speed based on 2D coordinate changes, and Y-axis speed based on 2D coordinates changes). By introducing another dimension in time (milliseconds), hand movement speed pattern can be easily identified to analyse acquired neurological impairments associated with motor symptoms (i.e. slowered movement) such as in Parkinson's disease. A longer trajectory within a shorter period shown in the right hand 3D trajectory (green Diagram in Fig. 8) indicates faster hand movement.

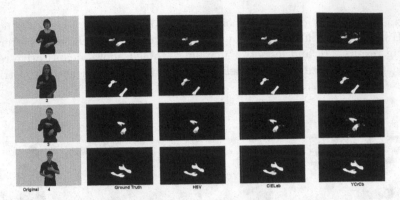

Fig. 6. Different colour models and their associated error map. From left to right column: original image, ground truth, HSV, CIELab, YCrCb and their associated error map (green and magenta pixels). Green pixels indicate False Negatives and magenta pixels indicate False Positives (Color figure online)

Table 1. Jaccard and dice scores for signing images

Signing image	Colour models	Dice score	Jaccard score
1	HSV	**0.87508**	**0.77914**
	CIELab	0.86467	0.7616
	YCrCb	0.70551	0.55474
2	HSV	0.78313	0.64356
	CIELab	**0.88988**	**0.8016**
	YCrCb	0.88428	0.79256
3	HSV	**0.88683**	**0.83695**
	CIELab	0.83965	0.72362
	YCrCb	0.70207	0.54591
4	HSV	0.90132	0.82037
	CIELab	**0.90239**	**0.82215**
	YCrCb	0.86992	0.76978

Figure 10 are the 2D plot (x-axis vs. time and y-axis vs. time) comparing hand movement tracking in a Deaf individual with Mild Cognitive Impairment (MCI) and a healthy individual. It shows that the MCI signer's trajectory resembles a straight line rather than the up and down trajectory characteristic of a healthy individual, indicating that the MCI signer produced more static poses/pauses during signing. Moreover, the X and Y trajectory lines of the signer with MCI are closer to each other as a result of a limited sign space envelope.

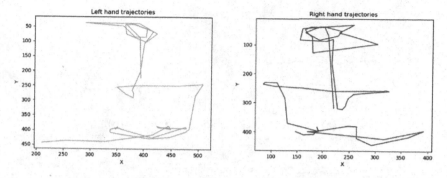

Fig. 7. 2D real time hand tracking trajectory

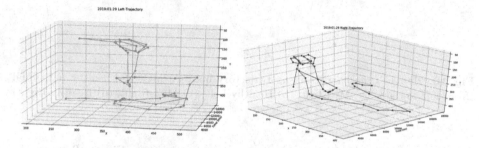

Fig. 8. 3D real time hand tracking trajectory (Color figure online)

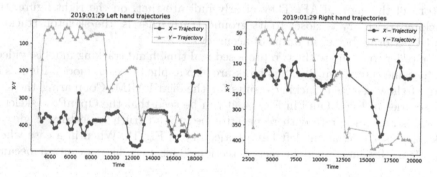

Fig. 9. 2D real time hand tracking trajectory over time

3.3 Comparisons Between Two Tracking Models

In order to compare the two proposed real time hand tracking models, firstly data from real time web camera capturing were analyzed. The hand tracking trajectory obtained from the tracking model was then compared with its ground truth as collected by the Polhemus magnetic tracker. As shown in Fig. 11, two receivers of the magnetic tracker are attached to both wrists and used to track the ground truth trajectory at the same time as the tracking model performs tracking. So far we are measuring the ground truth and its trajectory data obtained

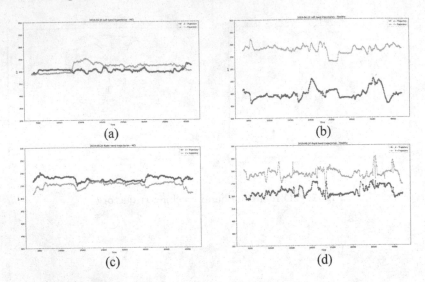

Fig. 10. 2D real time hand tracking trajectory between MCI and healthy individual (from left to right: (a) MCI-Left, (b) Healthy-Left, (c) MCI-Right, (d) Healthy-Right)

with each tracking model individually in a qualitative way. Figure 12 shows the differences between the tracking models and the ground truth in the hand trajectory of the sign DEAF. They clearly indicate that, on the right figure, the tracking trajectory is closer to its ground truth: that is, the skeleton tracking model performs better in terms of accuracy.

In order to compare the two proposed real time hand tracking models, videos of the same signs from BSL Signbank are also applied to each model. Figures 13 and 14 show selected tracking results for the sign FARM. Comparing the sign trajectories in Fig. 13 and in Fig. 14, it can be seen that the OpenPose skeleton model is more accurate with respect to the ground truth trajectory. Figure 15 takes a closer look at the left hand trajectory of Fig. 13. When in a case where the Haar classifier failed in face detection. This may have occurred because

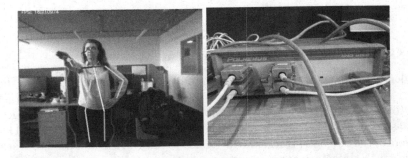

Fig. 11. Ground truth data collection setup

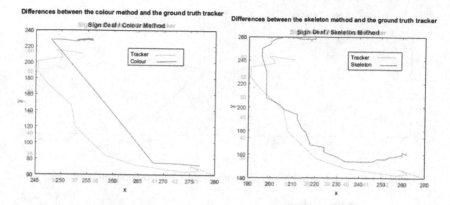

Fig. 12. Differences between trajectory obtained from tracking models and its ground truth

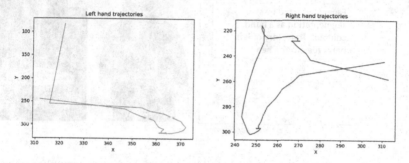

Fig. 13. 2D sign tracking trajectory from colour filtering model

prominent black features are missing or because the image is very bright, meaning that the mask could not be drawn on the face as no face detection bounding box was returned. The face was then detected by skin segmentation and was sorted as a hand contour due to its size. At the same time, when two hands were joined together, they were segmented and taken as a single hand contour. Consequently, the left hand tracking trajectory in Fig. 15 is incorrectly connected to the head as highlighted in the blue box. Similarly, when a face is partially occluded or turnws to the side, a Haar classifier will fail in detection, and the skin segmentation model will generate an inaccurate trajectory. Despite the above mentioned drawbacks, the skin segmentation model is easy to implement, and performs a relatively fast and accurate tracking result under low operating system requirements.

CNN-based part detection using the OpenPose skeleton model is not influenced by the colour of the background and participants' clothing. This makes it more robust in hand tracking. Figure 14 shows that details of changes in hand movement are also well tracked. This is because the model uses the wrist joint for motion trajectory. Unlike the contour centroid that can be shifted as the gesture or posture changes, the PAF of the wrist joint is relatively stable.

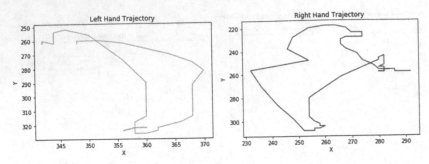

Fig. 14. 2D sign tracking trajectory from openpose skeleton model

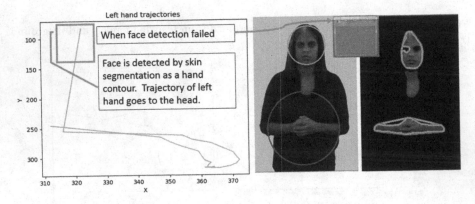

Fig. 15. Analysis of 2D left hand sign tracking trajectory from colour filtering model

The performance of the second model relies highly on the system's computational capability. In order to have a better knowledge of its tracking performance in speed, we applied the model in Windows with both a CPU and a GPU. The system specifications are listed in Table 2. The processing speed of the web camera input is slower than the video input. However, with utilisation of the GPU, overall performance speeds are greatly improved. In conclusion, the second model has significantly enhanced tracking accuracy and is more robust in tracking. To obtain the best performance especially for web camera capture, system computational capability plays a key role. As low processing capability causes loss of tracking points, this will add errors to real time trajectory tracking.

Table 2. OpenPose tracking model performance between GPU and CPU

	System specifications	Average FPS
CPU	Win7, Intel Core CPU@3.00 GHz, RAM 8 GB	1.2 (video)
		0.9 (web camera)
GPU	2 NVIDIA GeForce GTX 1080Ti, Win10,	60 (video)
	Intel Core CPU@3.30 GHz, RAM 16 GB	23 (web camera)

3.4 Discussion

In this project, we needed to decide on what would be the most promising approach to pursue in order to maximise the probability of extracting hand gestures and their trajectories that were as accurate as possible. Not investigating and comparing the main approaches in this context, would have been detrimental to the quality of the follow-up process of extraction of high level features from the related sign envelope. This, in turn, would have affected the quality of interpretation of the information provided by the sign envelope in relation to one that might be potentially atypical. In that sense, the main two approaches selected for comparison: skin colour filtering and pose (skeleton), can be viewed as representatives of larger families of algorithmic approaches: image processing techniques versus pre-trained machine learning (ML) models.

The experiments and comparisons verified that the pre-trained, ML based model is superior to the image processing one, in several aspects: simplicity in setting up the experiment, simplicity in capturing hand gesture trajectories as it is less sensitive to background and environment, increased speed and accuracy of measurement. Even if pre-trained ML-based models for skin colour filtering are used, which has not been the case in our methodology and comparisons, the pose (skeleton) based approach retains its superiority with regard to simplicity in setting up the experiment and simplicity in capturing hand gesture trajectories as it is less sensitive to background and environment.

As far as accuracy is concerned, ground truth data other than the video-recorded signs have been used. In order to increase trustworthiness in our methodological approach and comparisons, tracking data from real participants articulating signs have also been captured as test data to be used for verifying the hypothesis that the pose (skeleton), pre-trained ML-based approach delivers more accurate hand trajectories. Accuracy is defined as the closest possible trajectory to the one captured by the hand tracker. This is also based on the assumption that a tracker's data trajectories are close to real signs; hence, the use of this data set as ground truth data. In that sense, the captured trajectories from image processing and the skin colour filtering approach significantly deviated from the tracker data based trajectories. As pointed out previously, even if such a pre-trained ML-based, skin colour filtering system does exist or will be developed, it is unlikely that better accuracy will be achieved, and if so, it would be at the expense of complexity and intrinsic vulnerability to errors. It is also worth mentioning that the tracker data have been captured twice, once with each approach. Hence, for the sake of fairness, we decided to use only the tracker data corresponding to either the skin colour filtering or pose (skeleton) approach, respectively. An average of the two trajectories could also be drawn and used as common ground truth data, however, no significant difference with the results and comparisons will be observed.

Finally, first preliminary comparisons with real patient data has confirmed the significance of this methodological approach and the comparison results in identifying the approach delivering most accurate hand trajectories possible.

Although at a very early stage, it appears that there is difference between healthy Deaf individuals and those with early evidence of mild cognitive impairment.

4   Conclusions

Two types of real time hand movement trajectory tracking models have been introduced with the aim of enhancing dementia screening in ageing deaf signers of BSL. In the first model, hand sign trajectory is tracked by implementing skin colour filtering to track hand blob trajectories based on contour centroids. As an image can be presented in a number of different colour space models, multiple colour space filtering models with multi-colour thresholds (HSV/YCrCb/Lab) for skin segmentation are also addressed, to perform relatively accurate and fast hand tracking with low platform requirements. The second model is based on the OpenPose library for real time multi-person keypoint detection. The hand movement trajectory is obtained via wrist joint motion trajectory. It provides enhanced tracking accuracy and more robust tracking. To obtain the best performance, system computational capability plays an important role and as such the implementation has been performed on both CPU and GPU architectures. Based on the differences in patterns obtained from facial and trajectory motion data, further research work will implement machine learning and deep neural network models (CNN/LSTM/Hybrid) for the incremental improvement of dementia recognition rates. The final screening toolkit will be trained and validated against behavioural cognitive screening tests designed for users of BSL. As the proposed system focuses on analysing the sign space envelope and facial expression of BSL signers using normal 2D videos without requiring any ICT/medical facilities setup, the proposed system will be more economical, simpler, more flexible, and more adaptable.

5   Funding

This work has been supported by the Dunhill Medical Trust Grant RPGF1802\ 37 UK.

References

1. Cao, Z., Simon, T., Wei, S.E., Sheikh, Y.: Realtime multi-person 2D pose estimation using part affinity fields. In: Proceedings of the IEEE Conference on Computer Vision and Pattern Recognition (CVPR), pp. 7291–7299. IEEE Press, Honolulu (2017). https://doi.org/10.1109/CVPR.2017.143
2. Cao, Z., Hidalgo, G., Simon, T., Wei, S.E., Sheikh, Y.: OpenPose: realtime multi-person 2D pose estimation using part affinity fields. In: arXiv preprint arXiv:1812.08008 (2018)
3. Kleinberger, T., Becker, M., Ras, E., Holzinger, A., Müller, P.: Ambient intelligence in assisted living: enable elderly people to handle future interfaces. In: Stephanidis, C. (ed.) UAHCI 2007. LNCS, vol. 4555, pp. 103–112. Springer, Heidelberg (2007). https://doi.org/10.1007/978-3-540-73281-5_11

4. Urwyler, P., Stucki, R., Rampa, L., Müri, R., Mosimann, U., Nef, T.: Cognitive impairment categorized in community-dwelling older adults with and without dementia using in-home sensors that recognise activities of daily living. In: Scientific Reports, vol. 7, 42084 (2017)
5. Banerjee, T., Keller, J.M., Popescu, M., Skubic, M.: Recognizing complex instrumental activities of daily living using scene information and fuzzy logic. In: Computer Vision and Image Understanding, vol. 140, pp. 68–82 (2015)
6. Negin, F., Cogar, S., Bremond, F., Koperski, M.: Generating unsupervised models for online long-term daily living activity recognition. In: 3rd IAPR Asian conference on Pattern recognition (ACPR), pp. 186–190. IEEE Press, Kuala Lumpur (2015)
7. Sheriff, R.: Employing ICT in smart cities for the health and well-being of older people with dementia. In: RISUD Annual International Symposium (RAIS) - Smart Cities, Hong Kong (2016). https://doi.org/10.13140/RG.2.2.24997.29923
8. Enshaeifar, S., et al.: Health management and pattern analysis of daily living activities of people with dementia using in-home sensors and machine learning techniques. PLoS One **13**, e0195605 (2018). https://doi.org/10.1371/journal.pone.0195605
9. Singh, D., et al.: Human activity recognition using recurrent neural networks. In: Holzinger, A., Kieseberg, P., Tjoa, A.M., Weippl, E. (eds.) CD-MAKE 2017. LNCS, vol. 10410, pp. 267–274. Springer, Cham (2017). https://doi.org/10.1007/978-3-319-66808-6_18
10. Pellegrini, E., et al.: Machine learning of neuroimaging to diagnose cognitive impairment and dementia: a systematic review and comparative analysis. arXiv: 1804.01961 (2018)
11. Young, A., et al.: the genetic FTD initiative (GENFI), the Alzheimer's disease neuroimaging initiative (ADNI): uncovering the heterogeneity and temporal complexity of neurodegenerative diseases with subtype and stage inference. Nature Commun. **9**, 4273 (2018). https://doi.org/10.1038/s41467-018-05892-0
12. Negin, F., et al.: PRAXIS: towards automatic cognitive assessment using gesture. Expert Syst. Appl. **106**, 21–35 (2018)
13. Iarlori, S., Ferracuti, F., Giantomassi, A., Longhi, S.: RGBD camera monitoring system for Alzheimer's disease assessment using recurrent neural networks with parametric bias action recognition. In: Proceedings of the 19th World Congress the International Federation of Automatic Control (IFAC), Cape Town, pp. 3863–3868 (2014)
14. Atkinson, J.A., Marshall, J., Thacker, A., Woll, B.: When sign language breaks down: deaf people's access to language therapy in the UK. Deaf Worlds **18**, 9–21 (2002)
15. Liang, X., Angelopoulou, A., Woll, B., Kapetanios E.: Enhancing dementia screening in ageing deaf signers of British sign language via analysis of hand movement trajectories. In: Workshop of RSLondonSouthEast2019. Royal Society, London (2019)
16. British Sign Language Corpus Project. https://bslcorpusproject.org/
17. BSL SignBank. http://bslsignbank.ucl.ac.uk/
18. Viola, P., Jones, M.: Rapid object detection using a boosted cascade of simple features. In: Proceedings of the IEEE Conference on Computer Vision and Pattern Recognition (CVPR), pp. 511–518. IEEE Press, Kauai (2001) . https://doi.org/10.1109/CVPR.2001.990517
19. Chai, D., Ngan, K.: Face segmentation using skin-color map in videophone technology. IEEE Trans. Circ. Syst. Video Technol. **9**, 551–564 (1999). https://doi.org/10.1109/76.767122

20. Angelopoulou, A., et al.: Evaluation of different chrominance models in the detection and reconstruction of faces and hands using the growing neural gas network. J. Pattern Anal. Appl. **22**, 1–19 (2019). https://doi.org/10.1007/s10044-019-00819-x

21. OpenCV. https://opencv.org/

22. OpenPose. https://github.com/CMU-Perceptual-Computing-Lab/openpose

23. OpenPose in Tensorflow. https://github.com/ildoonet/tf-pose-estimation

24. O'Suilleabhain, P.E., Dewey, R.B.: Validation for tremor quantification of an electromagnetic tracking device. Mov. Disord. **16**, 265–271 (2001)

Commonsense Reasoning Using Theorem Proving and Machine Learning

Sophie Siebert[1]([✉])[iD], Claudia Schon[2][iD], and Frieder Stolzenburg[1][iD]

[1] Automation and Computer Sciences Department, Harz University
of Applied Sciences, Friedrichstr. 57–59, 38855 Wernigerode, Germany
{ssiebert,fstolzenburg}@hs-harz.de
[2] Institute for Web Science and Technologies, Universität Koblenz-Landau,
Universitätsstr. 1, 56070 Koblenz, Germany
schon@uni-koblenz.de
http://artint.hs-harz.de/, http://www.uni-koblenz.de/

Abstract. Commonsense reasoning is a difficult task for a computer to handle. Current algorithms score around 80% on benchmarks. Usually these approaches use machine learning which lacks explainability, however. Therefore, we propose a combination with automated theorem proving here. Automated theorem proving allows us to derive new knowledge in an explainable way, but suffers from the inevitable incompleteness of existing background knowledge. We alleviate this problem by using machine learning. In this paper, we present our approach which uses an automatic theorem prover, large existing ontologies with background knowledge, and machine learning. We present first experimental results and identify an insufficient amount of training data and lack of background knowledge as causes for our system not to stand out much from the baseline.

Keywords: Commonsense reasoning · Causal reasoning · Machine learning · Theorem proving · Large background knowledge

1 Introduction

Commonsense reasoning is the sort of everyday reasoning humans typically perform about the world [22]. It allows to derive knowledge about continuity and object permanence, e.g., if a person enters a room, then afterwards, the person normally will be in the room, if she has not left the room. People have knowledge about objects, events, space, time, and mental states and may use that knowledge. All this implicit background knowledge is part of everyday human reasoning and must be added to a cognitively adequate automated reasoning system.

The authors gratefully acknowledge the support of the German Research Foundation (DFG) under the grants SCHO 1789/1-1 and STO 421/8-1 *CoRg – Cognitive Reasoning*. A short and preliminary version of this paper appeared in [28].

Commonsense reasoning seems to be a task solved easily by humans. However, for a computer it is a rather difficult task, as the knowledge needed for this kind of problems often is huge and complex. Rather than relying on explicitly given facts, like e.g. geography, commonsense reasoning needs a broad understanding of the world on a very general level. This includes among others knowledge about physics, social interaction, cultural nuances, and basic objects present in the world. This kind of knowledge is so vast that it is difficult to explicitly formulate it in knowledge bases for computers. Problems resulting from this are incompleteness and inconsistency of the knowledge among others.

Error prone and inconsistent data are often addressed with machine learning techniques, e.g., in question-answering tasks. These algorithms perform very well on various tasks and are rather easily adaptable to multiple scenarios. However, they often lack explainability. Especially neural networks are more or less only a black box. While the behavior of a single neuron unit can be described easily, the sheer amount of computational entities leads to a complex overall behavior which is difficult to retrace and to understand. In addition, designing a neural network is often a process of guessing and trying out the right parameters. Also it is difficult to determine which inputs where important for certain processes. To maintain explainability one idea may be to make use of deductive reasoning techniques with theorem provers. They work in a deterministic way to derive facts from given statements, i.e., starting from a situation and a knowledge base, one can conduct all valid conclusions. However approaches relying solely on an automated theorem prover are facing problems with incomplete knowledge bases [5].

In this paper, we want to address commonsense reasoning by combining theorem proving with machine learning. Given a question-answering task in natural language (in English), we search in large knowledge bases for relevant information and feed both the task and the selected knowledge into an automated theorem prover which infers a logical model. This model contains additional facts that the theorem prover has been able to derive. However, it cannot be expected that the theorem prover alone is able to derive the answer to the question at hand. The derived facts can be fed into a machine learning algorithm, closing the gap to answer the commonsense reasoning task.

2   Related Works and Explainable AI

Explainable AI is artificial intelligence programmed to describe its purpose, rationale and decision-making process in a way that can be understood by the average person.[1] It is often discussed in relation to deep learning for applications like text analysis or object recognition in medical diagnosis systems or for autonomous cars.[2] In this context, *post-hoc* and *ante-hoc* analysis can be distinguished [17]: Models of the former type explain the given answer afterwards,

---

[1] http://whatis.techtarget.com/definition/explainable-AI-XAI,   accessed: 14-June-2019.

[2] http://heatmapping.org/slides/2018_MICCAI.pdf, accessed: 14-June-2019.

e.g., by inspecting a learned neural network, while in the other case the model itself is explanatory, e.g., a decision tree.

How can explainable AI be achieved for commonsense reasoning? For this, a hybrid approach that combines a logic-based method with machine learning seems to be the right way [9]. Logical reasoning alone does not seem to be sufficient to handle commonsense reasoning tasks. Here, recurrent networks with LSTM (long short-term memory) [15] are a promising strategy and often used. They achieve a success rate of up to 84% on commonsense reasoning benchmarks, e.g., SemEval [24] which comprises 1,000 questions with two possible answers that require commonsense knowledge for finding the correct answers. Most of the teams participating in this competition use neural network approaches in combination with ontological knowledge like ConceptNet [19,29].

A general problem of machine learning approaches is often that they need big data to learn the desired behavior. This problem can be addressed by unsupervised pretraining, also for improving natural language understanding. [7] and [25] report recent encouraging results on a variety of benchmarks, e.g., question answering, based on this procedure. So, from a behavioral point of view, pure machine learning approaches can solve commonsense reasoning tasks. However, there is no representation of a reasoning process and hence usually no explanatory component in these systems.

As already mentioned, the reason is that machine learning with deep learning neural networks works as a black box. Without further components, drawn conclusions can neither be explained nor corrected if necessary. In the worst case, the computed answers are biased and may be discriminating. A famous example for this was Amazon's recruiting engine that was not rating candidates in a gender-neutral way. The computer models were trained to vet applicants by observing patterns in résumés. But since as a matter of fact of statistics most came from men, male candidates were preferred.[3]

Thus, the benefit of explainable AI may be diverse: First, systems with an explicit (symbolic) knowledge representation may help to find the correct answer. An example for this is the reasoning capacity of theorem provers which employs inference rules on given facts and rules. Second, it allows us to understand and hence to evaluate and possibly to revise answers. Furthermore, artificial intelligence systems should not only provide explanations but should also be advisable by explanations to guide the search for answers and to avoid biases or discrimination.

Commonsense reasoning tasks require a vast amount of knowledge data. Hence reasoning techniques from the fields of deduction, logics, and nonmonotonic reasoning should be employed and combined with machine learning. There are two ways of combination: Machine learning can be used as a subsystem to improve the reasoning process of theorem provers [10]. But it is also possible to do it the other way round. This means, we learn also the argument leading to the conclusion and thus provide explanations only *a posteriori*. In the context

---

[3] www.theguardian.com/technology/2018/oct/10/amazon-hiring-ai-gender-bias-recruiting-engine.

of big data, both procedures (with deductive reasoning and with machine learning) may be problematic, because possibly the resulting longish explanations, e.g., a complete proof of an argument, may also not be helpful. In this paper, we attempt to combine theorem proving with machine learning on top of it and try to tackle commonsense reasoning problems.

There are already several approaches for extracting rules from neural networks in general [8]. Special approaches combine inductive logic programming and machine learning [11,12]. These neural-symbolic learning systems start with a set of logical rules, encoded in neural networks. Then in addition, more knowledge is incorporated from examples into the system. Finally, a modified rule set can be extracted from the improved learned network. But in this context it is assumed that the input is given as logic-based representation. For general natural-language question answering or text comprehension, which we want to address here, this does not hold, however.

```
Premise: The man broke his toe. What was the CAUSE of this?
Alternative 1: He got a hole in his sock.
Alternative 2: He dropped a hammer on his foot.

Premise: The pond froze over for the winter. What happened as a RESULT?
Alternative 1: People skated on the pond.
Alternative 2: People brought boats to the pond.
```

Fig. 1. Problems 273 and 13 from the Choice of Plausible Alternative challenge.

3   Basic Methods and System Architecture

The objective of our project is to answer commonsense reasoning question-answering tasks like COPA [27] or the Story Cloze Test [21]. In general, each of these commonsense reasoning tasks consists of several parts of textual input: the premise describing a situation, a question about the situation together with n sentences describing an alternative answer from which the solution has to be selected. In our project, we currently focus on the COPA challenge.

The task is to determine the answer candidate which has a stronger (causal) relationship to the premise. Our approach to tackle these benchmarks is based on a combination of symbolic and subsymbolic methods: Knowledge represented in ontologies shall be used as background knowledge to perform inferences with the help of an automated theorem prover. The result of these inferences is then evaluated using machine learning, more precisely neural networks, to find answers.

3.1   Benchmarks

To get a better understanding of the project, we first describe the commonsense reasoning benchmarks used to evaluate our implementation. They form the input and consist of a question or situation and a set of possible answers. Currently we focus on COPA (Choice of Plausible Alternatives) [27] and the Story Cloze Test [21] which uses the ROCStories Corpora.

Problems in the COPA challenge (see Fig. 1) consist of a premise and two alternative answers, each given in natural language. The corpus is equally divided into two categories, marked by the question: *cause* and *result*. The *cause* category requires backward causal reasoning, while the *result* category requires forward causal reasoning. All together there are 1,000 tasks which are split into 500 training and 500 test tasks.

The Story Cloze Test has a similar structure. It also has two answers per task, however, the situation part is longer there, see Fig. 2. It is based on the ROCStories Corpora of 98,159 five-sentence stories. 3,744 of these stories were crafted into the Story Cloze Test, by taking the first four sentences of a ROC-Story to describe a situation, and the last sentence for the correct answer, i.e., the most plausible continuation of the story. The wrong answer is a new element to complete the test.

```
Premise: Karen was assigned a roommate her first year of college. Her
roommate asked her to go to a nearbycity for a concert. Karen agreed
happily. The show was absolutely exhilarating.
Alternative 1: Karen became good friends with her roommate.
Alternative 2: Karen hated her roommate.
```

Fig. 2. A Story Cloze Test example.

3.2 The System

Based on the example of the COPA challenge, we now describe the structure of our system and provide details of our approachwhich is implemented in the system CoRg – Cognitive Reasoning (see Fig. 3). As mentioned before, each problem in the COPA challenge consists of a premise and two alternatives. Since all three are given in natural language and our aim is to use an automated theorem prover, the first step of our systems transforms the input into first-order logic formulae. This is achieved using KNEWS [2], a tool that performs semantic parsing, word sense disambiguation, and entity linking. Predicate symbols used in the formulae created by KNEWS in most cases correspond to words (e.g., nouns, verbs and adjectives) of the original text. Word sense disambiguation in our case considers the predicate names of the formulae consisting of the so-called synset IDs of WordNet [20], a lexical-semantic network of the English language. Synset IDs group words with similar meaning for which short definitions, examples and also relations to other synsets are given. They correspond to the predicate symbols occurring in the first-order logic formulae and are used to determine the synonyms, hyponyms and hypernyms of the symbols from WordNet, providing us with a small lexical knowledge base related to the original text. Figure 4 presents the first-order logic formula for the premise of the COPA example 13 in Fig. 1 created by KNEWS.

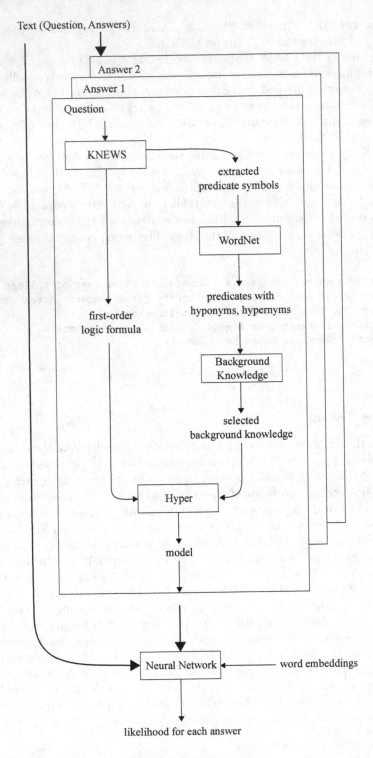

Text (Question, Answers)

Answer 2

Answer 1

Question

KNEWS

extracted
predicate symbols

WordNet

predicates with
hyponyms, hypernyms

first-order
logic formula

Background
Knowledge

selected
background knowledge

Hyper

model

Neural Network ◄─── word embeddings

likelihood for each answer

Fig. 3. The CoRg system.

$$\exists x \big(pond(x) \wedge \exists y, z, v (r1to(y,x) \wedge r1Theme(y,z) \wedge r1Actor(y,v) \wedge bring(y) \wedge boat(z) \wedge people(v)) \big)$$

Fig. 4. First-order logic formula for the first alternative of the COPA example 13 presented in Fig. 1 by KNEWS. To increase readability, the WordNet synset IDs determined by KNEWS for the words *pond* (09420266-n), *bring* (01441539-v), *boat* (02861626-n) and *people* (07958392-n) were replaced by their natural language identifier.

The symbols as well as the related gathered lexical information from WordNet are used to select relevant information from large first-order logic knowledge bases like SUMO [23], Adimen-SUMO [1], ResearchCyc [18] and YAGO [30]. Our system currently only uses Adimen-SUMO but we plan to integrate other ontologies and knowledge graphs such as ConceptNet [19,29]. Section 3.3 provides details on the selection of background knowledge. All gathered background knowledge together with the logical representation of the natural language is fed into the automated theorem prover Hyper [3] which performs inferences resulting in a (possibly partial) model. This is done separately for each answer and the premise of a problem resulting in $n+1$ (partial) models for each task, with n being the number of answer candidates of the benchmarks. In the case of problems from the COPA challenge, this process leads to the construction of three (partial) models.

These models represent the inferences performed by the theorem prover and are fed into a neural network to come to a decision. Each COPA task is split into two training examples such that the model of each answer is paired with the model of the respective answer candidate. The neural network calculates a likelihood that the presented answer indeed fits to the situation it is paired with. The alternative with the higher likelihood is assumed to be the answer of the system. In case of multiple choice questions with n alternatives, the answer with the highest likelihood of being the right answer is selected.

3.3 Background Knowledge

Solving a reasoning task, humans naturally use their broad knowledge about the world. This background knowledge contains knowledge about physical relationships (e.g., a vehicle can overtake another vehicle only if it is faster) but also general knowledge (like the fact that dogs like bones). Since this kind of background knowledge goes far beyond statistical correlations on texts, we aim to use background knowledge which is represented in ontologies and knowledge graphs. Furthermore, we only use already existing knowledge bases and refrain from manually creating knowledge bases for the commonsense reasoning tasks. Currently WordNet and Adimen-SUMO are used as background knowledge.

As described in the previous section, the natural language text of a COPA problem is translated into first-order logic formulae using the KNEWS system, which furthermore performs word sense disambiguation by providing a WordNet synset ID for each noun, verb or adjective occurring in the text. As mentioned before, WordNet contains relations between synsets. Relations that are particularly interesting for our purposes are the hyper- and hyponym relation between synsets. From these relations we generate background knowledge in the form of first-order logic formulae. For example, the facts that *lake* is a hypernym of *pond*, *fishpond* is a hyponym of *pond*, and *pond* and *pool* are synonymous are translated into the following formulae:

$$\forall x \ (pond(x) \rightarrow lake(x))$$
$$\forall x \ (fishpond(x) \rightarrow pond(x))$$
$$\forall x \ (pond(x) \leftrightarrow pool(x))$$

Since this knowledge generated from WordNet has only a taxonomic character, these formulae are supplemented by parts of Adimen-SUMO. Adimen-SUMO is a large ontology consisting of axioms and individuals.

The problem that the Adimen-SUMO symbols do not coincide with the symbols of the KNEWS output in general is solved using the Adimen-SUMO WordNet mapping. This mapping specifies for each WordNet synset Adimen-SUMO symbols that are equivalent or belong to a subclass of the synset. From this information, we generate bridging formulae. Since the generation of these formulae corresponds to the generation of formulae from hypernyms and synonyms described above, we refrain from a more precise description.

Because of the huge size of Adimen-SUMO it cannot be used by a theorem prover as a whole. Therefore we use selection techniques to select a subset of the knowledge provided by Adimen-SUMO that is relevant for the respective COPA problem under consideration. The selection technique we currently use is a relevance-based selection called SInE (Sumo INference Engine) [16] and is broadly used by first-order logic theorem provers.

In a preprocessing step, SInE computes some information about the knowledge base: For each symbol s, the number of occurrences in the whole knowledge base, denoted by $occ(s)$, is determined. Next, a *triggers* relation between symbols and formulae is defined. For each formula F and symbol s occurring in F, $triggers(s, F)$ is true iff for all symbols s' occurring in F $occ(s) < t \cdot occ(s')$ for some $t \in \mathbb{R}$. For $t = 1$, this means that each formula is triggered by the symbol with the fewest occurrences. For $t > 1$, the formula F is triggered by exactly those symbols from F that occur at most t times more frequently than the rarest symbol in F. The parameter t is called *tolerance*.

The selection of formulae suitable for a problem is performed by computing the so-called d-relevance as follows: Every symbol occurring in the problem and the formulae created using WordNet is set be 0-relevant. For a symbol s, which is d-relevant, all formulae F with $triggers(s, F)$ are d-relevant. Furthermore, all symbols occurring in a d-relevant formula become $d + 1$-relevant. Given a problem, a knowledge base and $d \in \mathbb{N}$, this selection extracts the subset of the knowledge base consisting of all formulae which are d-relevant for the problem. More or less background knowledge is selected depending on the selection of the parameter d. In the following, the parameter d is referred to as *recursion depth*.

Recursion depth and tolerance are two parameters that we can vary when generating background knowledge. In addition, we can generate WordNet formulae for the text under consideration or not. Table 1 gives an overview of the different parameters for the selection of background knowledge. In future work, we also plan to add further sources for background knowledge like ResearchCyc, Yago and ConceptNet.

Table 1. Parameters for background knowledge.

Parameter	Value
Integrate WordNet	True or false
Recursion depth (SInE)	1–5
Tolerance (SInE)	1–5

3.4 Using an Automated Theorem Prover

After background knowledge for the premise and both alternatives has been gathered, the automated theorem prover is called. We use the theorem prover Hyper [3] as it is not only able to provide proofs of unsatisfiability but also constructs models for satisfiable problems. These models consist of a set of facts that can be inferred from the knowledge base (see Fig. 5 for an example). Since this inferred knowledge is an important input for the following machine learning step, it is essential for us that the theorem prover is able to deliver this output. In addition, Hyper outputs the formulae used for the performed inferences which we want to use in future work to generate explanations.

The application of Hyper in our system has limitations: Even if multiple sources are used for background knowledge, the background knowledge still does not contain all important information for all problems. Therefore, we cannot expect the theorem prover to construct a complete inference chain to one of the alternatives but only a few inferences useful for the problem. Therefore, these are used as an input for our machine learning component.

4 Employing Machine Learning

Our system as described until now deduces a logical model for each premise and both of the alternative answer candidates. In those models, in addition to the original natural language text, facts are derived and can be used for further processing. As the overall goal is to answer the commonsense reasoning benchmark, we want to determine the likelihood for each of the answers belonging to the respective premise. For this, we use neural networks, as they proved their suitability in various other commonsense reasoning and text processing tasks (cf. Sect. 2). Thus, in this section, we will present the machine learning part of our system. This includes preprocessing the models and the explanation of the neural network architecture.

4.1   Preprocessing

Given a commonsense reasoning task, the natural language input usually needs to be encoded in a suitable manner. For neural networks, the input consists of the natural language text itself, as well as additional information like word embeddings, part-of-speech embeddings, and other features.

In contrast to other approaches, in our system we additionally face the challenge to process a logical model. While the natural language text in our context mostly is rather short and its meaning depends on the word order, this does not hold for a logical model derived by the theorem prover (Hyper). The logical model can contain up to thousands of lines. Furthermore, the derived facts are order-independent, as they do not form sentences but rather single statements on their own. Due to the size it is not reasonable to fed the whole logical model into the network. Therefore we apply numerous preprocessing steps to it which we shall explain now. To start, we depict representative parts of a logical model in Fig. 5.

$$winter(sK0).$$
$$pond(sK1).$$
$$n1froze(sK2).$$
$$p\_d\_subclass(c\_Freezing, c\_StateChange).$$
$$p\_d\_instance(sK2, c\_Freezing).$$
$$p\_d\_instance(sK5(sK2), c\_Cooling).$$
$$p\_d\_instance(sK0, c\_WinterSeason).$$
$$p\_d\_instance(sK0, c\_SeasonOfYear).$$

Fig. 5. Excerpt of a model produced by the automated theorem prover Hyper for the formula representing the sentence *The pond froze over for the winter.* (cf. Fig. 1) together with the gathered background knowledge.

While preprocessing, we dismiss all structural information of the model and extract only the predicate and function symbols, like *winter*, *pond*, or *SeasonOfYear*. First, we replace the special characters (),. with spaces, so that we can process the single elements. For each of those elements we apply some normalization, as they are later linked to word embeddings. This means: We drop all skolem constants (like $sK0$) and skolem functions (like $sK5$). We further drop the prefixes from the background knowledge base which in case of Adimen-SUMO is $p\_d\_$ and $c\_$ as well as other underscore character variants. Often this preprocessing step causes two lines of the logical knowledge to become identical, e.g., both $(sk3(sk8(bird)))$ and $sk4(bird)$ are transformed into *bird*. In this case, we delete one of the duplicates. The remaining elements are usually common words and are looked up in the word embedding. If there is no corresponding entry, they are either not a real word or we cannot map the word to a numerical representation. This makes them useless for further processing. Thus we delete them.

This procedure results in a sequence of words which can be interpreted as text, although it is not a grammatically correct sentence in any form. The size of this text is greatly reduced in comparison to the original logical model and mostly does not exceed 100 words which is a reasonable size to put into the network. Optionally, we can transform the remaining words into a set, getting rid of duplicates. As the model consists of derived facts, ordering or duplicate words should not influence the outcome. This again reduces the input size. In the future we could also get rid of the meta-predicates from the background knowledge, like *subclass* or *instance*. As long as we do not integrate structure information, these words do not help deciding for an answer candidate, as they appear in both alternative models.

4.2   The Neural Network Architecture

In the neural network, we implement an attentive reader approach [31] making use of the framework Keras.[4] The input of the neural network consists of question-answer pairs, i.e., each task consisting of n answers is transformed into n training examples. We use a multi-input neural network, where the question and the answer are first separately encoded in a single network, and later on merged into a calculation core. The output is the likelihood that the given answer fits to the corresponding question. The general structure is shown in Fig. 6 and explained now.

Each training example input consists of a premise P, an alternative A and a classification $y \in \{(0, 1), (1, 0)\}$. P and A are a list of word indices. The first step of the picture covers the preprocessing as explained in the previous section. The second step is the integration of word embeddings to enable similarity calculations among identifiers. For each word in P we use a word embedding vector from ConceptNet Numberbatch. Word embeddings map words into a high-dimensional numerical space, so that similar words have a short distance. In this project we use the state-of-the-art word embedding ConceptNet Numberbatch [29]. It outperforms other word embeddings like word2vec[5] and gloVe[6] in several benchmarks and reduce the bias of prejudices.[7] In our work, we reimplemented the built-in Keras embedding layer, so we can adjust the weight of the single words. We can choose between a constant weighting and a frequency weighting. In the future we may consider TF-IDF weighting [26] as well.

The third step is the encoding of the word embeddings information using a BiLSTM (bidirectional long short term memory) layer h^P. Analogously, this is done for the words in the answer candidate A. An LSTM can remember and forget previous information, providing a context of a current word into its previous text. In a bidirectional LSTM, the input is read both forward and backwards. Analogous to the forward-read input, the backward-read input provides a con-

---

[4] www.keras.io, accessed: 22-April-2019.
[5] https://github.com/tmikolov/word2vec, accessed: 21-June-2019.
[6] https://nlp.stanford.edu/projects/glove/, accessed: 21-June-2019.
[7] github.com/commonsense/conceptnet-numberbatch, accessed: 22-April-2019.

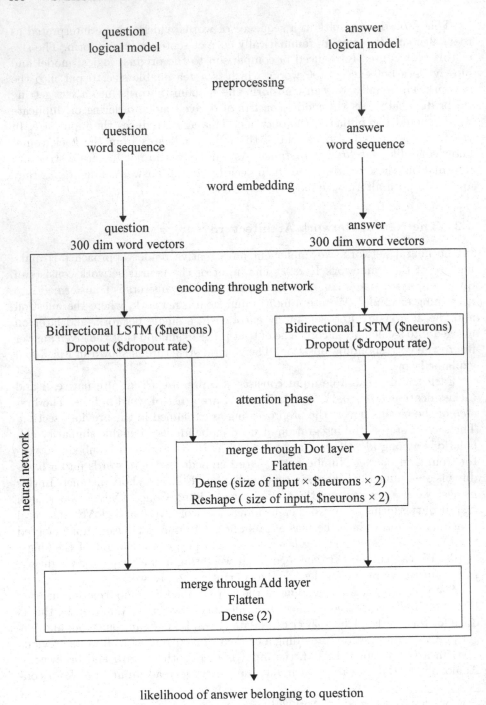

Fig. 6. The neural network. The terms marked with $ are hyperparameters, which we use to tune the network. They are varied, such that the resulting network is optimized.

text into its following text, i.e., a look-ahead. Bidirectional LSTMs therefore embed a word into its context for previous and succeeding information.

Next, the encoded text passages are merged together using a dot product $Dot_P^A(h^P, h^A) = \sum_{i=1}^{N} h_i^P \cdot h_i^A$. This process is the attention phase of the network [31], calculating the shared features between the texts. It is followed by a fully connected so-called *dense* layer, to generate an answer embedding in the context to the question $Att_P^A(Dot_P^A) = \sum_{i=1}^{N} Dot_P^{A_i} \cdot w_i$. In the last step, both the merged core network as well as the input question network is again merged using addition $Add_P^A(h^P, Att_P^A) = \sum_{i=1}^{N} h_i^P + Att_P^A$. A two-neuron dense layer with softmax activation $y(Add_P^A) = \sigma(\sum_{i=1}^{N} Add_P^{A_i} \cdot w_i)$ assigns an output in the shape of $y^* \in [0, 1]^2$, with $|y^*| = 1$ and describes a likelihood.

Throughout the network we use various layers to reshape our tensors as well as dropout and kernel constraint measures to avoid overfitting. We also changed the number of neurons of the LSTMs and dense layers. We use categorical cross entropy as loss, adamax as optimizer and accuracy as metric [13]. Currently, we can feed both the logical model and the natural language as is into the neural network. For future experiments, we want to integrate both approaches into one network.

During experiments we varied multiple parameters, see Table 2. First, one can change the general network structure. We propose an attentive reader approach [31] with LSTMs. In our first attempts, we also did experiments with a simple feed-forward net and an LSTM approach without the attentive reader model. However, they did not score well on our task. Second, we can choose between natural language and a logical model as input. Third, we can choose the embedding weights. In our current experiments we choose the input to be a set as it kept the input small and thus saves computation time. The following parameters are standard neural network hyperparameters [4] and were evaluated in the current experiments.

Table 2. Parameters for the neural network.

Parameter	Value
Network structure	feed-forward, LSTM, attentive reader
Data type	natural language, logical model
Embedding weights	sequence, set, frequency
Neurons	2–100 in first layer
Dropout rate	0.1–0.5
Optimizer	adam, adamax
Learning rate	0.0005–0.002
Kernel constraint	1.0–10.00

5 First Ideas Towards Explainability

The architecture of our system allows us to generate explanations for the decisions made. Since this has not yet been implemented, we will illustrate our idea briefly with an example. Let us take another look at COPA example 13 from Fig. 1. The theorem prover Hyper derives the facts given in Fig. 5 from the formula representation of the premise of this problem together with the selected background knowledge. Machine learning finds out that the facts in this model point more in the direction of model generated for alternative one than the model belonging to alternative two and therefore makes the decision for alternative one. With the help of a word embedding we determine the symbols in the model of the premise that point most in the direction of alternative one. Because of the similarity of *skate* to *winter*, *freezing* and *cooling* these symbols are recognized as relevant for the decision. In the computed model, the following facts (among others) contain these symbols:

$$n1froze(sK2).$$

$$p\_d\_instance(sK2, c\_Freezing).$$

$$p\_d\_instance(sK5(sK2), c\_Cooling).$$

To derive these facts, Hyper used (among others) the following two formulae:

$$\forall x((n1froze(x)) \rightarrow (p\_d\_instance(x, c\_Freezing))) \tag{1}$$

$$\forall x(p\_d\_instance(x, c\_Freezing) \rightarrow \\ \exists y(p\_d\_instance(y, c\_Cooling) \land p\_subProcess(y, x))) \tag{2}$$

From the second formula, it is possible to generate an explanation that freezing involves a subprocess of cooling. We are aware that this is not yet an explanation for the correctness of alternative one. The reason for this is the fact that the background knowledge we currently use contains mainly taxonomic knowledge and does not adequately represent commonsense knowledge and reasoning in the strict sense. Therefore, we do not find any formulae in it that associates a frozen lake with winter sports. We hope to solve this problem by adding more sources of background knowledge such as ConceptNet. However, the basic idea for generating explanations remains the same even after the background knowledge has been extended.

By neural-symbolic learning systems [11] (cf. Sect. 2) it is possible to encode the information including background knowledge more explicitly. After the learning phase it is possible to extract the actual rules from the neural network. They may yield the basis for human-understandable explanations.

6 Evaluation

In this section, we describe our evaluation and the achieved results on the COPA challenge as well as first experiments with the Story Cloze Test. Concerning the latter, the logical models are not yet conducted, thus we have only experiments with the natural language and machine learning part without using the deduction part of the system.

6.1   Cross Evaluation

The COPA challenge specifies 1,000 problems, 500 for training and 500 for testing. We evaluated our system using stratified 10-fold cross-evaluation, splitting the training set ten times into 450 training and 50 validation examples. As our training examples for the neural network are crafted pairing the premise with each of the answers, we have 900 training and 100 validation examples. After processing both training examples of a pair through the network, we got a likelihood for each answer belonging to the respective premise. The answer of a pair which got the higher likelihood is assumed to be the answer and chosen by the system.

The goal of the cross evaluation is to identify the parameters which might lead to a good performance on the test set. In Fig. 7 we show the results on the cross-evaluation for different selected parameters. The green plots refer to the training set, while the blue plots show the respective validation set. Each data point refers to a cross-evaluation such that each boxplot represents a 10-fold cross-evaluation. The y-axis corresponds to the accuracy of our predictions and the x-axis gives the value of the respective parameter. The left plot describes different selection methods for the background knowledge, resulting in a different amount of knowledge available for the theorem prover. This in turn leads to bigger or smaller conducted logical models. *no bgk* refers to not using knowledge from Adimen-SUMO, but still from WordNet. *rec* refers to the recursion depth, while *tol* stands for the tolerance of the SInE selection. The right plot presents the amount of neurons used in the first starting layer in the neural network. The amount of neurons in the following layers are dependent on that number, as seen in Fig. 6.

The results on the training sets are very good. They range from 82.6% to 100% with a mean of 98.4% and a low derivation of 3.6%. The accuracy on the validation sets ranges from 30% to 72% with a mean of 50.2% and a derivation of 7.7%. The experiments on the parameters we do not present here behave similarly. The accuracy of our system on the test set is not good, with an average of only 50%. Nevertheless we can observe a few indicators. For instance, concerning the background knowledge integrating more information does not improve the performance in our setting, however it lowers the variance of the performance.

In the right diagram one can see that choosing two or four neurons in the starting layer, the network cannot learn the training data well, with scores of 82% to 98% and not 100%. Simultaneously, the accuracy of the validation data does not decrease in comparison to more neurons. This indicates that fewer neurons are better in generalization, probably because they do not overfit so easily. In addition there are models, which score on up to 70% on the validation data, indicating that our approach is theoretically capable of learning the right parameters.

As the training set in general is almost perfectly learned, we assume to have an overfitting problem. We tried to tackle this problem with constraints on the weights and dropout layers as presented in Table 2, however we did not get better results. We believe that we can tackle our problems with more training

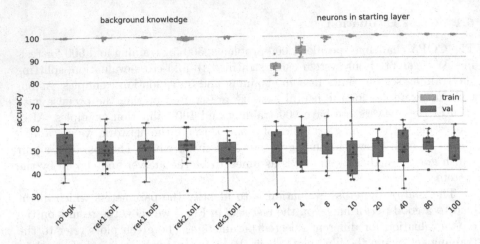

Fig. 7. Results on the different parameters *background knowledge* and *neurons in the starting layer*. The green boxplots present the performance on the training set, while the blue boxplots present the performance on the validation set. Note that the total number of neurons of the whole network is several order of magnitudes higher than the number of neurons in the starting layer. (Color figure online)

data, as the overall number of neurons in the tiniest possible network in the attentive reader approach is still one million. To tackle this size with 1,000 training examples seems to be impossible.

6.2 Performance and Discussion

Using the validation set and a fixed random seed, we identified the most promising parameters for both the background knowledge and the neural network, as described in Sect. 6.1 to calculate a best model. However, we did not yet conducted an exhaustive search of all parameter combinations. We choose the input to be a set erased of duplicates, a SInE selection recursion depth of 2, a SInE selection tolerance parameter of 1, 20 neurons in the starting layer, 40 epochs, a dropout rate of 0.4, and a learning rate of 0.002. With those parameters and a varying seed, we calculated an ensembled model out of 7 models. We repeated this 30 times to get stable results ranging from 49.30% to 56.10% with a mean of 52.51%, i.e., only slightly above chance, and a low variance of 1.49%.

The baseline given for the COPA challenge is a PMI approach (Pointwise Mutual Information), as described in [6], scoring 58.8% [27]. When the challenge came out, first approaches scored 65.4% [14] in 2011, while the state-of-the-art approach from OpenAI currently scores 78.6% [25]. The latter uses an approach solely based on neural networks with pretraining techniques and an enormous amount of training data and thus lacks explainability. With the same setup as before we also did experiments using only the natural language as it is as input, neither with background information nor a logical model. On COPA we achieved then results ranging from 53.80% to 57.00% with a mean of 55.57%, and a variance of 0.85%. On the Story Cloze Test we achieved better results ranging from 69.05% to 71.51% with a mean of 70.17%, and a variance of 0.43%. The baseline here is 59.5% [21] which is an improvement of 17.93%. However, OpenAI already scores 86.5% [25].

So far, our results do not reach the performance of state-of-the-art systems. Nevertheless we can identify tendencies. The model approach on the COPA set with 52.51% accuracy works slightly worse in comparison to the natural language only approach with 55.57% accuracy. But this might be not a general trend. Also, the Story Cloze Test using natural language with 70.17% scores better than the COPA set using natural language with 55.57%. This might be either due to the bigger data set of 3,744 problems in comparison to 1,000 problems in COPA, or to the longer description part of four sentences instead of one. Taking a look at the OpenAI results indicate that indeed the Story Cloze Tests are easier to tackle, because they also score 7.9% better on this problems.

7 Summary

In this paper, we present first experiences in combining automated theorem provers with machine learning methods to solve commonsense reasoning problems. This combination is motivated by the fact that approaches based solely on machine learning cannot provide explanations for the decisions made by the system. The use of background knowledge in the form of ontologies suggests that this is achievable in our system. Finally we present an idea of how explanations can be generated.

Unfortunately, our first experiments did not lead to good results: We obtain an accuracy of 52.51% on the COPA test set, thus our approach is hardly better than guessing. Nevertheless, the accuracy on the training set is close to 100%. We believe that this mainly is caused by too few training data. As already said in Sects. 3.1 and 6.2, we are currently integrating the Story Cloze Test. For now, we only use them as natural language, but we are soon processing them into logical models. They consist of 3,744 problems and are therefore three times as many training examples as with the COPA challenge. They are an additional benchmark and can be used as a pretraining set for the COPA tasks, as they are similar in structure and inference. In addition, we consider to make use of unsupervised pretraining on continuous text (cf. [25]).

Another problem is the quality of the background knowledge. The background knowledge we are currently using contains mostly taxonomic knowledge which is possibly only of little help in the area of commonsense reasoning. Hence, in future work, we plan to integrate further sources of background knowledge like ConceptNet. They provide knowledge graphs representing factual knowledge as triplets of the form (s, p, o) (subject – predicate – object). For this, the machine learning procedure shall also be refined to deal with the structural information in knowledge graphs.

References

1. Álvez, J., Lucio, P., Rigau, G.: Adimen-SUMO: reengineering an ontology for first-order reasoning. Int. J. Semant. Web Inform. Syst. (IJSWIS) **8**(4), 80–116 (2012)
2. Basile, V., Cabrio, E., Schon, C.: KNEWS: using logical and lexical semantics to extract knowledge from natural language. In: Proceedings of the European Conference on Artificial Intelligence (ECAI) (2016)
3. Bender, Markus, Pelzer, Björn, Schon, Claudia: System description: E-KRHyper 1.4. In: Bonacina, Maria Paola (ed.) CADE 2013. LNCS (LNAI), vol. 7898, pp. 126–134. Springer, Heidelberg (2013). https://doi.org/10.1007/978-3-642-38574-2_8
4. Bengio, Y.: Practical recommendations for gradient-based training of deep architectures. In: Montavon, G., Orr, G.B., Müller, K.-R. (eds.) Neural Networks: Tricks of the Trade. LNCS, vol. 7700, pp. 437–478. Springer, Heidelberg (2012). https://doi.org/10.1007/978-3-642-35289-8_26
5. Bos, J.: Is there a place for logic in recognizing textual entailment? Perspect. Semant. Represent. Text. Inference **9**, 27–44 (2013)
6. Church, K.W., Hanks, P.: Word association norms, mutual information, and lexicography. Comput. Linguist. **16**(1), 22–29 (1989)
7. Devlin, J., Chang, M., Lee, K., Toutanova, K.: BERT: pre-training of deep bidirectional transformers for language understanding. CoRR - Computing Research Repository abs/1810.04805, Cornell University Library (2018). http://arxiv.org/abs/1810.04805
8. Diederich, J., Tickle, A.B., Geva, S.: Quo vadis? Reliable and practical rule extraction from neural networks. In: Koronacki, J., Ras, Z.W., Wierzchon, S.T., Kacprzyk, J. (eds.) Advances in Machine Learning I. Studies in Computational Intelligence, vol. 262, pp. 479–490. Springer, Heidelberg (2010). https://doi.org/10.1007/978-3-642-05177-7_24. Dedicated to the Memory of Professor Ryszard S. Michalski
9. Doran, D., Schulz, S., Besold, T.R.: What does explainable AI really mean? A new conceptualization of perspectives. In: Besold, T.R., Kutz, O. (eds.) Proceedings of the First International Workshop on Comprehensibility and Explanation in AI and ML 2017, Co-located with 16th International Conference of the Italian Association for Artificial Intelligence (AI*IA 2017). CEUR Workshop Proceedings, vol. 2071. CEUR-WS.org, Bari (2018). http://ceur-ws.org/Vol-2071/CExAIIA_2017_paper_2.pdf
10. Furbach, U., Schon, C., Stolzenburg, F., Weis, K.H., Wirth, C.P.: The RatioLog project: rational extensions of logical reasoning. KI **29**(3), 271–277 (2015). https://doi.org/10.1007/s13218-015-0377-9
11. d'Avila Garcez, A.S., Broda, K., Gabbay, D.M.: Symbolic knowledge extraction from trained neural networks: a sound approach. Artif. Intell. **125**(1–2), 155–207 (2001). https://doi.org/10.1016/S0004-3702(00)00077-1
12. d'Avila Garcez, A.S., Zaverucha, G.: The connectionist inductive learning and logic programming system. Appl. Intell. **11**(1), 59–77 (1999). https://doi.org/10.1023/A:1008328630915
13. Goodfellow, I., Bengio, Y., Courville, A.: Deep Learning. Adaptive Computation and Machine Learning. MIT Press, Cambridge (2016). http://www.deeplearningbook.org
14. Gordon, A.S., Bejan, C.A., Sagae, K.: Commonsense causal reasoning using millions of personal stories. In: Twenty-Fifth AAAI Conference on Artificial Intelligence (2011)

15. Hochreiter, S., Schmidhuber, J.: Long short-term memory. Neural Comput. **9**(8), 1735–1780 (1997). https://doi.org/10.1162/neco.1997.9.8.1735
16. Hoder, K., Voronkov, A.: Sine qua non for large theory reasoning. In: Bjørner, N., Sofronie-Stokkermans, V. (eds.) CADE 2011. LNCS (LNAI), vol. 6803, pp. 299–314. Springer, Heidelberg (2011). https://doi.org/10.1007/978-3-642-22438-6_23
17. Holzinger, A.: Explainable AI (ex-AI). Inform. Spekt. **41**(2), 138–143 (2018). https://doi.org/10.1007/s00287-018-1102-5. Aktuelles Schlagwort, in German
18. Lenat, D.B.: CYC: a large-scale investment in knowledge infrastructure. Commun. ACM **38**(11), 33–38 (1995)
19. Liu, H., Singh, P.: ConceptNet - a practical commonsense reasoning tool-kit. BT Technol. J. **22**(4), 211–226 (2004)
20. Miller, G.A.: WordNet: a lexical database for English. Commun. ACM **38**(11), 39–41 (1995)
21. Mostafazadeh, N., Roth, M., Louis, A., Chambers, N., Allen, J.: LSDSem 2017 shared task: the story cloze test. In: Proceedings of the 2nd Workshop on Linking Models of Lexical, Sentential and Discourse-level Semantics, pp. 46–51 (2017)
22. Mueller, E.T.: Commonsense Reasoning, 2nd edn. Morgan Kaufmann, San Francisco (2014)
23. Niles, I., Pease, A.: Towards a standard upper ontology. In: Proceedings of the International Conference on Formal Ontology in Information Systems, pp. 2–9. ACM (2001)
24. Ostermann, S., Roth, M., Modi, A., Thater, S., Pinkal, M.: SemEval-2018 task 11: machine comprehension using commonsense knowledge. In: Proceedings of the 12th International Workshop on Semantic Evaluation, pp. 747–757 (2018)
25. Radford, A., Narasimhan, K., Salimans, T., Sutskever, I.: Improving language understanding by generative pre-training. Technical report Open AI (2018). http://openai.com/blog/language-unsupervised/
26. Ramos, J., et al.: Using TF-IDF to determine word relevance in document queries. In: Proceedings of the First Instructional Conference on Machine Learning, Piscataway, NJ, USA, vol. 242, pp. 133–142 (2003)
27. Roemmele, M., Bejan, C.A., Gordon, A.S.: Choice of plausible alternatives: an evaluation of commonsense causal reasoning. In: AAAI Spring Symposium: Logical Formalizations of Commonsense Reasoning, pp. 90–95 (2011)
28. Siebert, S., Stolzenburg, F.: CoRg: commonsense reasoning using a theorem prover and machine learning. In: Benzmüller, C., Parent, X., Steen, A. (eds.) Selected Student Contributions and Workshop Papers of LuxLogAI 2018. Kalpa Publications in Computing, vol. 10, pp. 20–26. EasyChair (2019). Deduktionstreffen 2018, Luxembourg. https://doi.org/10.29007/lt5p
29. Speer, R., Chin, J., Havasi, C.: ConceptNet 5.5: an open multilingual graph of general knowledge. In: AAAI Conference on Artificial Intelligence, pp. 4444–4451 (2017). http://aaai.org/ocs/index.php/AAAI/AAAI17/paper/view/14972
30. Suchanek, F.M., Kasneci, G., Weikum, G.: YAGO: a large ontology from Wikipedia and WordNet. Web Semant. **6**(3), 203–217 (2008). https://doi.org/10.1016/j.websem.2008.06.001
31. Tan, M., Santos, C.D., Xiang, B., Zhou, B.: LSTM-based deep learning models for non-factoid answer selection. CoRR - Computing Research Repository abs/1511.04108, Cornell University Library (2015). http://arxiv.org/abs/1511.04108

Author Index

Printed in the United States
By Bookmasters